Environmental Oncology

Eric H. Bernicker
Editor

Environmental Oncology

Theory and Impact

Springer

Editor
Eric H. Bernicker
Neal Cancer Center
Houston Methodist Hospital
Houston, TX, USA

ISBN 978-3-031-33749-9 ISBN 978-3-031-33750-5 (eBook)
https://doi.org/10.1007/978-3-031-33750-5

This Springer imprint is published by the registered company Springer Nature Switzerland AG
The registered company address is: Gewerbestrasse 11, 6330 Cham, Switzerland

Contents

Part I

Biology

Molecular Mechanisms of Environmental Oncogenesis

1

Kenneth S. Ramos and Abeer A. I. Hassanin

Overview

Cancer is a term broadly used to describe a variety of conditions associated with uncontrolled division of aberrant cells and the ability to invade surrounding tissues and migrate to distant sites. Although cancer continues to rank as the second-leading cause of mortality worldwide [1], advances in cancer detection and treatment now support increasing survival rates for a variety of cancers. Because cancer cells may develop anywhere in the body, many different types of cancer are recognized, and the precise term used to describe them often depends on the location and type of tissue involved. The availability of molecular phenotyping is now giving rise to new taxonomies that are based on mutation profiles.

Carcinogenesis is a multi-step/multi-pathway process that involves diverse molecular and cellular changes that can be broadly classified into three phases: initiation, promotion, and progression [2–4]. Initiation includes spontaneous or carcinogen-induced genetic mutations associated with dysregulation of biochemical signaling pathways involved in cellular proliferation, survival, and differentiation. Carcinogen-induced mutations can be impacted by metabolic and DNA repair functions [5]. During the promotion stage, actively proliferating preneoplastic cells accumulate and transition from a premalignant state to an invasive state that can break down the surrounding tissue matrix. During progression, additional genetic

K. S. Ramos (✉)
Center for Genomic and Precision Medicine, Texas A&M Institute of Biosciences and Technology, Texas Medical Center, Houston, TX, USA
e-mail: kramos@tamu.edu

A. A. I. Hassanin
Center for Genomic and Precision Medicine, Texas A&M Institute of Biosciences and Technology, Texas Medical Center, Houston, TX, USA

Department of Animal Wealth Development, Faculty of Veterinary Medicine, Suez Canal University, Ismailia, Egypt

E. H. Bernicker (ed.), *Environmental Oncology*,
https://doi.org/10.1007/978-3-031-33750-5_1

and phenotypic alterations appear that afford rapid tumor growth and increased invasive and metastatic alterations that allow cancer cells to migrate to distant sites [6, 7].

Cancer Etiology

There is no single cause of cancer. Genetic factors account for several malignancies including prostate cancer (42%), breast cancer (27%), and colon cancer risk (35%) [8]. Inherited genes, such as BRCA1 and BRCA2 [9, 10] have been associated with breast cancer. Likewise, FAP-associated with colon cancer [8]. When genetic factors are paired with other factors, either voluntary or involuntary, the combination can greatly increase a person's cancer risk. Lifestyle choices, such as smoking and heavy drinking [11], sexual behavior [12, 13], high consumption of red meat [14], or low consumption of fiber [15, 16], indoor and outdoor air pollution [17, 18], and exposure to chemical substances and radiation [19–21] are broadly recognized as major risk factors.

Genetic Causes of Cancer

Cancer is caused by alterations in genes involved in the regulation of cellular growth and differentiation. Genes are DNA segments that regulate how cells produce the proteins and other products the body requires to maintain homeostasis. An abnormal change in a gene is called a mutation. Gene mutations can be either inherited or acquired; with inherited mutations present in the egg or sperm cell at the time of conception and acquired mutations appearing during development. Given that every cell in the developing embryo derives from the initial fertilized egg, germ line mutations are present in every cell of the embryo and may be transmitted to future generations. On the other hand, acquired mutations, also known as somatic mutations, are not inherited from a parent's egg or sperm, and are not passed to future generations. These concepts are important when considering that there are many more acquired mutations than inherited mutations, and that most tumors are caused by somatic mutations [22].

There are mainly three types of genes that regulate cellular proliferation and may lead to the development of cancer, namely, oncogenes, tumor suppressor genes, and DNA repair genes (Table 1.1). Proto-oncogenes enable cells to grow or divide when activated and normally function as on/off switches involved in cellular proliferation. Oncogenes, the mutant form of proto-oncogenes, are constantly "on" and allow cells to divide uncontrollably. A change in the nucleotide sequence of a proto-oncogene that results in constitutive expression is a major part of the onco-mutation process. Three of the best characterized oncomutations include activating mutations (such as those seen in the K-H- or N-ras proto-oncogenes or EGFR) and gene amplifications (such as those seen in ErbB2/HER2 or Myc) as well as viral promoter insertions (such as those seen with LCK and E2a genes) [23–28]. K-ras mutations are common in lung adenocarcinoma and non-small cell lung carcinomas, but uncommon in small cell lung carcinomas. This proto-oncogene is also often mutated in colorectal cancers [29]. Myc gene overexpression is also often seen in both small cell and non-small cell lung carcinomas [30–38]. Normally occurring genes termed tumor suppressors or anti-oncogenes, inhibit cell growth and division, correct DNA errors, and favor apoptosis when required [39]. When these genes undergo mutation, they become inactive, leading to uncontrolled cell growth and often malignant transformation. Four primary mechanisms by which tumor suppressor genes may become defective have been identified. These include direct

Table 1.1 Oncogenes, tumor suppressor genes, and DNA repair proteins and their roles, mutations, and cancer associations

	Gene name	Protein function	Mechanism of action	Cancer
Oncogenes	RAS	GTPase	Point mutation	Colon, lung, pancreas, thyroid carcinomas, and melanoma
	EGFR	Transmembrane protein with cytoplasmic kinase activity	Gene amplification/ point mutations	Squamous cell carcinoma, non-small cell lung cancer
	HER2 (ERBB2)	Receptor	Gene amplification/ Overexpression	Breast, esophagus, and stomach cancers
			Insertion/Duplication	Lung cancer
	MYC	Transcription factor	Gene amplification/ Deregulated activity	Skin, brain, and lung carcinomas
	RNS	Guanosine Triphosphate Signal Transduction	Point mutations	Pancreas, colon, and lung carcinomas
	LCK	Protein tyrosine kinases (PTKs)	Translocation / overexpression	Leukemia, brain, breast, colorectal, and prostate cancers
	E2a	Transcription factor	Deregulated activity	Colorectal, breast cancers and leukemia
	ABL	Nonreceptor tyrosine kinase activity	Translocation t (9:22)	Myelogenous leukemia (CML) and acute lymphocytic leukemia (ALL)
	ALK	Transmembrane tyrosine kinase	Gene rearrangement / Overexpression	Non-small cell lung cancer
	TRK	Tyrosine receptor kinase	NTRK genes fusion	Colon and thyroid cancers
	BRAF	(MAPK/ERK) signaling pathway	Point mutation	Melanoma, colorectal cancer, and myeloma
	BCL-2	Regulate cell death (apoptosis)	t(14;18)(q32;q21) translocation	Follicular lymphoma
	AKT3	Key regulator of PI3K-AKT-mTOR pathway	MAGI3-AKT3 fusion	Breast cancer

(continued)

Table 1.1 (continued)

	Gene name	Protein function	Mechanism of action	Cancer
Tumor Suppressor genes	TP53	Regulate the cell cycle and repair damaged DNA	Germline mutations	Li-Fraumeni (LFS) and Li-Fraumeni-like syndromes (LFL)
			Somatic mutations including Missense substitutions, frameshift insertions and deletions, and nonsense mutations	50% of all cancers including bladder, breast, brain, head and neck, liver, lung, colorectal, and ovarian cancers.
	BRCA	Homologous recombination DNA repair	Missense, nonsense mutations, and frameshift deletion mutations.	Breast and ovarian cancers. Male breast and prostate cancers, and pancreatic cancer
	PTEN	Regulate the cell cycle, control cell growth and division	Deletion mutations	Prostate, uterine, and brain cancers
	APC	Controlling the Wnt pathway and regulation of intracellular β-catenin levels	Germline mutations	Familial adenomatous polyposis (FAP)
			Somatic mutations including point and frameshift mutations	Colorectal, stomach and pancreatic cancers
DNA repair proteins	MMR	Post-replication repair	Germline and somatic mutations in (MMR) genes including MLH1, MSH2, MSH6, and PMS2	Stomach, kidney, bladder, ovarian, colorectal, and uterine cancers

mutations/gene silencing mutations; gene deletion; loss of heterozygosity; and promoter hypermethylation [40–44]. DNA repair genes correct errors that might occur during DNA replication. Mutated DNA repair genes can no longer correct mutations in oncogenes or tumor suppressor genes leading to genomic instability. For instance, mutant BRCA1 and BRCA2 tumor suppressor genes may raise the risk of breast and ovarian cancers [45]. Mutations in the BRCA2 gene also increase the risk of male breast and prostate cancers. Men and women with BRCA2 mutations have a slightly greater risk of pancreatic cancer [46–48].

The tumor suppressor gene TP53 controls cell growth and division, directs other genes to repair damaged DNA, and regulates apoptosis. If DNA damage is not repaired, TP53 halts cell division and the cell undergo programmed cell death. When TP53 is mutated, cells with damaged DNA proliferate out of control [49]. TP53 mutations occur in about 50% of all cancers and have historically been recognized as the genome's guardian angel [50–53]. Cell division is less likely to result in DNA mistakes when DNA mismatch repair (MMR) genes including MLH1, MSH2, MSH6, and PMS2 are active. Mutations in these genes have been found in

Lynch syndrome patients [54]. Increased risk of stomach, kidney, bladder, and ovarian colorectal and uterine cancers have also been linked to these mutations [55]. The tumor suppressor gene adenomatous polyposis coli (APC) controls cell growth on colonic epithelial cells [56]. Patients with familial adenomatous polyposis (FAP) carry mutations in this gene and have a markedly increased risk of colorectal cancers [57, 58]. In addition, small intestine and pancreatic cancers are more likely to occur in people with APC mutations. HER2 (human epidermal growth factor receptor 2) is a cell surface protein that promotes cell growth of epithelial cells [59]. Mutated HER2 genes play a major role in several cancers including breast, esophagus, and stomach [60]. In several cancers, excess production of HER2 proteins is responsible for tumor growth [59].

The BCR-ABL fusion gene (also known as the Philadelphia chromosome) is created when fragments of chromosomes 9 and 22 exchange locations. Bone marrow cells carrying the fusion gene produce increased levels of tyrosine kinase. This molecular deficit is associated with the development of chronic myelogenous leukemia (CML) [61]. The BCR-ABL fusion gene is found in ~95% of patients with CML and approximately 25% of adults with acute lymphocytic leukemia (ALL). Mutations in the epidermal growth factor receptor (EGFR) and the anaplastic lymphoma kinase (ALK) have also been found in certain non-small cell lung cancer patients [62]. Cancers of the thyroid, colon, lung, and brain have been linked to gene fusions of the neurotrophic tyrosine receptor kinase (NTRK) [63, 64]. Overexpression or constitutive activation of tropomyosin receptor kinases (TRKs) is achieved through fusions involving any of the three NTRK genes (NTRK1, NTRK2, and NTRK3), and this can promote oncogenesis [65]. Therefore, Malignancies caused by a fusion of genes in the NTRK family are often referred to as "TRK fusion cancers" [66–70]. Lastly, BRAF gene mutations are seen in around half of all melanomas [71] and contribute to constitutive MAPK pathway signaling [72].

Most occurrences of melanoma are caused by mutations in BRAF, which leads to its hyperactivation and makes melanoma dependent on the MAPK oncogenic signal pathway along with additional alterations like NRAS and KRAS. It is estimated that the BRAF V600E mutation accounts for almost 50% of melanoma cases [72]. The above-mentioned gene mutations can occur in several different ways including insertions, deletions, duplications, translocations, and inversions. Follicular lymphoma, for instance, frequently displays the t(14;18) (q32;q21) translocation, which joins the BCL2 gene on chromosome 18 with the transcriptional enhancer of the IgH locus on chromosome 14 [73–76]. Furthermore, breast cancers frequently exhibit MAGI3-AKT3 translocations, accompanied by MAGI3 hemizygous deletions, resulting in the loss of function of the tumor suppressor gene PTEN and activation of the oncogene AKT3 [77]. Other examples include the frequent translocation in Burkitt lymphoma, t(8;14) (q24;q32), which fuses MYC with an immunoglobulin heavy chain [78]; and gene fusions involving the RAF family of serine/threonine protein kinases in pediatric low-grade astrocytomas [79].

Environmental Causes of Cancer

It is now broadly recognized that gene-environment interactions account for most cancers [80]. Physical, chemical, and biological agents in the environment are often responsible for the mutation spectrum highlighted above (Figure 1.1). Not surprisingly, numerous cancers have been associated with outdoor, indoor, and food contaminants. Of note are the polycyclic aromatic hydrocarbon (PAH) family of carcinogens that is present in various forms of environmental pollution and linked to lung, skin, bladder, and gastrointestinal cancers [81–84].

Indoor and Outdoor Pollutants Table 1.2 lists some of the best recognized indoor and outdoor pollutants. Of note are the PAHs, indoor and outdoor contaminants generated from the incomplete combustion of organic matter. Among them, benzo[a]pyrene and benzo[a]antracene are two of the most hazardous chemicals identified [101]. Benzo[a]pyrene, a prototypical PAH, is highly genotoxic (IARC Group 1), and one of the most prevalent lung cancer-causing environmental carcinogens [102]. The weight-of-the-evidence categorization for benz[a]antracene classifies this agent as a probable human carcinogen when inhaled or consumed [103]. Exposure to indoor air pollutants such as volatile organic compounds or pesticides increases the risk of leukemia, lymphoma, brain cancers, Wilm's tumors, Ewing's sarcoma, and germ cell tumors [104–107]. Cancers have also been linked to long-term exposures to chlorinated drinking water products, with N-nitroso compounds

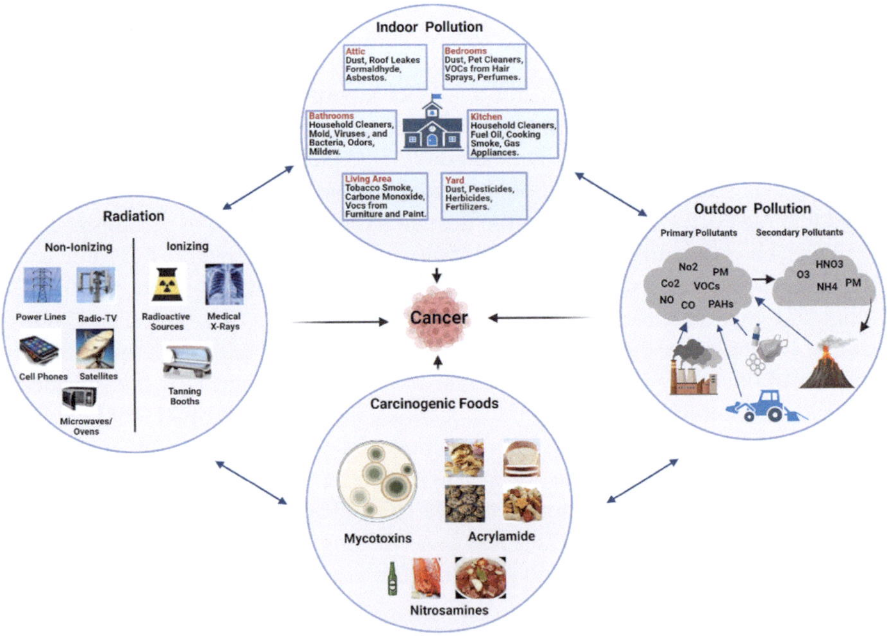

Fig. 1.1 Environmental risk factors linked with cancer development

in drinking water linked to increased risk of lymphoma, leukemia, colorectal cancer, and bladder cancers [108].

Carcinogenic Foods Mutagenic and carcinogenic chemicals, either natural compounds or those generated during food preservation and processing, can be present in food products. Most of these agents are genotoxins that covalently attach to DNA leading to mutations during the process of DNA repair and resulting in mutant

Table 1.2 Best Recognized indoor and outdoor pollutants and their sources

	Name of the pollutant	Source
Indoor Pollutants	Polycyclic aromatic hydrocarbons (PAHs)	Cigarette smoking, exhaust, cooking, heating, and household cleaning [85].
	Asbestos	Construction materials including roofing shingles, ceiling and floor tiles, paper products, and asbestos cement products. friction products including automobile clutches, automobile brakes, and transmission components, heat-resistant textiles, packaging, gaskets, coating [86–93].
	Carbon monoxide	Leaky chimneys and furnaces, unvented kerosene, and gas heater, tobacco smoke, outdated or poorly adjusted and maintained combustion devices, back drafting from furnaces, gas water heaters, wood stoves and fireplaces, gas stoves and generators and other gasoline powered equipment (e.g., boilers, furnaces) [94].
	Formaldehyde	Produced as a by-product of combustion and other natural processes and utilized in the production of a wide variety of construction supplies and household objects [95]. Formaldehyde is released into the air by a variety of consumer products, including some hair smoothing and straightening products, cleaning chemicals, glues, and adhesives [95].
	Lead	Utilized in paint, gasoline, water pipelines, and indoor operations such as soldering and stained-glass production [96].
	Nitrogen Dioxide	Combustible appliances that do not have proper exhaust venting, such as gas stoves, poorly installed vented appliances, welding, tobacco smoke, and kerosene heaters [86–93].
	Pesticides	Polluted soil or dust that floats in or is tracked in from the outdoors, stored pesticide containers, domestic surfaces that gather and release pesticides, products used in and around the house to control insects, termites, rodents, fungi, microorganisms [86–93].
	Indoor Particulate Matter	Preparation of foods, combustion activities such as the use of candles, fireplaces, unvented space heaters, kerosene heaters, and cigarettes [86–93].
	Indoor Volatile Organic Compounds	Household products such as paints, paint removers, and solvents, wood-preserving agents, aerosol sprays, cleaning and disinfecting agents, pesticides, and air fresheners [86–93, 97].

(continued)

Table 1.2 (continued)

	Name of the pollutant	Source
Outdoor pollutants	PAHs	Anthropogenic emission sources, which can be split into four categories: industrial, mobile, household, and agricultural emission sources. Also, natural emission sources, such as volcanic eruptions, natural forest fires, and lightning-caused moorland fires, are minor or insignificant [97].
	Carbon Monoxide (CO)	Fossil fuel burning from vehicles and heavy machinery [86–93].
	Ozone (O3)	Harmful industrial chemicals and fuel combustion [86–93].
	Nitrogen Oxides (Nox)	Cars, power plants, and other fuel-burning applications [99].
	Sulfur Dioxide (SO2)	Combustion of fuel in industries and power plants, as well as ships and heavy-equipment-equipped vehicles. Also constituting a natural source of SO2 emissions are volcanoes [86–93].
	Outdoor Volatile Organic Compounds such as Benzene, 1,3-Butadiene	Most benzene emissions come from gasoline cars. distillation, refining, and evaporation from gasoline production. 1,3-Butadiene is also primarily produced by automobiles, specifically through the combustion of gasoline and diesel fuel [100].
	Particulate matters (PM2.5)	Exhaust from vehicles, power plants, woodstoves, and wildfires, as well as certain industries [86–93].

proteins with altered functions [109]. Mold, primarily Penicillium, Asparagillus, and Fusarium, produce mycotoxins which can be extremely carcinogenic. Aflatoxin, ochratoxin, fumonisin, and cearalenone are among the 400 mycotoxins now thought to be responsible for the development of breast, liver, esophageal, and prostate cancers [110, 111]. Nitrosamines are carcinogenic chemicals found in pickled pork, smoked salmon, beer, soy sauce, and other foods [112–114]. Several studies have linked nitrosamines to stomach, esophageal, nasopharynx, urinary bladder, and breast cancers [115–119]. Fresh vegetables, tea, smoked sausage, and ham can be contaminated with PAHs, as do baked goods, fried foods, and smoked meats, all these previously linked with significant mutagenesis potential [120, 121]. Furthermore, acrylamide, one of the best recognized human carcinogens, can be generated naturally and be present in carbohydrate-containing meals such as potato chips; crackers, bread, and cookies when cooked or baked at high temperatures (over 120 °C) [122–124]. Human exposure to acrylamide is thought to occur mostly through inhalation of cigarette smoke and consumption of foods that have been subjected to high heat processing [125, 126]. Acrylamide has been labeled a "probable human carcinogen" by the International Agency for Research on Cancer [127]. even though it is considered "likely to be carcinogenic to humans" by the US Environmental Protection Agency [128].

Radiation Ionizing and nonionizing radiation may account for up to 10% of all cancer cases [108]. This category includes radiation from radioactive chemicals and

UV as well as pulsed electromagnetic fields. Leukemia, lymphoma, thyroid cancers, skin cancers, sarcomas, lung, and breast carcinomas are all examples of cancers that may be caused by radiation. Age, physiology, synergistic interactions between radiation and carcinogens, and genetic susceptibility are key variables associated with radiation-induced cancers in humans. The first reports on the dangers of ionizing radiation appeared in the mid-20th century, in reference to studies of children who had survived the atomic bombings of Hiroshima and Nagasaki, as well as children who had undergone prenatal Radioisotope Thermoelectric Generator (RTG). In these studies, leukemia and thyroid cancers were shown to have an elevated risk [129]. However, ionizing radiation may also cause cancers of the chest (lung) and bladder [109]. Ionizing radiation comes from a variety of sources, such as radioactive materials and radiation-producing devices. Materials with radioactive properties can be either found in nature (like uranium and radium) or created artificially in an accelerator. In addition, X-ray machines, computerized tomography [CT], and fluoroscopy generate radiation [130].

Radon and radon decay products inside homes and workplaces (such as mines) are frequent sources of ionizing radiation shown to increase lung and stomach cancers in humans [131, 132]. Of concern is that the X-rays utilized for diagnostic or therapeutic reasons emit radiation that may increase the risk of breast cancer in females exposed to chest irradiation, with most cases occurring before the age of 40 [133]. Pregnant mothers who have had diagnostic radiography have an increased chance of having a child that develops cancer. Likewise, children who have been exposed to X-rays are more likely to acquire liver cancer, bone cancer, or leukemia, with radiation dosage and number of exposures recognized as the main factors that influence risk [109]. Moreover, childhood radiation treatment can induce cancers of the gastrointentinal tract [133].

The ultraviolet (UV) rays of sunlight can also be carcinogenic [134]. Approximately 5% of the UV light that reaches the Earth's surface is UVB (ultraviolet shortwave) radiation with a wavelength between 280 and 315 nm, with the remaining being UVA (ultraviolet long wave radiation) with a wavelength between 315–400 nm [135]. UVB radiation is roughly 1000 times more mutagenic than UVA and primarily responsible for the direct mutagenic and local immunosuppressive effects seen in skin. Excessive exposure to UV radiation increases the incidence of pigmentary and non-pigmentary skin cancers—melanoma, squamous cell, and basal cell carcinoma [136, 137]. Upon exposure to UV radiation, DNA suffers oxidative damage because of free radical formation and the generation of reactive oxygen species [138, 139]. Reactive oxygen species (ROS) cause oxidative stress, which is in turn linked to a variety of human diseases [140]. This stress is caused by oxygen-free radicals at high concentrations which damage cell structures, lipids, proteins, and DNA. All the DNA's building blocks—the purine and pyrimidine bases and the deoxyribose backbone—are susceptible to damage from hydroxyl radicals (OH°) [138, 139]. On one hand, UV rays disrupt the pathways leading to the death of damaged cells, and on the other, encourages the proliferation of altered, immature cells. UV increases formation of thymidine dimers and/or thymine dimers which damage the DNA of epidermal cells [137]. The International Agency for Research on Cancer (IARC) identified solarium lights as a risk factor for skin cancer in 2009 [141, 142], as the UVA radiation dosages released by tanning lamps exceed the amounts to which the skin is exposed to when exposed to sunlight, and this may deplete cellular defense systems.

Electromagnetic Fields High-voltage power lines, transformers, electric train engines, and other electrical appliances emit electromagnetic fields (EMFs). A monograph of the International Programme on Electromagnetic Fields, the World Health Organization stated that a magnetic field of 50/60 Hz is a possible carcinogen [143]. However, others have refuted these conclusions indicating that extremely low-frequency magnetic fields (ELF MFs) solely have not revealed any cancer-causing effects. In keeping with the latter conclusion, genotoxicity investigations have not found impacts from MFs alone. However, several animal research and in vitro investigations have revealed that ELF MFs augment the effects of carcinogenic or mutagenic chemicals [144]. While the jury is still out, several reports have appeared suggesting that EMF exposure increases the risk of leukemia, brain tumors, and breast cancer [145]. Workers at a New York telecommunications business and women living within 300 meters of power lines in Sweden were found to have a 6.5-fold increased risk of breast cancer, with the relative risk associated with exposure to fields with an induction value greater than 0.01T [109]. Children within 200 meters of high-voltage power lines showed a 69% relative risk of leukemia, while those between 200 and 600 meters away showed a 23% risk. On another controversial front, a recent meta-analysis of epidemiologic data found that everyday mobile phone usage for 10 years or more increased the incidence of brain cancers [108]. This is a hotly debated area for which additional studies are sorely needed.

Lifestyle and Cancer

In most cases, the root causes of cancer are not genetic and instead, dominated by lifestyle variables such as nutrition, smoking, and alcohol intake. According to the IARC) GLOBOCAN project, the worldwide incidence of cancer may rise to 22.2 million worldwide by 2030, with much of this increase involving lifestyle-related cancers tightly linked to economic and demographic influences [146]. Smoking, alcohol consumption, diet, obesity, physical inactivity [147–150], and infectious agents, are among the most impactful lifestyle choices influencing cancer incidence and death.

Smoking

At least 14 types of cancer have been linked to smoking; with the most affected organs including the lungs, oropharynx, larynx, and esophagus, tissues directly exposed to air pollutants, as well as the pancreas, kidney, and urine bladder, tissue exposed to circulating toxins and toxicants. The proportion of cancers that may be attributed to tobacco smoking has been estimated to range from 25 to 35% for liver, kidney, and exocrine pancreas; 65% for cancers of the oral cavity, pharynx, and esophageal cavity, and 85% for lung cancer [151] with 87% of deaths from lung cancer directly linked to the tobacco carcinogen metabolite, benzopyrene diol epoxide [152]. Numerous compounds and certain carcinogens present in tobacco have a variety of cancer-causing effects, including DNA damage and epigenetic alterations, as well as inflammation [153, 154]. There is a strong relationship between cigarette smoking and inflammation, specifically related to cancers linked to activation of the pro-inflammatory marker NF-κB [155, 156].

Alcohol

Globally, it has been estimated that 3.5% of cancer deaths can be attributed to alcohol use [157]. Cancers of the upper aerodigestive tract, including cancers of the oral cavity, pharynx, hypopharynx, larynx, and esophagus carry excess relative risks (RRs) for heavy drinkers compared to nondrinkers of 5.1, 2.73 and 5.05, respectively, as well as cancers of the liver, with a RR of 2.1, pancreas, 1.2, colorectal, 1.4, lung, RR 1.2 and breast, 1.6 [158–161]. Other studies have suggested that up to 60% of women's alcohol-related cancers are breast cancers [157]. Alcohol usage causes 4% of newly diagnosed breast cancer cases in the US [162] with each extra 10 g/day of alcohol increasing breast cancer risk by 7.1% [163]. Heavy alcohol use (greater than 50–70 g/day) also increases the risk of hepatocellular carcinoma (HCC) and colorectal cancer [160, 164, 165]. There are different mechanisms by which alcohol stimulates carcinogenesis; one linked to metabolism. Toxic free radicals are created when alcohol is metabolized in the liver and these radicals are thought to mediate damage to DNA and proteins, disrupt folate metabolism, and cause loss of proliferative control. Alcohol can also promote carcinogenesis by inducing cytochrome P-4502E1, which increases free radical load and activates numerous procarcinogens present in alcoholic beverages [81], and activates the NF-κB proinflammatory pathway [166]. Alcohol and smoking can strongly interact, with evidence that entry of the carcinogen benzo-a-pyrene to the esophagus is increased by ethanol [167]. Moreover, for upper aerodigestive cancers, the Population Attributable Risk (PAR) for smoking or drinking was 72%, with PARs of 33%, 4%, and 35% for smoking alone, alcohol alone, and synergistic unfavorable effects of both [168].

Diet

In 2007, the World Cancer Research (WCR) indicated that poor diet and insufficient physical exercise are responsible for 35% of all cancer cases globally. In fact, human exposure to most carcinogens occurs via the oral route and originates from foods, food additives, or cooking, including chemicals such as nitrates, nitrosamines, pesticides, and dioxins. Heavy consumption of red meat has long been recognized as a risk factor for various cancers, including gastrointestinal, colorectal [169–171], prostate [172], bladder [173], breast [174], gastric [175], pancreatic, and oral [176]. The process of cooking meat produces carcinogenic heterocyclic amines, while charcoal heating and/or smoke curing meat produces cancer-causing carbon compounds such as PhIP (2-amino-1-methyl-6-phenyl-imidazo[4,5-b] pyridine), a mutagen believed to be responsible for ~20% of the total mutagenicity found in fried beef [177]. Long-term exposure to food additives such as nitrite preservatives and azo dyes has been shown to cause cancer [178]. Breast [179] and prostate [180] cancers have been linked to bisphenol A (BPA) from plastic food containers and cancers of the bladder, kidney, liver, and lungs may be exacerbated by arsenic ingestion [181]. Foods high in saturated fat (e.g., butter), trans fat (e.g., margarine), refined sugars, and wheat (like white bread) have also been associated with various forms of cancer [182, 183].

Obesity

Modernization coupled with a Western diet and changing lifestyles have been associated with a rise in the prevalence of obesity [16]. In countries with a high human development index, the burden of obesity-related cancers is greater (5.3%) than in countries with a low human development index (1%), with 3.6% of all cancer cases worldwide linked to obesity (women, 5.4%, males, 1.9%) [184, 185]. Obesity has been linked with a higher risk of 13 types of cancer including adenocarcinoma of the esophagus, breast (postmenopausal), colon and rectum, uterus, gallbladder, upper stomach, kidneys, liver, ovaries, pancreas, thyroid, meningioma, and myeloma [186]. J-shaped correlations with BMI have been reported for gastric and oral cavity cancers as well as lung cancer [187]. BMI-cancer risk associations appear to be heterogeneous when correlated with smoking status and absent in nonsmokers, suggesting that smoking-related residual confounding may bias the relationship between BMI and cancer risk [187]. Other studies have shown that neurochemicals, hormones (insulin like growth factor-1 (IGF-1), insulin, and leptin, sex steroids), adiposity, insulin resistance, and inflammation are common denominators for obesity and cancer [188]. Both obesity and cancer have been related to signaling pathways such the IGF/insulin/Akt signaling pathway, the leptin/JAK/STAT signaling system, and other inflammatory cascades. However, it remains unclear whether inhibiting these signaling cascades might lessen the cancer risk associated with obesity [81].

Physical Activity

Identification of the specific impacts of physical activity on cancer risk is difficult to achieve due to difficulties in the accurate assessment of individual physical activity patterns. Despite this limitation, several studies have demonstrated a decrease in cancer risk among persons who report high activity levels [189–191]. Cancers in which obesity is a major risk factor such as endometrial, pancreatic, colon, and esophageal cancers exhibit inverse relationships with increasing levels of physical activity [192–195]. However, clear inverse relationships between physical activity and the incidence of cancers not obesity related cancers, such as bladder and stomach cancers, have also been reported [196, 197], and suggest that the benefits of physical exercise may extend beyond obesity itself [198]. The advantages afforded by physical activity in primary and secondary cancer prevention can be explained by the strong influence of physical activity on inflammation, immunological function, insulin sensitivity, and levels of adiponectin, leptin, and sex hormones [199, 200]. These complex relationships indicate that it is rather difficult to separate cancer-protective pathways from obesity-related metabolic changes and their relative contributions to cancer outcomes.

Infectious Agents

Infections are thought to be the cause of 16.1–17.8% of all cancer cases worldwide [201–203]. Most infection-associated cancers are viral in nature, with several viruses including human papillomavirus (HPV), Epstein-Barr virus (EBV), Kaposi's sarcoma-associated herpes virus (KSAV), human T-lymphotropic virus 1 (HTLV-1), HIV, HBV, and HCV, linked to cervical, anogenital, skin, liver, and nasopharyngeal cancers, among others. Approximately 50–80% of primary hepatocarcinoma cancer cases are linked to HBV [204–207]. Human papillomavirus types 16 and 18 are responsible for more than 90% of squamous and 75% of cervical adeno carcinomas, respectively [208, 209]. Leukemia, lung cancer, and lymphoma have been linked to the emergence of human Merkel cell polyoma virus (HTLV-1), human T-lymphotropic retrovirus type 1 (HTLV-1), and human immunodeficiency virus (HIV) types 1 and 2. On the other hand, some parasites, such as *Opisthorchis viverrini* and *Schistosoma haematobium*, and bacteria, such as *Helicobacter pylori*, may also have a role in the development of cancer [108]. Cholangiocarcinomas and hepatocellular carcinomas in south-eastern Thailand and southern China have been associated with *Opisthorchis viverinni* and *Clonorchis sinensis* [209–211]. Infections with *Helicobacter pylori*, *C. pneumoniae*, and *Chlamydia trachomatis* have been linked to stomach, lung, and cervical cancers, respectively. Moreover, *Salmonella typhi* may cause cholangiocarcinoma, while *Streptococcus bovis* has been linked to colorectal cancer. Lyme disease spirochete infections have been linked to lymphoma [212]. The inflammation that accompanies an infection is believed to be a key contributor to cancer risk, and the inflammatory marker NF-κB has been shown to be activated by both cancer-causing viruses and *Helicobacter pylori* [213, 214].

Environmental and Lifestyle Influences by Cancer Type

According to the World Health Organization, 35% of all cancer deaths worldwide are attributable to modifiable lifestyle risk factors, including smoking and alcohol consumption in low, medium, and high income countries, infections, parasites, overexposure to UV light, tanning using devices that emit radiation, environmental tobacco, smoking, dietary factors, hormone replacement therapy, and ionizing radiation [215–220]. Key features for some of these cancers are discussed below.

Brain Adult primary central nervous system CNS cancers occur in 7 per 100,000 person-years [221]. These cancers generally have a poor prognosis, with 5.4 deaths per 100,000 person-years [221]. Brain tumors are the most common solid tumors found in children and are associated with high mortality [222]. Tumors that develop within the brain itself or in its immediate surroundings are referred to as primary brain tumors. Primary tumors are classified as benign or malignant and/or glial or non-glial (originating from nerves, blood vessels, or glands) [223]. In contrast, a metastatic brain tumor is a tumor that originated in another organ and makes its way to the brain, typically through the circulation [223]. Ionizing radiation is regarded as the primary environmental risk factor associated with brain tumors, notably gliomas, meningiomas, and nerve sheath tumors, especially in children exposed to high-dose radiotherapy [224, 225]. Of 10,834 individuals who received low dose cranial and cervical irradiation for tinea capitis (mean dose to neural tissue: 1.5 Gy), the RR of developing a brain tumor was estimated to be 6.9; with glioma risk estimated at 2.6 [226]. Children and adolescents exposed to head CT scans have an excess RR of brain cancer as high as 23 per Gy [227, 228]. Exposure to non-ionizing radiation also plays a prominent role in brain tumor development. Shintani et al. showed that 20 years after the atomic bomb detonations in Hiroshima and Nagasaki, the incidence of meningioma in adults rose from 5 per 100,000 to 15 per 100,000 [229] and this relationship correlated with the dose of radiation received.

Brain carcinogenesis may also be influenced by environmental factors such as N-nitroso compounds, pesticides, and arguably, radiofrequency electromagnetic radiation [230]. Exogenous and endogenous sources contribute equally to N-nitroso compound (NOC) exposure. Exogenous sources include cigarettes smoke, cosmetics, car interiors, and cured foods, rubber items (pacifiers, bottle nipples) and some medications (antihistamines, diuretics, oral hypoglycemic agents, antibiotics, tranquilizers, opioids) [230]. The stomach produces NOCs endogenously relying on the presence of NOC precursors, stomach pH, microbes, and other physiological conditions [231]. The link between maternal diet and childhood brain tumor risk has been examined in several studies [232], but it was difficult to draw solid conclusions due to the study design limitations, including recall and selection biases. Data on pesticide exposures and brain tumors are inconsistent, with a significant number of studies in a variety of occupational settings [233–237] showing no link between occupational status and the development of brain tumors [230]. Radiofrequency (RF) radiation is a non-ionizing form of radiation released by mobile phones and their base stations, radio and TV transmission, Wi-Fi, satellite communications, microwave ovens and radar. IARC identified RF as a probable carcinogen [238]. Electromagnetic fields (EMFs) do not induce chromosomal damage, but they are thought to promote tumor growth [239, 240]. No substantial evidence to date has linked EMF exposure from hair dryers, electric shavers, laptops, TVs, and microwave ovens to CNS malignancies [241]. However, long-term users may be at risk for acoustic neuroma [230].

Bladder The American Cancer Society estimates for bladder cancer incidence in the United States in 2022 revealed that approximately 81,180 new cases of bladder cancer are diagnosed each year (about 61,700 in men and 19,480 in women), with a total of 17,100 deaths (about 12,120 in men and 4980 in women) [242]. Bladder cancer affects elderly persons, with 9 out of 10 patients above 55 years old, with an average age of 73 at the time of diagnosis. Overall, males have a 1:27 risk of developing this cancer throughout their lifetime. The likelihood of a woman developing bladder cancer is around 1:89 [242]. However, this estimate can be influenced by a variety of risk factors, with smoking of cigarettes identified as the most significant risk factor [243, 244]. It has been shown that current cigarette smokers are three times more likely to get bladder cancer than non-smokers, with a strong association established between the number of years smoked and the number of cigarettes smoked per day [243, 245, 246]. A total of 23% of all female bladder cancers can be attributed to cigarette smoking, whereas in males, cigarette smoking is responsible for 50% of all bladder cancers [247]. The RR of bladder cancer associated with tobacco use is 3.0 [248], black tobacco (pungent, relatively crude air-cured tobacco) consumers face a two- to three-fold greater risk than Virginia tobacco (flue-cured tobacco) users [249].

Several compounds present in smoke, such as 2-naphthylamine and 4-aminobiphenyl, are linked to bladder cancer development [247]. Arsenic in drinking water is classified as a bladder carcinogen by IARC. There is strong evidence for a link between bladder cancer and arsenic in drinking water at concentrations of 300–500 g/l [250, 251]. However, no increases in risk have been found at arsenic concentrations below 200 g/l [252], except for smokers [253]. Other arsenic sources in air, food, job-related, and cigarettes may contribute to the risk of bladder cancer [250, 251, 253]. Arsenic indirectly inhibits sulfhydryl-containing enzymes and interferes with cellular metabolism to elicit cytotoxicity, genotoxicity, and suppression of antioxidant enzymes [254]. The p53 protein, on the other hand, may have a role in the development of bladder cancer, with p53 gene mutations at codon 175 and transitions at points 9 and 10 found in bladder tumors from patients exposed to arsenic for extended periods in Taiwan [253]. There was no indication, however, that arsenic exposure was linked to increased incidence of p53 mutations or immunopositivity of p53 protein in bladder cancer in a South American cohort [255], and this inconsistency may be accounted for by genetic variations in DNA repair that differentially influence arsenic carcinogenesis [256].

Water chlorination by-products such as trihalomethanes (THMs) and haloacetic acids produced when chlorine reacts with organic compounds may be a risk factor for the development of bladder cancer [257]. Aromatic amines including 2-naphthylamine, 4-aminobiphenyl and benzidine are among the most notable risk factors for bladder cancer [258–261]. Occupational exposure to 2-naphthylamine exposure raised the incidence of bladder cancer in English and Welsh workers by a factor of 200 [262]. Moreover, 19 incidences of bladder cancer were found among 171 rubber workers owing to 4-aminobiphenyl [263]. Benzidine, used in dye and rubber manufacture, is regarded as the most bladder-damaging carcinogenic aromatic amine present in cloth, paper, and leather dyes, rubber-compounding agent, and plastic films [264]. A total of 92 of 331 German employees occupationally exposed to benzidine before 1967 had bladder cancer [263]. In a Chinese cohort study, benzidine exposure raised bladder cancer risk by 35-fold [265]. Another study on rubber workers found that o-toluidine exposure increased the incidence of bladder cancer [266]. Multiple mechanisms contribute to o-toluidine's carcinogenicity including oxidative DNA damage, chromosomal damage, mutagenicity, and cytotoxicity [267, 268]. There are three main types of bladder cancers [269] including urothelial carcinoma, squamous cell carcinoma and adenocarcinoma. Around 90% of bladder malignancies are caused by urothelial carcinoma, also known as transitional cell carcinoma (TCC), involving the lining cells of the urinary system (urothelial cells) [269]. Squamous cell carcinoma originates from bladder irritation and inflammation and accounts for about 4% of all bladder tumors [269]. Lastly, adenocarcinomas arise from glandular cells and account for roughly 2% of all cases [269].

Blood Leukemia is the most common type of cancer among children and adolescents worldwide, accounting for 36.1% of all cases in those aged 0–14, and 15.4% in those aged 15–19 [270]. Of all cancers in children, leukemia accounts for about 27% of cases in the United States, with the highest rates seen in White Hispanic children and the lowest rates seen in Black children [270]. A total of 31% of all children's cancers in Ireland and France, 35% in Shanghai, China and 33% in Germany are blood cancers that arise from the transformation of a blood pluripotent stem cell or, in rare instances, a mature bone marrow cell [271]. Hematological neoplasms are disorders with varying cytology, histology, epidemiology, etiology, and prognosis [272]. The four main types of leukemia include Acute Lymphocytic Leukemia (ALL), the most prevalent kind accounting for one-fourth of all cases of leukemia, Acute Myeloid Leukemia (AML), Chronic Lymphocytic Leukemia (CLL), and Chronic Myeloid Leukemia (CML). Ionizing radiation is the only environmental risk factor shown to be strongly associated with ALL and AML [272], with a greater chance of cancer development from radiation exposure during childhood than later in life [273–277]. Ionizing radiation sources include medicinal and man-made environmental radiation (e.g., nuclear weapons testing) [278].

An increase in leukemia cases among Hiroshima and Nagasaki bomb survivors was reported by Folley in 1952 [279], with individual risk increasing with exposure time, radiation dosage, and age. Moreover, a 50% increased risk of leukemia in children has been associated with maternal X-ray exposure during pregnancy. On the other hand, there are currently no conclusive findings concerning the relationship between natural background radiation including radon and gamma radiation exposure and childhood cancers (including leukemia) [280]. Non-ionizing radiation also is a potential risk factor for ALL, with exposure to power frequency magnetic fields larger than 0.4 mT shown to raise the incidence of childhood leukemia by a factor of two [281]. A 1.7-fold increase in the risk of leukemia was discovered in children who lived near the 0.3mT source [282]. Children exposed to very low frequency magnetic fields larger than 0.3 microT had an increased chance of developing ALL compared to children exposed to levels below 0.1 microT [283]. Daniels and Zahm linked pesticide usage to leukemia in studies showing that the use of pesticides in the home and garden three months before pregnancy, throughout pregnancy, and during childhood, was linked to an increased risk of blood cancers [284–287].

Lymphoblastic and myeloblastic acute leukemia in children have been linked to maternal insecticide, herbicide, and fungicide use. Acute leukemia has also been linked to paternal use of pesticides [288]. In conclusion, environmental elements may interact synergistically to cause blood cancers. This may explain why epidemiological research discovered a weak link between leukemia and risk factors [272]. In the context of infections and leukemia there are different possible hypotheses, including Kinlen's "population mixing" theory [289] and Greaves' "delayed infection" idea [290]. Both concepts suggest that ALL develops from an aberrant response to infections. Hauer et al. [291] recently presented three additional models of ALL development, all of which point to infection-induced immunological abnormalities as the cause of leukemia.

Breast Breast cancer (BC) is caused by a combination of genetic, endocrine, and environmental factors, with 60% of BC caused by environmental carcinogens (physical, chemical, and biological) [292, 293], together with inheritance, reproductive life, and exposure age may raise breast cancer risk [294]. Worldwide, breast cancer (BC) is the most frequent non-skin cancer among women. It is estimated that in Western nations, the average age at which women are diagnosed with breast cancer is 60–70 years old, whereas in Asia, the average age is 40–50 years [295]. When carcinomas originate in the breast, they are frequently adenocarcinomas involving cells in the milk duct or the glands in the breast lobule that manufacture milk [296]. Ductal carcinoma in situ (DCIS), also known as intraductal carcinoma or in situ carcinoma of the breast, begins in the duct but has not spread to other parts of the breast. Any form of breast cancer that has progressed into neighboring tissues is referred to as invasive (or infiltrating) breast cancer [296]. Divalent metals, such as cadmium, copper, cobalt, nickel, lead, mercury, tin, and chromium, have been shown in the human breast cancer cell line MCF-7 to activate estrogen receptor (ER-α) and to drive cell proliferation [294]. The link between cadmium and breast cancer has been stated in many studies; with creatinine-adjusted urine cadmium levels over 0.58 g/g doubling breast cancer risk [297]. Cadmium levels in malignant breast tumors and healthy breast tissue vary significantly [294, 298, 299]. Antila et al. found high cadmium contents (3.2–86.9 g/g) in breast cancer patients' tissue [300]. Cadmium forms a high-affinity complex with ER-α hormone-binding domain, making it act like steroidal estrogen in breast cancer cells [301, 302]. This interaction stimulates the development of side branches and alveolar buds in the mammary gland, as well as induction of casein, whey acidic protein, PgR, and C3 [303].

It is well-known that ionizing radiation causes cancer of the breast [304]. Increased breast cancer risk has been documented after acute radiation exposure from the atomic bombs in Japan [305], and following high cumulative doses associated with the treatment of various disorders and frequent diagnostic radiography testing [306, 307]. Moreover, radiologic technicians who had daily low-dose radiation exposures for years show an increased risk of breast cancer. The radiation dose-response to breast cancer is linear [308], with radiation doses up to 40 gray (Gy) increasing breast cancer risk [309], and a 1 to 3 Gy low-dose threshold for breast carcinogenesis [306] taking into consideration the impact of age and age at the time of exposure on breast cancer risk, with a greater risk for those treated before age 20 [304, 310] compared to exposures occurring after age 40 [311]. Exposure to electromagnetic fields in the power frequency range of 50–60 Hz may raise the risk of BC by suppressing nocturnal melatonin synthesis, as melatonin has been suggested to be a preventive factor against BC by influencing estrogen levels [312]. BC mortality and incidence rates inversely correlate with the total average sunlight energy [313], which might be partially explained by the proposed role of vitamin D as an inhibitor of breast epithelial cell proliferation through the nuclear vitamin D receptor [314, 315]. Early exposure to persistent endocrine disrupting chemicals (EDCs) including dioxins, PCBs and OCPs may change breast development or hormone response and enhance adult breast cancer risk [316]. For example, hormone replacement therapy (HRT) may increase tumor development via estrogen- or progesterone-mediated mechanisms [317, 318]. Dioxins that bind to the aryl hydrocarbon receptor (AhR), a transcription factor which affects gene expression, enzyme metabolism, and hormone signaling pathways can lead to the promotion of BC [319]. One possible mechanism by which AhR-activating toxicants cause BC is through to involve changes in the expression profile of cytokines and immune modulatory enzymes [320]. The AhR repressor protein (AhRR) has been shown to inhibit both normal AhR signaling and the production of inflammatory cytokines [321, 322].

Women who were exposed in utero to the synthetic estrogen diethylstilbestrol (DES) exhibited an increased risk of BC beyond the age of 40 [323]. Breast cancer risk was shown to be 2.5 times higher in DES-exposed daughters over the age of 40 [324]. Women over the age of 40 who were exposed to DES had a nearly 2-fold greater risk for BC, while women over the age of 50 were projected to have a 3-fold increased risk [325]. Furthermore, studies have shown that mothers treated with DES have a slightly increased chance of developing BC [326–329]. Also, terminal end bud and ductal development may be increased by DES during puberty [330–332], providing a biological explanation for the higher risk of BC.

Colon Cancer of the colon is a disease of the large intestine that can occur at any age, mostly affecting persons over the age of 50 [333]. It is worth noting that the incidence of colon cancer is increasing in younger individuals and that the age of screening was recently lowered to 40 years of age by the United States Prevention

Services Task Force [334]. Both men and women are at risk for colon cancer, and this cancer is the third most frequent cancer diagnosed in both sexes, and the second most common cause of cancer mortality in the US [22]. Colon cancer often starts with the formation of noncancerous (benign) clusters of cells called polyps on the inside of the colon, with some of these polyps developing into colon cancer over time. Colorectal cancer (CRC) is a term that combines colon cancer with rectal cancer, to include cancers that originate in the rectum. Only 10% of those who develop colon cancer carry germ-line mutations. As such, colon cancer is recognized as an environmental disease in which multiple exposures (internal and external) are combined to define disease risk [335]. Environmental risk factors of CRC include red, processed, and grilled meat diets, preexisting disorders (obesity, inflammatory bowel diseases, type 2 diabetes), smoking, and alcohol consumption [336]. Red and processed meats are linked to increased CRC risk in a dose-dependent manner [337], with each 100 g/day of red and processed meats increasing risk by 12% [338]. High levels of heme iron in red meat may contribute to the increased risk of CRC [339]. Free iron may take part in Fenton reactions, resulting in elevated levels of reactive oxygen species and lipid peroxidation products in the gastrointestinal system that may cause significant cell damage [340]. Processed meats may raise CRC risk because they also include N-nitroso compounds as byproducts of the curing process, PAHs generated from meat smoking, and heterocyclic amines formed upon heating [339]. Intriguingly, the carcinogenic chemicals detected in these meats are also widespread in other environmental sources, including cigarette smoke, dust, and vehicle engines, showing that persons exposed to high concentrations of these environmental contaminants may also have an increased risk of CRC [335].

High fat intake also raises the risk of CRC via increased bile acid secretion and microbial development of genotoxic secondary bile acids which disrupt cell membranes, create ROS and reactive nitrogen species that cause DNA damage [341]. Overweight and obesity are also well-known risk factors for CRC [342], with an 18% increase in cancer risk per each 5 unit rise in BMI [343]. Smoking has long been recognized as a significant risk factor for the occurrence and mortality from colon cancer [344], with greater impact on the formation of serrated polyps [345] or tumors with microsatellite instability than other CRC subtypes [346]. Tobacco use is associated with a 15% higher risk of colon cancer and this relationship shows a clear dose-response [347]. CRC and alcohol intake are also linked [348], with heavy drinkers more likely to get CRC than moderate drinkers [349], and a 7% increase in risk with each daily consumption of 10 g of ethanol [338]. Age, gender, amount, and quality of alcohol consumed as well as pattern of consumption have all been demonstrated to influence risk [348, 350]. Alcohol-induced CRC is tightly connected to metabolic pathways and mostly attributable to the production of carcinogenic acetaldehyde which interacts with DNA bases to create DNA adducts that cause genetic mutations [351]. *Enterotoxigenic Bacteroides fragilis* [352], *Fusobacterium nucleatum* [353, 354], and *Escherichia coli* strain NC101 are among the invasive pathogenic bacteria associated to CRC) [355]. *Fusobacterium nucleatum* promotes an immunocompromised, pro-inflammatory microenvironment that promotes tumor growth [353, 356–358]. However, due to a lack of prospective data, it is controversial whether this bacterium is an opportunistic invader prospering in the hypoxic environment of growing tumors or a causative agent. Rather than the proportional abundance of any bacterial species, the interactive microbial balance, strain balance, and overall bacterial function may be more essential to CRC development [359].

Kidney Kidney cancer is caused by abnormal proliferation of modified cells within renal parenchymal tissue. This cancer is more frequent among individuals who are 65–74-years of age, with more than 131,000 people dying each year from kidney cancer throughout the world [360, 361]. The risk factors for kidney cancer include lifestyle risk factors (cigarette use, obesity, alcohol intake, physical activity, and diet); environmental and occupational exposures (trichloroethylene and aristolochic acid); genetic risk factors and others [362, 363]. As defined by the American Cancer Society [364, 365], renal cancer comes in several forms, including renal cell carcinomas, transitional cell carcinomas, Wilms tumors, and renal sarcomas, with renal cell carcinoma (RCC) recognized as the most common type. The tubule lining is a common site for the development of malignant cells in the kidney. These cells have the potential to accumulate into a solid mass that eventually causes obstruction. RCC can affect either one or both kidneys [366, 367]. A common site for the development of transitional cell carcinoma (TCC) (also known as urothelial cancer or renal pelvis carcinoma) is at the junction of the ureters and the kidney. Under the microscope, tumor cells may start to resemble bladder cancer cells more than kidney cancer cells [366, 367]. Wilms tumor is a form of childhood kidney cancer that typically affects children between the ages of 3 and 4. Wilms tumors account for many pediatric kidney malignancies, with a higher risk for African American kids compared to White kids [366, 367]. Renal sarcoma is an extremely uncommon cancer that begins in the connective tissue. Unlike renal cell carcinomas and transitional cell carcinomas, this is a form of soft tissue sarcoma [366, 367].

Cigarette smoking has been linked to an increased risk of RCC [154, 368]. Smokers with 20 or more pack years show a 30% increased risk of developing RCC compared to nonsmokers [369]. Ever-smoking individuals show an increased RCC risk by 1.38 with a clear dose-response association and men's risk higher (1.50) than women's (1.27) [370]. Chronic tissue hypoxia from carbon monoxide exposure and smoking-related diseases, such as chronic obstructive pulmonary disease, has been linked to an increased risk of RCC in smokers [371]. Moreover, a tobacco-specific N-nitrosamine was shown to increase DNA damage in peripheral blood cells of renal cell carcinoma patients [372]. RCC patients were shown by Zhu et al. [373] to have higher rates of chromosome 3p deletions in their peripheral blood lymphocytes following exposure to benzo[a]pyrene diol epoxide, a common carcinogen found in tobacco smoke. The risk of RCC was observed to rise in a dose-response manner for overweight and obese adults at baseline [374–377], with a 24% increase for men and 34% for women for every 5 kg/m2 increase in BMI [378]. Several mechanisms may impact renal cell cancer in obese people, including chronic tissue hypoxia, insulin resistance and hyperinsulinemia, altered endocrine milieu and adipokine production, obesity-induced inflammatory response, lipid peroxidation and oxidative stress [379]. Physical exercise reduces body weight, blood pressure, chronic inflammation, and oxidative stress [380–382] which may lower RCC risk. High intake of animal-based fat and protein has been linked to increased risk of kidney cancer. Some epidemiological studies linked acrylamide, a Group 2A "probable" human carcinogen [127] present in high quantities in fried and baked foods [383] with RCC. On the other hand, a 32% reduction in the incidence of RCC was associated with a high consumption of fruits and vegetables [384]. In the NIH-AARP diet and health study, alcohol consumption was shown to be protective against RCC, with the RR of RCC reduced from 0.75 to 0.71 with light (0–5 g/d) to moderate (15–30 g/d) alcohol use in both men and women [385]. Occupational exposure to a variety of industrial compounds has been associated with an increased risk of renal cancer. Trichloroethylene (TCE), a common environmental pollutant and metal degreaser [386] and a Group 2A "probable" human carcinogen, increases renal cell cancer risk with increasing levels of exposure [387]. Workers exposed to high amounts of TCE at a German manufacturing facility had a roughly 8-fold increased risk of kidney cancer [388] when compared to a reference Danish population. Furthermore, many other chemicals have been postulated as possible causes of RCC, including lead, arsenic, pesticides, and benzene [389–394].

Liver The incidence and mortality of liver cancer have risen worldwide [395–397]. Liver cancer is the 6th most prevalent cancer in the world, the 5th most prevalent cancer among males, and the 9th among women. Several liver cancer subtypes have been identified, with 75–85% of all cases classified as hepatocellular carcinoma (HCC), also known as hepatoma. Hepatocellular carcinoma develops from hepatocellular cells [366, 367, 398]. Cholangiocarcinoma, also known as bile duct cancer, is a malignant tumor that forms in the tiny tubes (bile ducts) of the liver involved in the transport bile to the gallbladder. Cholangiocarcinoma is a rarer form of liver cancer, making up 10–20% of all liver tumors. When bile ducts in the liver become malignant, it is called intrahepatic bile duct cancer. However, cancer of the bile ducts outside the liver is called extrahepatic [366, 367]. About 1% of all liver malignancies are angiosarcomas, also known as hemangiocarcinomas. Angiosarcomas begin in the blood vessels of the liver and grow fast. Liver metastases, or secondary liver cancer, occur when cancer originates in another part of the body and spreads to the liver [366, 367]. The incidence of secondary liver cancers far exceeds that of initial liver tumors [399]. Although uncommon, hepatoblastoma is the most frequent liver cancer in children, affecting the lobes of the liver [366, 367].

According to ACS an estimated 30,520 persons (20,420 men and 10,100 women) will die from liver cancer [364, 365]. HCC is the third-leading cause of cancer-related death globally. Alcohol consumption, pesticides, aflatoxin, vinyl chloride, arsenic, and PAHs are among the main risk factors for liver cancer [400, 401]. Approximately 15% to 30% of HCC cases may be attributable to alcohol usage [402]. Over the course of a person's lifetime, consuming >600,000 ml of alcohol raises the risk of developing HCC [403], and more than twice as many people who have had high levels of alcohol consumption and HBV or HCV carry an elevated risk of developing HCC compared to those with moderate levels of alcohol use and no chronic viral infection [404]. Pesticide exposures in agriculturally concentrated regions raises the incidence of HCC in farmers via genotoxic and immunotoxic pathways, in addition to deficits in hormonal and tumor-promoting signaling [405]. Aflatoxins, especially Aflatoxin B1 (AFB1), are potent hepatocarcinogens found mostly in sub-Saharan Africa, Southeast Asia, and China as the tropical and subtropical climates in these regions enhance *Aspergillus* development [406]. Dietary Aflatoxin B1 is a powerful mycotoxin produced by *Aspergillus* spp. [407]. Upon intake, the hepatic cytochrome P-450-dependent monooxygenase system converts Aflatoxin B1 to a potent AFB1-8,9 oxide [408] that causes DNA damage [398]. Pesticides, including dichlorodiphenyltrichloroethane (DDT) have been linked to hepatocarcinogenesis via mechanisms of oxidative stress, genotoxicity, and immunotoxicity [409, 410]. Air pollution includes PAHs and particulate matter (PM) from natural and anthropogenic sources that have been classified as Group 1 human carcinogens [411, 412]. There is a strong correlation between exposure to PM and NOx and liver cancer risk [413–415]. For residents of the Taiwanese Penghu Islands, higher levels of PM2.5 exposure were linked to increased risk of HCC development [414]. High blood BaP levels have also been correlated with HCC risk [416], with HCC patients showing greater serum BPDE-albumin adducts and GSTP (detoxification gene) hypermethylation than controls [417]. Oral ingestion of polluted drinking water is the principal route of exposure to hepatotoxic microcystins (MCs), the most reported cyanotoxins in eutrophic freshwaters [401, 418]. The cyanobacteria (e.g., Microcystis, Anabaena) found in eutrophic freshwaters are the primary source of the hepatotoxic microcystins which belong to the class of cyclic heptapeptides and are referred to as hepatotoxins because the liver is their primary target [418]. Long-term exposure to low-dose MCs may also cause liver cancer. However, more relevant large-scale, population-based case-control studies and cohort studies are required to validate the causal link between MC exposure and hepatotoxicity [419].

Pancreas Pancreatic cancer is the 10th most frequent cancer and the 7th highest mortality rate worldwide, with ~460.000 new cases of identified each year [409] and a 5-year survival rate of only 9% [420]. Exocrine pancreatic cancer, which includes adenocarcinomas and neuroendocrine pancreatic cancers are the two main kinds of pancreatic cancer [421]. The exocrine gland and ducts of the pancreas are the origin of exocrine pancreatic cancer. More than 95% of all pancreatic cancers are caused by exocrine pancreatic tumors which include adenocarcinomas, also known as ductal carcinomas, and are recognized as the most common kind of pancreatic cancer, making up more than 90% of all cases. Pancreatic ductal carcinoma forms in the ductal epithelium [422]. The cells responsible for making pancreatic enzymes are also a potential source of adenocarcinoma. This condition, known as acinar cell carcinoma, accounts for 2% of all exocrine malignancies [423]. An extremely rare form of nonendocrine pancreatic cancer, squamous cell carcinoma, develops in the pancreatic ducts and is composed entirely of squamous cells, which are not normally found in the pancreas, many cases are not identified until after metastasis has occurred, giving this cancer type a very poor prognosis [424, 425]. Adenosquamous carcinoma is an extremely uncommon kind of pancreatic cancer that accounts for just 1% to 4% of all exocrine pancreatic malignancies. This tumor type shares characteristics with both ductal adenocarcinoma and squamous cell carcinoma [423, 426]. Colloid carcinomas, which account for 1% to 3% of exocrine pancreatic malignancies, are an uncommon kind that typically arises from a benign cyst called an intraductal papillary mucinous neoplasm (IPMN) [423, 427]. Neuroendocrine tumors (NETs) of the pancreas arise in the endocrine gland of the pancreas, which is responsible for secreting the hormones insulin and glucagon into the bloodstream to control blood sugar levels. Rare neuroendocrine tumors, also called endocrine or islet cell tumors, account for less than 5% of all instances of pancreatic cancer [423, 428, 429].

The environmental pancreatic cancer risk factors, that is the non-genetic factors giving rise to cancer of the pancreas include cigarette smoking, heavy alcohol drinking, diet, obesity, and low physical activity. Tobacco use is a well-known contributor to the development of pancreatic cancer. Smoking is responsible for 20% to 25% of pancreatic cancer cases [430, 431], with only a few years or only a few smokes per day causing a dramatic rise in risk [432]. The risk of pancreatic cancer is believed to be reduced to the level of never smokers 15–20 years after quitting smoking [433, 434]. While no direct linkage between pancreatic cancer and passive smoking has been demonstrated, a positive association between maternal smoking and pancreatic cancer in the offspring has been reported [435]. This may be due to prolonged exposure to passive smoke in utero and early in life. There is no correlation between smoking and mutations in cancer-associated genes such as KRAS and TP53 in pancreatic tissues [436, 437]. However, pancreatic tissues from smokers exhibit a greater total mutational burden than those from nonsmokers [436]. NNK interacts with pancreatic cells through β-adrenergic and nAChR receptors [438], causing the activation of cyclooxygenase 2, epidermal growth factor receptor, and extracellular signal-regulated kinase (ERK) [439, 440]. NNK and other smoke constituents promote pancreatic ductal cell proliferation and prevent apoptosis through activation of AKT and AMP kinase [441]. Smoking components not only contribute to the onset of pancreatic carcinogenesis and its progression, but also play a role in the metastatic progression and chemotherapy resistance [442]. Excessive alcohol intake has shown a positive association with pancreatic cancer, with a daily intake of 30 grams of alcohol increasing the risk of pancreatic cancer ~20% [443]. No significant link has been detected with light to moderate drinking (4 drinks/day) [430]. Ethanol may induce pancreatic carcinogenesis via a direct effect exerted by acetaldehyde, a recognized carcinogen capable of forming protein and DNA adducts [444, 445]. East Asians may be more susceptible to pancreatic cancer [446], as 30–50% of the East

Asian population carries the ALDH2*2 allele that reduces enzyme activity leading to poor acetaldehyde metabolism [447]. Consumption of red and processed meats is also associated with a higher risk of pancreatic cancer [433, 448, 449], possibly due the high amounts of mutagenic heterocyclic amines and PAHs formed during high-temperature cooking [449, 450].

On the other hand, fruits, vegetables, and other plant-based diets lower pancreatic cancer risk [433, 451–454]. Plant-based diets that include fruits, vegetables, whole grains, and nuts contain phytochemicals (carotenoids, phenolics, alkaloids, nitrogen-containing compounds, and organosulfur compounds) as well as dietary fibers shown to exert potent anti-cancer properties [455, 456]. The anti-cancer activities of these phytochemicals are mediated via different mechanisms that include antioxidant and anti-inflammatory activity, suppression of cell proliferation, development, and invasion, and promotion of DNA damage repair [455]. In fact, an inverse association between dietary fiber intake and pancreatic cancer risk has been established [455], with every 10 grams of daily fiber intake associated with a 12% reduction in risk [457]. Obesity and pancreatic cancer are linked by unknown molecular pathways, however. inflammation and hormone imbalance are believed to be contributing factors [458]. While inconclusive evidence exists about the prevention of pancreatic cancer with physical activityy, regular physical exercise may control obesity, which is a recognized risk factor for pancreatic cancer [446]. Poor oral health, particularly periodontal disease and tooth loss have also been linked to an increased risk of pancreatic cancer [459–461]. An association between oral bacteria such as *P. gingivalis* and *A. actinomycetemcomitans* and pancreatic risk has been proposed [462, 463]. Two possible mechanisms for this association proposed suggest that oral infection may increase systemic inflammation and consequently pancreatic inflammation [464, 465] and oral pathogens may be circulated to the pancreas where they cause persistent inflammation [446].

Occupational exposures to chlorinated hydrocarbon compounds such as those found in diesel exhaust or aluminum manufacturing, as well as other airborne pollutants have been associated with increased pancreatic cancer risk [466]. Concerning heavy metals, cadmium has received special attention given its ubiquitous presence in diet, cigarette smoke, air, soil, and water [467]. Cadmium exposure has been linked to an elevated risk of pancreatic cancer among those who were not exposed to cadmium in the workplace [32, 33]. The main ways in which the "general" population can be exposed to cadmium is through diet or with tobacco use [467]. It is widely believed that cigarette smoking is a major contributor to cadmium exposure in smokers. For smokers, the chance of acquiring pancreatic cancer is around double that of nonsmokers [468].

Gastric Cancer Gastric cancer (GC) is a fast-growing mucinous adenocarcinoma of the stomach. Although GC can begin in any of the stomach's five layers, the mucosa is the most likely site of malignancy [469]. Adenocarcinoma is the most frequent form of stomach cancer, making up between 90 and 95% of all cases. It starts in the mucosa and then spreads to the other four layers. The other types include, carcinoid tumors, which originate in the hormone-producing cells of the stomach and rarely metastasize, cancer of the skin-like cells that line the digestive tract (squamous cell carcinoma), the extremely rare neuroendocrine tumors known as small cell carcinomas, frequently arise in tandem with other tumor types, leiomyosarcomas that develop in the stomach's smooth muscle cells, and tumors of the gastrointestinal stroma (GIST) which originate in the stomach's interstitial cells of Cajal [469]. GC is the 5th most common cancer in the world and the 3rd most common cause of death worldwide, with a specific regional pattern of distribution [470]. Globally, a higher incidence of GC is seen in Eastern Asia, Eastern Europe, and South America, compared to Northern Europe, Africa, and North America [470, 471]. GC incidence and mortality vary greatly across the same country and even within a single state or region. It is probable that environmental, genetic, and infectious factors contribute to these geographical variations. Environmental factors including water, soil, and air [472]. In Zanhuang County, Hebei Province, China, and Eastern Transylvania [473], drinking pond, river, or shallow groundwater was found to enhance GC incidence [474], whereas low GC incidence was reported in certain regions of France with deep groundwater [475]. It has been proposed that long-term exposure to extremely salty drinking water damages the stomach mucosa and may set the stage for cancer development [476]. GC mortality tends to rise in locations with high levels of water pollution, which is believed to be a major contributing cause [477]. Some pollutants in water have been linked to increased GC incidence. For example, an increase in the number of cases may be associated with organic matter contamination [478]. Heavy metal pollution in water has also been studied as a contributing factor, given that the ability of GC cells to penetrate and metastasize is increased by heavy metals [479]. Cadmium [480, 481], lead [480, 481], arsenic [482], and chromium [483] have been shown to be positively associated with GC incidence and mortality. Furthermore, the presence of nitrate and nitrite in drinking water may lead to increased incidence and mortality of GC [484–488]. Nitrate and nitrite can be transformed into nitrosamines in the stomach, metabolites with strong carcinogenic activity [485]. Numerous investigations revealed a striking negative association between water hardness and GC, as calcium and magnesium ions in hard water are apparently protective [489, 490], while softer water facilitates the dissolution of hazardous compounds from pipes into drinking water [488] Soil type and pollution have a great impact on GC, with the incidence of GC prevalent in peat soil regions with poor soil drainage and crops that are deficient in nutrients and copper [491, 492]. GC prevalence is lower in places with reasonably adequate drainage, such as those with porous subsoils containing sandstone and limestone [473, 475]. The incidence of GC is higher in areas with calcareous or acidic sandy soils [493]. GC mortality was found to be low in red soil locations, but high in decaying organic matter or flooded soils [491, 492].

Pollution in the air has been linked to increased risk of GC, both in terms of incidence and mortality, primarily due to inhalation of suspended particles that enter the stomach. A significant association was found between PM2.5 and GC incidence, with an increase of 5.0 mg/m3 PM2.5 raising the GC risk by 38% [494], possibly owing to sulfur (S) present in the particles [495]. Radiation exposure is also a risk factor for GC [496]. In uranium mines, even modest levels of radiation have been linked to GC [497, 498]. Moreover, increased GC morbidity has been observed in volcanic locations and linked to high levels of radiation in volcanic soil [499]. Studies on survivors of the atomic bombs in Hiroshima and Nagasaki, Japan, found that ionizing radiation raised GC risk and cancer death rates [500, 501]. Diet, and in particular salt, is a significant modifiable risk factor for GC [502, 503]. People with high and moderate salt consumption had a 40–70% higher risk of cancer than those with low salt intake [504]. High salt concentrations affect the mucosal barrier of the stomach causing inflammation and atrophy [505, 506]. CagA expression, an H. Pylori virulence factor and GC risk factor, is significantly elevated by high salt content in the diet [507]. Also, salt may increase the toxic effects of N-nitroso compounds [508, 509]. Cigarette smoke is also associated with GC, with the risk of GC estimated to be 1.5–1.8 times higher for smokers than nonsmokers and to follow a dose-dependent relationship [510–513]. Activation of nicotinic acetylcholine receptors, production of DNA adducts, promotion of tumor angiogenesis, and induction of cell proliferation are all possible mechanisms for smoking associated GC [514, 515]. Moreover, smoking is a risk factor for chronic inflammation in the GI tract through alterations in mucosal cell proliferation, induction of immunological dysfunction, and increases in the risk of bacterial or viral infection, all of which contribute to carcinogenesis [514]. Heavy alcohol use raises GC risk by up to 65% [516–519]. The mechanisms behind the increased risk include N-nitrosodimethylamine (NDMA), a carcinogen present in several alcoholic drinks, especially beer [516]. As noted earlier, alcohol intake also produces carcinogenic acetaldehyde [519]. The cellular damage caused by alcohol may facilitate carcinogenic chemicals to penetration across cellular and subcellular membranes [519]. Finally, heavy alcohol use may cause stomach inflammation [518].

Lung Worldwide, lung cancer is the main cause of cancer-related death, with over 2 million diagnoses and 1.8 million deaths annually [470]. Although the 5-year survival rate for lung cancer has climbed from 21% in 2014 to 25% in 2018, a jump that specialists have called "amazing progress," lung cancer remains the leading cause of cancer death in the United States. This is largely because most lung cancers are not recognized until they are at an advanced or metastatic stage, and most of them (75%) are discovered after the disease has spread. WHO classifies lung cancer into two broad histological subtypes: non-small cell lung cancer NSCLC, which accounts for 85% of cases, and small-cell lung cancer (SCLC), which makes up the remaining 15% [520]. Three subtypes of NSCLC have been identified, lung adenocarcinoma (LUAD), lung squamous carcinoma (LUSC) which together account for 40% and 25% of the cases, respectively, and large cell carcinoma (LCC), which account for 10% of cases [521].

At 11.6%, lung cancer was the most common cancer in 2018 [522], with smoking being directly responsible for 80% of cases [522, 523], The risk of developing lung cancer is elevated for those who smoke even infrequently, or on a light daily regular basis. The danger grows with the number of cigarettes smoked daily and the length of time a person has been a smoker [524]. Eighty-nine thousand cases of lung cancer were attributed to outdoor air pollution [525]. Radon is the second major cause of lung cancer in the United States, after smoking. In its natural state, radon can be found in a variety of environments, including soil, water, and rocks and can accumulate in the air inside a building if allowed to enter through cracks or holes. Radon is responsible for roughly 21,000 cases of lung cancer per year, according to the Environmental Protection Agency (EPA) in the United States. Radon exposure increases the risk of lung cancer, but the risk is greatest for smokers. The Environmental Protection Agency (EPA) has found that more than 10% of radon-related lung cancer deaths occur in adults who have never smoked cigarettes [524]. Asbestos, arsenic, diesel exhaust, and some types of silica and chromium are all examples of pollutants found in some workplaces that raise lung cancer risk [524]. On the other hand, 36,000 cases were attributed to the use of solid fuels for cooking and heating, which is more common in developing countries [525, 526], Many carcinogens, including benzene, carbon monoxide, formaldehyde, and PAHs, can be released into the air when coal is burned for house heating or cooking. Indoor coal combustion is a leading cause of cancer in humans, classified as a Group 1 carcinogen [527]. Furthermore, an estimated 21,000 lung cancer cases were attributed to passive smoking [528]. Exposure to secondhand tobacco smoke or e-cigarette aerosol has been linked to an increased risk of developing lung cancer [529].

Concluding Remarks

Environmental and lifestyle factors have been recognized as primary contributors to about 50% of the worldwide cancer burden (IARC), and to significantly modify most of the remaining cancer burden (IARC) [530]. Carcinogens are substances or environmental elements that cause cancer in humans by chemical, physical, or biological interactions [531]. In fact, a substance's carcinogenicity status does not guarantee that it will cause cancer in humans. The likelihood that someone exposed to a carcinogen would get cancer depends on many factors, such as the extent and length of the exposure and the person's own genetic susceptibility. Subpopulations, such as those who work in industries where they may be exposed to carcinogens, face a higher risk of developing cancers. Genetic and epigenetic studies have shown that environmental stressors damage DNA or modify the epigenome in ways that allow cells to escape normal regulatory control and to develop tumors [532]. The lens of precision medicine is crucial to the development of genetic and epigenetic based therapies in the future. Epigenetic changes are among the most often identified cancer abnormalities and these alterations are in principle amenable to pharmacological intervention. Understanding the interaction between these molecular changes may help determine which mutations produce exploitable vulnerabilities.

References

1. GBD 2016 Causes of Death Collaborators. Global, regional, and national age-sex specific mortality for 264 causes of death, 1980–2016: a systematic analysis for the Global Burden of Disease Study 2016. Lancet (London, England). 2017;390(10100):1151–210. https://doi.org/10.1016/S0140-6736(17)32152-9. www.thelancet.com.

2. Alberts B, Bray D, Johnson A, Lewis J, Raff M, Roberts K, Walter P. Essential cell biology. 4th ed. W.W. Norton & Company; 2013. https://doi.org/10.1201/9781315815015.
3. Pitot HC, Goldsworthy T, Moran S. The natural history of carcinogenesis: implications of experimental carcinogenesis in the genesis of human cancer. J Supramol Struct Cell Biochem. 1981;7:133–46.
4. Stewart BW. Mechanisms of carcinogenesis: from initiation and promotion to the hallmarks. In: Baan RA, Stewart BW, Straif K, editors. Tumour site concordance and mechanisms of carcinogenesis. International Agency for Research on Cancer; 2019. http://www.ncbi.nlm.nih.gov/books/NBK570326/.
5. Dixon K, Kopras E. Genetic alterations and DNA repair in human carcinogenesis. Semin Cancer Biol. 2004;14(6):441–8. https://doi.org/10.1016/j.semcancer.2004.06.007.
6. Compton C. Cancer initiation, promotion, and progression and the acquisition of key behavioral traits. In: Compton C, editor. Cancer: the enemy from within. Springer International Publishing; 2020. p. 25–48. https://doi.org/10.1007/978-3-030-40651-6_2.
7. Siddiqui IA, Sanna V, Ahmad N, Sechi M, Mukhtar H. Resveratrol nanoformulation for cancer prevention and therapy: resveratrol nanoformulations for cancer. Ann N Y Acad Sci. 2015;1348(1):20–31. https://doi.org/10.1111/nyas.12811.
8. Pomerantz MM, Freedman ML. The genetics of cancer risk. Cancer J. 2011;17(6):416–22.
9. Reid BM, Permuth JB, Sellers TA. Epidemiology of ovarian cancer: a review. Cancer Biol Med. 2017a;14(1):9–32. https://doi.org/10.20892/j.issn.2095-3941.2016.0084.
10. Reid BM, Permuth JB, Sellers TA. Epidemiology of ovarian cancer: a review. Pathologic classification of OC. Cancer Biol Med. 2017b;14:9–32.
11. Katzke VA, Kaaks R, Kühn T. Lifestyle and cancer risk. Cancer J. 2015;21(2):104–10. https://doi.org/10.1097/PPO.0000000000000101.
12. Haile ZT, Kingori C, Chavan B, Francescon J, Teweldeberhan AK. Association between risky sexual behavior and cervical cancer screening among women in Kenya: a population-based study. J Community Health. 2018;43(2):238–47. https://doi.org/10.1007/s10900-017-0410-z.
13. Hayes RB, Pottern LM, Strickler H, Rabkin C, Pope V, Swanson GM, Greenberg RS, Schoenberg JB, Liff J, Schwartz AG, Hoover RN, Fraumeni JF. Sexual behaviour, STDs and risks for prostate cancer. Br J Cancer. 2000;82(3):718–25. https://doi.org/10.1054/bjoc.1999.0986.
14. Farvid MS, Sidahmed E, Spence ND, Mante Angua K, Rosner BA, Barnett JB. Consumption of red meat and processed meat and cancer incidence: a systematic review and meta-analysis of prospective studies. Eur J Epidemiol. 2021;36(9):937–51. https://doi.org/10.1007/s10654-021-00741-9.
15. Bagot RC, Meaney MJ. Epigenetics and the biological basis of gene × environment interactions. J Am Acad Child Adolesc Psychiatry. 2010;49(8):752–71. https://doi.org/10.1016/j.jaac.2010.06.001.
16. Kopp TI, Vogel U, Andersen V. Associations between common polymorphisms in CYP2R1 and GC, vitamin D intake and risk of colorectal cancer in a prospective case-cohort study in Danes. PLoS One. 2020;15(2):e0228635. https://doi.org/10.1371/journal.pone.0228635.
17. Chen K-C, Tsai S-W, Shie R-H, Zeng C, Yang H-Y. Indoor air pollution increases the risk of lung cancer. Int J Environ Res Public Health. 2022;19(3):1164. https://doi.org/10.3390/ijerph19031164.
18. Turner MC, Andersen ZJ, Baccarelli A, Diver WR, Gapstur SM, Pope CA, Prada D, Samet J, Thurston G, Cohen A. Outdoor air pollution and cancer: an overview of the current evidence and public health recommendations. CA Cancer J Clin. 2020;70(6):460–79. https://doi.org/10.3322/caac.21632.
19. Parsa N. Telomerase: from aging to human cancers. In: Mehdipour P, editor. Telomere territory and cancer. Springer; 2013. p. 1–28. https://doi.org/10.1007/978-94-007-4632-9_1.
20. Sankpal UT, Pius H, Khan M, Shukoor MI, Maliakal P, Lee CM, Abdelrahim M, Connelly SF, Basha R. Environmental factors in causing human cancers: emphasis on tumorigenesis. Tumor Biol. 2012;33(5):1265–74. https://doi.org/10.1007/s13277-012-0413-4.

21. Filho APDR, Silveira MAD, Demarco NR, D'Arce LPG. Increased DNA damage, instability and cytokinesis defects in occupationally exposed car painters. In Vivo. 2019;33(6):1807–11. https://doi.org/10.21873/invivo.11672.
22. ACS. Cancer Facts & Figures. Atlanta: American Cancer Society; 2016.
23. Hobbs GA, Der CJ, Rossman KL. RAS isoforms and mutations in cancer at a glance. J Cell Sci. 2016;jcs.182873 https://doi.org/10.1242/jcs.182873.
24. Martín-Hernández J, Sørensen AB, Pedersen FS. Murine Leukemia virus Proviral insertions between the N- *ras* and *unr* genes in B-cell lymphoma DNA affect the expression of N- *ras* only. J Virol. 2001;75(23):11907–12. https://doi.org/10.1128/JVI.75.23.11907-11912.2001.
25. Mikkers H, Allen J, Berns A. Proviral activation of the tumor suppressor E2a contributes to T cell lymphomagenesis in EμMyc transgenic mice. Oncogene. 2002a;21:6559–66.
26. Mikkers H, Allen J, Knipscheer P, Romeyn L, Hart A, Vink E, Berns A. High-throughput retroviral tagging to identify components of specific signaling pathways in cancer. Nat Genet. 2002b;32(1):153–9. https://doi.org/10.1038/ng950.
27. Naab TJ, Gautam A, Ricks-Santi L, Esnakula AK, Kanaan YM, DeWitty RL, Asgedom G, Makambi KH, Abawi M, Blancato JK. MYC amplification in subtypes of breast cancers in African American women. BMC Cancer. 2018;18(1):274. https://doi.org/10.1186/s12885-018-4171-6.
28. Voronova AF, Adler HT, Sefton BM. Two lck transcripts containing different 5′ untranslated regions are present in T cells. Mol Cell Biol. 1987;7(12):4407–13. https://doi.org/10.1128/mcb.7.12.4407-4413.1987.
29. Dinu D, Dobre M, Panaitescu E, Bîrlă R, Iosif C, Hoara P, Caragui A, Boeriu M, Constantinoiu S, Ardeleanu C. Prognostic significance of KRAS gene mutations in colorectal cancer—preliminary study. J Med Life. 2014;7(4):581–7.
30. Cao J, Li D. Searching for human oncoviruses: histories, challenges, and opportunities. J Cell Biochem. 2018;119(6):4897–906. https://doi.org/10.1002/jcb.26717.
31. Chaturvedi AK, Engels EA, Pfeiffer RM, Hernandez BY, Xiao W, Kim E, Jiang B, Goodman MT, Sibug-Saber M, Cozen W, Liu L, Lynch CF, Wentzensen N, Jordan RC, Altekruse S, Anderson WF, Rosenberg PS, Gillison ML. Human papillomavirus and rising oropharyngeal cancer incidence in the United States. J Clin Oncol. 2011;29(32):4294–301. https://doi.org/10.1200/JCO.2011.36.4596.
32. Chen C, Xun P, Nishijo M, Sekikawa A, He K. Cadmium exposure and risk of pancreatic cancer: a meta-analysis of prospective cohort studies and case–control studies among individuals without occupational exposure history. Environ Sci Pollut Res. 2015a;22(22):17465–74. https://doi.org/10.1007/s11356-015-5464-9.
33. Chen H, Chen X-Z, Waterboer T, Castro FA, Brenner H. Viral infections and colorectal cancer: a systematic review of epidemiological studies: viral infections and colorectal cancer. Int J Cancer. 2015b;137(1):12–24. https://doi.org/10.1002/ijc.29180.
34. Chen K, Liao QL, Ma ZW, Jin Y, Hua M, Bi J, Huang L. Association of soil arsenic and nickel exposure with cancer mortality rates, a town-scale ecological study in Suzhou China. Environ Sci Pollut Res. 2015c;22(7):5395–404. https://doi.org/10.1007/s11356-014-3790-y.
35. Hussein HAM, Okafor IB, Walker LR, Abdel-Raouf UM, Akula SM. Cellular and viral oncogenes: the key to unlocking unknowns of Kaposi's sarcoma-associated herpesvirus pathogenesis. Arch Virol. 2018;163(10):2633–43. https://doi.org/10.1007/s00705-018-3918-3.
36. Lau L, Gray EE, Brunette RL, Stetson DB. DNA tumor virus oncogenes antagonize the cGAS-STING DNA-sensing pathway. Science. 2015;350(6260):568–71. https://doi.org/10.1126/science.aab3291.
37. Yeo-Teh N, Ito Y, Jha S. High-risk human papillomaviral oncogenes E6 and E7 target Key cellular pathways to achieve oncogenesis. Int J Mol Sci. 2018;19(6):1706. https://doi.org/10.3390/ijms19061706.
38. Yuan H, Krawczyk E, Blancato J, Albanese C, Zhou D, Wang N, Paul S, Alkhilaiwi F, Palechor-Ceron N, Dakic A, Fang S, Choudhary S, Hou T-W, Zheng Y-L, Haddad BR, Usuda Y, Hartmann D, Symer D, Gillison M, et al. HPV positive neuroendocrine cervical cancer

cells are dependent on Myc but not E6/E7 viral oncogenes. Sci Rep. 2017;7(1):45617. https://doi.org/10.1038/srep45617.
39. Oliveira AM, Ross JS, Fletcher JA. Tumor suppressor genes in breast cancer. Pathol Patterns Rev. 2005;124(suppl_1):S16–28. https://doi.org/10.1309/5XW3L8LU445QWGQR.
40. Dunford A, Weinstock DM, Savova V, Schumacher SE, Cleary JP, Yoda A, Sullivan TJ, Hess JM, Gimelbrant AA, Beroukhim R, Lawrence MS, Getz G, Lane AA. Tumor-suppressor genes that escape from X-inactivation contribute to cancer sex bias. Nat Genet. 2017;49(1):10–6. https://doi.org/10.1038/ng.3726.
41. Gao W, Li W, Xiao T, Liu XS, Kaelin WG. Inactivation of the PBRM1 tumor suppressor gene amplifies the HIF-response in VHL $^{-/-}$ clear cell renal carcinoma. Proc Natl Acad Sci. 2017;114(5):1027–32. https://doi.org/10.1073/pnas.1619726114.
42. Li L, Xu J, Qiu G, Ying J, Du Z, Xiang T, Wong KY, Srivastava G, Zhu X-F, Mok TS, Chan AT, Chan FK, Ambinder RF, Tao Q. Epigenomic characterization of a p53-regulated 3p22.2 tumor suppressor that inhibits STAT3 phosphorylation via protein docking and is frequently methylated in esophageal and other carcinomas. Theranostics. 2018;8(1):61–77. https://doi.org/10.7150/thno.20893.
43. Luchini C, Veronese N, Yachida S, Cheng L, Nottegar A, Stubbs B, Solmi M, Capelli P, Pea A, Barbareschi M, Fassan M, Wood LD, Scarpa A. Different prognostic roles of tumor suppressor gene *BAP1* in cancer: a systematic review with meta-analysis: different prognostic roles of bap1 in cancer. Genes Chromosom Cancer. 2016;55(10):741–9. https://doi.org/10.1002/gcc.22381.
44. Roa I, de Toro G, Fernández F, Game A, Muñoz S, de Aretxabala X, Javle M. Inactivation of tumor suppressor gene pten in early and advanced gallbladder cancer. Diagn Pathol. 2015;10(1):148. https://doi.org/10.1186/s13000-015-0381-2.
45. Alsop K, Fereday S, Meldrum C, deFazio A, Emmanuel C, George J, Dobrovic A, Birrer MJ, Webb PM, Stewart C, Friedlander M, Fox S, Bowtell D, Mitchell G. *BRCA* mutation frequency and patterns of treatment response in *BRCA* mutation–positive women with ovarian cancer: a report from the Australian Ovarian Cancer Study Group. J Clin Oncol. 2012;30(21):2654–63. https://doi.org/10.1200/JCO.2011.39.8545.
46. Couch FJ, Farid LM, DeShano ML, Tavtigian SV, Calzone K, Campeau L, Peng Y, Bogden B, Chen Q, Neuhausen S, Shattuck-Eidens D, Godwin AK, Daly M, Radford DM, Sedlacek S, Rommens J, Simard J, Garber J, Merajver S, Weber BL. BRCA2 germline mutations in male breast cancer cases and breast cancer families. Nat Genet. 1996;13(1):123–5. https://doi.org/10.1038/ng0596-123.
47. Easton D, Thompson D, McGuffog L, et al. Breast cancer linkage consortium cancer risks in BRCA2 mutation carriers. J Natl Cancer Inst. 1999;91:1310.
48. Tai YC, Domchek S, Parmigiani G, Chen S. Breast cancer risk among male BRCA1 and BRCA2 mutation carriers. JNCI J Natl Cancer Inst. 2007;99(23):1811–4. https://doi.org/10.1093/jnci/djm203.
49. Williams AB, Schumacher B. P53 in the DNA-damage-repair process. Cold Spring Harb Perspect Med. 2016;6(5):a026070. https://doi.org/10.1101/cshperspect.a026070.
50. Aubrey BJ, Kelly GL, Janic A, Herold MJ, Strasser A. How does p53 induce apoptosis and how does this relate to p53-mediated tumour suppression? Cell Death Different. 2018;25(1):104–13. https://doi.org/10.1038/cdd.2017.169.
51. Kastenhuber ER, Lowe SW. Putting p53 in context. Cell. 2017;170(6):1062–78. https://doi.org/10.1016/j.cell.2017.08.028.
52. Levine AJ. p53: 800 million years of evolution and 40 years of discovery. Nat Rev Cancer. 2020;20(8):471–80. https://doi.org/10.1038/s41568-020-0262-1.
53. Oren M. p53: not just a tumor suppressor. J Mol Cell Biol. 2019;11(7):539–43. https://doi.org/10.1093/jmcb/mjz070.
54. Heinen CD. Mismatch repair defects and Lynch syndrome: the role of the basic scientist in the battle against cancer. DNA Repair. 2016;38:127–34. https://doi.org/10.1016/j.dnarep.2015.11.025.

55. Pećina-Šlaus N, Kafka A, Salamon I, Bukovac A. Mismatch repair pathway, genome stability and cancer. Front Mol Biosci. 2020;7:122. https://doi.org/10.3389/fmolb.2020.00122.
56. Yamazaki D, Hashizume O, Taniguchi S, Funato Y, Miki H. Role of adenomatous polyposis coli in proliferation and differentiation of colon epithelial cells in organoid culture. Sci Rep. 2021;11(1):3980. https://doi.org/10.1038/s41598-021-83590-6.
57. Groden J, Thliveris A, Samowitz W, Carlson M, Gelbert L, Albertsen H, Joslyn G, Stevens J, Spirio L, Robertson M, Sargeant L, Krapcho K, Wolff E, Burt R, Hughes JP, Warrington J, McPherson J, Wasmuth J, Le Paslier D, et al. Identification and characterization of the familial adenomatous polyposis coli gene. Cell. 1991;66(3):589–600. https://doi.org/10.1016/0092-8674(81)90021-0.
58. Nishisho I, Nakamura Y, Miyoshi Y, Miki Y, Ando H, Horii A, Koyama K, Utsunomiya J, Baba S, Hedge P, Markham A, Krush AJ, Petersen G, Hamilton SR, Nilbert MC, Levy DB, Bryan TM, Preisinger AC, Smith KJ, et al. Mutations of chromosome 5q21 genes in FAP and colorectal cancer patients. Science. 1991;253(5020):665–9. https://doi.org/10.1126/science.1651563.
59. Iqbal N, Iqbal N. Human epidermal growth factor receptor 2 (HER2) in cancers: overexpression and therapeutic implications. Mol Biol Int. 2014;2014:1–9. https://doi.org/10.1155/2014/852748.
60. Gerson JN, Skariah S, Denlinger CS, Astsaturov I. Perspectives of HER2-targeting in gastric and esophageal cancer. Expert Opin Investig Drugs. 2017;26(5):531–40. https://doi.org/10.1080/13543784.2017.1315406.
61. Kang Z-J, Liu Y-F, Xu L-Z, Long Z-J, Huang D, Yang Y, Liu B, Feng J-X, Pan Y-J, Yan J-S, Liu Q. The Philadelphia chromosome in leukemogenesis. Chin J Cancer. 2016;35(1):48. https://doi.org/10.1186/s40880-016-0108-0.
62. Ess SM, Herrmann C, Frick H, Krapf M, Cerny T, Jochum W, Früh M. Epidermal growth factor receptor and anaplastic lymphoma kinase testing and mutation prevalence in patients with advanced non-small cell lung cancer in Switzerland: a comprehensive evaluation of real world practices. Eur J Cancer Care. 2017;26(6):e12721. https://doi.org/10.1111/ecc.12721.
63. Eggert A, Grotzer MA, Ikegaki N, Liu X, Evans AE, Brodeur GM. Expression of the neurotrophin receptor TrkA down-regulates expression and function of angiogenic stimulators in SH-SY5Y neuroblastoma cells. Cancer Res. 2002;62(6):1802–8.
64. Rajan N, Elliott R, Clewes O, Mackay A, Reis-Filho JS, Burn J, Langtry J, Sieber-Blum M, Lord CJ, Ashworth A. Dysregulated TRK signalling is a therapeutic target in CYLD defective tumours. Oncogene. 2011;30(41):4243–60. https://doi.org/10.1038/onc.2011.133.
65. Gatalica Z, Xiu J, Swensen J, Vranic S. Molecular characterization of cancers with NTRK gene fusions. Mod Pathol. 2019;32(1):147–53. https://doi.org/10.1038/s41379-018-0118-3.
66. Albert CM, Davis JL, Federman N, Casanova M, Laetsch TW. TRK fusion cancers in children: a clinical review and recommendations for screening. J Clin Oncol. 2019;37(6):513–24. https://doi.org/10.1200/JCO.18.00573.
67. Federman N, McDermott R. Larotrectinib, a highly selective tropomyosin receptor kinase (TRK) inhibitor for the treatment of TRK fusion cancer. Expert Rev Clin Pharmacol. 2019;12(10):931–9. https://doi.org/10.1080/17512433.2019.1661775.
68. Penault-Llorca F, Rudzinski ER, Sepulveda AR. Testing algorithm for identification of patients with TRK fusion cancer. J Clin Pathol. 2019;72(7):460–7. https://doi.org/10.1136/jclinpath-2018-205679.
69. Solomon JP, Linkov I, Rosado A, Mullaney K, Rosen EY, Frosina D, Jungbluth AA, Zehir A, Benayed R, Drilon A, Hyman DM, Ladanyi M, Sireci AN, Hechtman JF. NTRK fusion detection across multiple assays and 33,997 cases: diagnostic implications and pitfalls. Mod Pathol. 2020;33(1):38–46. https://doi.org/10.1038/s41379-019-0324-7.
70. Lassen U, Albert CM, Kummar S, et al. Larotrectinib efficacy and safety in TRK fusion cancer: an expanded clinical dataset showing consistency in an age and tumor agnostic approach. Ann Oncol. 2018;29(Suppl. 9):ix23–7.

71. Cheng L, Lopez-Beltran A, Massari F, MacLennan GT, Montironi R. Molecular testing for BRAF mutations to inform melanoma treatment decisions: a move toward precision medicine. Mod Pathol. 2018;31(1):24–38. https://doi.org/10.1038/modpathol.2017.104.
72. Alqathama A. BRAF in malignant melanoma progression and metastasis: potentials and challenges. Am J Cancer Res. 2020a;10(4):1103–14.
73. D'Achille P, Seymour JF, Campbell LJ. Translocation (14;18)(q32;q21) in acute lymphoblastic leukemia: a study of 12 cases and review of the literature. Cancer Genet Cytogenet. 2006;171(1):52–6. https://doi.org/10.1016/j.cancergencyto.2006.07.005.
74. Tsai AG, Lu H, Raghavan SC, Muschen M, Hsieh C-L, Lieber MR. Human chromosomal translocations at CpG sites and a theoretical basis for their lineage and stage specificity. Cell. 2008;135(6):1130–42. https://doi.org/10.1016/j.cell.2008.10.035.
75. Gu K, Chan WC, Hawley RC. Practical Detection of t(14;18)(*IgH/BCL2*) in Follicular Lymphoma. Arch Pathol Lab Med. 2008;132(8):1355–61. https://doi.org/10.5858/2008-132-1355-PDOBIF.
76. Osborne CS. Molecular pathways: transcription factories and chromosomal translocations. Clin Cancer Res. 2014;20(2):296–300. https://doi.org/10.1158/1078-0432.CCR-12-3667.
77. Banerji S, Cibulskis K, Rangel-Escareno C, Brown KK, Carter SL, Frederick AM, Lawrence MS, Sivachenko AY, Sougnez C, Zou L, Cortes ML, Fernandez-Lopez JC, Peng S, Ardlie KG, Auclair D, Bautista-Piña V, Duke F, Francis J, Jung J, et al. Sequence analysis of mutations and translocations across breast cancer subtypes. Nature. 2012;486(7403):405–9. https://doi.org/10.1038/nature11154.
78. Lawson ARJ, Hindley GFL, Forshew T, Tatevossian RG, Jamie GA, Kelly GP, Neale GA, Ma J, Jones TA, Ellison DW, Sheer D. *RAF* gene fusion breakpoints in pediatric brain tumors are characterized by significant enrichment of sequence microhomology. Genome Res. 2011;21(4):505–14. https://doi.org/10.1101/gr.115782.110.
79. Dalla-Favera R, Bregni M, Erikson J, Patterson D, Gallo RC, Croce CM. Human c-myc onc gene is located on the region of chromosome 8 that is translocated in Burkitt lymphoma cells. Proc Natl Acad Sci. 1982;79(24):7824–7. https://doi.org/10.1073/pnas.79.24.7824.
80. Simonds NI, Ghazarian AA, Pimentel CB, Schully SD, Ellison GL, Gillanders EM, Mechanic LE. Review of the gene-environment interaction literature in cancer: what do we know?: gene-environment interaction literature review. Genet Epidemiol. 2016;40(5):356–65. https://doi.org/10.1002/gepi.21967.
81. Anand P, Kunnumakara AB, Sundaram C, Harikumar KB, Tharakan ST, Lai OS, Sung B, Aggarwal BB. Cancer is a preventable disease that requires major lifestyle changes. Pharm Res. 2008;25(9):2097–116. https://doi.org/10.1007/s11095-008-9661-9.
82. Boada LD, Henríquez-Hernández LA, Navarro P, Zumbado M, Almeida-González M, Camacho M, Álvarez-León EE, Valencia-Santana JA, Luzardo OP. Exposure to polycyclic aromatic hydrocarbons (PAHs) and bladder cancer: evaluation from a gene-environment perspective in a hospital-based case-control study in the Canary Islands (Spain). Int J Occup Environ Health. 2015;21(1):23–30. https://doi.org/10.1179/2049396714Y.0000000085.
83. Boffetta P, Jourenkova N, Gustavsson P. Cancer risk from occupational and environmental exposure to polycyclic aromatic hydrocarbons. Cancer Causes Control. 1997;8(3):444–72. https://doi.org/10.1023/A:1018465507029.
84. Jameson CW. Polycyclic aromatic hydrocarbons and associated occupational exposures. In: Baan RA, Stewart BW, Straif K, editors. Tumour site concordance and mechanisms of carcinogenesis. Lyon (FR): International Agency for Research on Cancer, vol. 165). Chapter 7. (IARC Scientific Publications; 2019.
85. Harrison R, Komulainen H, et al. Polycyclic aromatic hydrocarbons. In: WHO guidelines for indoor air quality: selected pollutants. Geneva: World Health Organization; 2010. p. 6.
86. U.S. Enivronmental Protection Agencey (EPA). Asbestos' impact on indoor air quality. 2022a. https://www.epa.gov/indoor-air-quality-iaq/asbestos-impact-indoor-air-quality.
87. U.S. Enivronmental Protection Agencey (EPA). Indoor air quality (IAQ): nitrogen dioxide's impact on indoor air quality. 2022b. https://www.epa.gov/indoor-air-quality-iaq/nitrogen-dioxides-impact-indoor-air-quality.

88. U.S. Enivronmental Protection Agencey (EPA). Indoor air quality (IAQ): pesticides' impact on indoor air quality. 2022c. https://www.epa.gov/indoor-air-quality-iaq/pesticides-impact-indoor-air-quality.
89. U.S. Enivronmental Protection Agencey (EPA). Indoor air quality (IAQ): sources of indoor particulate matter (PM). 2022d. https://www.epa.gov/indoor-air-quality-iaq/sources-indoor-particulate-matter-pm.
90. U.S. Enivronmental Protection Agencey (EPA). Carbon monoxide (CO) pollution in outdoor air: basic information about carbon monoxide (CO) outdoor air pollution. 2022e. https://www.epa.gov/co-pollution/basic-information-about-carbon-monoxide-co-outdoor-air-pollution.
91. U.S. Enivronmental Protection Agencey (EPA). Sulfur dioxide (SO_2) pollution. 2022f. https://www.epa.gov/so2-pollution/sulfur-dioxide-basics.
92. U.S. Enivronmental Protection Agencey (EPA). Particle pollution and your patients' health. 2022g. https://www.epa.gov/pmcourse/what-particle-pollution.
93. U.S. Enivronmental Protection Agencey (EPA). Ozone pollution and your patients' health. 2022h. https://www.epa.gov/ozone-pollution-and-your-patients-health/what-ozone.
94. U.S. Enivronmental Protection Agencey (EPA). What is carbon monoxide? 2021. https://www.epa.gov/indoor-air-quality-iaq/what-carbon-monoxide.
95. National Institute of Environmental Health Sciences (NIH). Environmental agents: formaldehyde. 2022. https://www.niehs.nih.gov/health/topics/agents/formaldehyde/index.cfm.
96. National Institute of Environmental Health Sciences (NIH). Environmental agents: lead. 2022. https://www.niehs.nih.gov/health/topics/agents/lead/index.cfm.
97. U.S. Enivronmental Protection Agencey (EPA). Indoor air quality (IAQ): volatile organic compounds' impact on indoor air quality. 2022. https://www.epa.gov/indoor-air-quality-iaq/volatile-organic-compounds-impact-indoor-air-quality.
98. Abdel-Shafy HI, Mansour MSM. A review on polycyclic aromatichydrocarbons: source, environmental impact, effect on human health and remediation. Egypt J Pet. 2016;25(1):107–23. https://doi.org/10.1016/j.ejpe.2015.03.011.
99. Thurston GD. Outdoor air pollution: sources, atmospheric transport, and human health effects. In: International Encyclopedia of public health. Elsevier; 2017. p. 367–77. https://doi.org/10.1016/B978-0-12-803678-5.00320-9.
100. Smith J, Petersen J, Gray S. Benzene, 1,3 butadiene and other volatile organic compounds in Auckland. Prepared by NIWA for Auckland Regional Council. Auckland Regional Council Technical Report 2009/048; 2009.
101. IARC Working Group on the Evaluation of Carcinogenic Risks to Humans. Tobacco smoke and involuntary smoking, vol. 83. IARC Monographs on the Evaluation of Carcinogenic Risks to Humans/World Health Organization, International Agency for Research on Cancer; 2004. p. 1–1438.
102. Wang Z, Yang P, Xie J, Lin H-P, Kumagai K, Harkema J, Yang C. Arsenic and benzo[a]pyrene co-exposure acts synergistically in inducing cancer stem cell-like property and tumorigenesis by epigenetically down-regulating SOCS3 expression. Environ Int. 2020;137:105560. https://doi.org/10.1016/j.envint.2020.105560.
103. Soni MG. Benz[a]anthracene. In: Encyclopedia of toxicology. Elsevier; 2005. p. 250–1. https://doi.org/10.1016/B0-12-369400-0/00123-X.
104. Bassil KL, Vakil C, Sanborn M, Cole DC, Kaur JS, Kerr KJ. Cancer health effects of pesticides: systematic review. Can Fam Physician. 2007;53(10):1704–11.
105. Hall C, Heck JE, Ritz B, Cockburn M, Escobedo LA, von Ehrenstein OS. Prenatal exposure to air toxics and malignant germ cell tumors in young children. J Occupational Environ Med. 2019;61(6):529–34. https://doi.org/10.1097/JOM.0000000000001609.
106. Shrestha A, Ritz B, Wilhelm M, Qiu J, Cockburn M, Heck JE. Prenatal exposure to air toxics and risk of Wilms' tumor in 0- to 5-year-old children. J Occup Environ Med. 2014;56(6):573–8. https://doi.org/10.1097/JOM.0000000000000167.
107. Zhang Y, Chen D, Shi R, Kamijima M, Sakai K, Tian Y, Gao Y. Indoor volatile organic compounds exposures and risk of childhood acute leukemia: a case-control study in Shanghai. J

Environ Sci Health. Part A, Toxic/Hazardous Substances Environ Eng. 2021;56(2):190–8. https://doi.org/10.1080/10934529.2020.1861903.

108. Belpomme D, Irigaray P, Hardell L, Clapp R, Montagnier L, Epstein S, Sasco AJ. The multitude and diversity of environmental carcinogens. Environ Res. 2007;105:414–29.
109. Lewandowska AM, Rudzki M, Rudzki S, Lewandowski T, Laskowska B. Environmental risk factors for cancer—review paper. Ann Agric Environ Med. 2019;26(1):1–7.
110. Peterson E, De P, Nuttall R. BMI, diet and female reproductive factors as risks for thyroid cancer: a systematic review. PLoS One. 2012;7(1):e29177. https://doi.org/10.1371/journal.pone.0029177.
111. Szkaradkiewicz A. Microbes and oncogenesis. Contemporary Oncol. 2003;7(2):96–101.
112. Gushgari AJ, Halden RU. Critical review of major sources of human exposure to N-nitrosamines. Chemosphere. 2018;210:1124–36. https://doi.org/10.1016/j.chemosphere.2018.07.098.
113. Izquierdo-Pulido M, Barbour JF, Scanlan RA. N-nitrosodimethylamine in Spanish beers. Food Chem Toxicol. 1996;34(3):297–9. https://doi.org/10.1016/0278-6915(95)00116-6.
114. Iko Afé OH, Kpoclou YE, Douny C, Anihouvi VB, Igout A, Mahillon J, Hounhouigan DJ, Scippo M. Chemical hazards in smoked meat and fish. Food Sci Nutr. 2021;9(12):6903–22. https://doi.org/10.1002/fsn3.2633.
115. Dias EP, da Rocha ML, de Oliveira Carvalho MO, da Fonte de Amorim, L. M. Detection of Epstein-Barr virus in recurrent tonsillitis. Braz J Otorhinolaryngol. 2009;75(1):30–4. https://doi.org/10.1016/S1808-8694(15)30828-4.
116. Gankhuyag N, Lee K-H, Cho J-Y. The role of nitrosamine (NNK) in breast cancer carcinogenesis. J Mammary Gland Biol Neoplasia. 2017;22(3):159–70. https://doi.org/10.1007/s10911-017-9381-z.
117. Gurski RR, Schirmer CC, Kruel CR, Komlos F, Kruel CDP, Edelweiss MI. Induction of esophageal carcinogenesis by diethylnitrosamine and assessment of the promoting effect of ethanol and N-nitrosonornicotine: experimental model in mice*. Dis Esophagus. 1999;12(2):99–105. https://doi.org/10.1046/j.1442-2050.1999.00010.x.
118. Jakszyn P, González CA, Luján-Barroso L, Ros MM, Bueno-de-Mesquita HB, Roswall N, Tjønneland AM, Büchner FL, Egevad L, Overvad K, Raaschou-Nielsen O, Clavel-Chapelon F, Boutron-Ruault M-C, Touillaud MS, Chang-Claude J, Allen NE, Kiemeney LA, Key TJ, Kaaks R, et al. Red meat, dietary nitrosamines, and heme iron and risk of bladder cancer in the european prospective investigation into cancer and nutrition (EPIC). Cancer Epidemiol Biomark Prev. 2011;20(3):555–9. https://doi.org/10.1158/1055-9965.EPI-10-0971.
119. Ward MH, Pan WH, Cheng YJ, Li FH, Brinton LA, Chen CJ, Hsu MM, Chen IH, Levine PH, Yang CS, Hildesheim A. Dietary exposure to nitrite and nitrosamines and risk of nasopharyngeal carcinoma in Taiwan. Int J Cancer. 2000;86(5):603–9.
120. Chybicka A. Influence of mother's diet during pregnancy on cancer incidence in children. Prevention activities. Pol Med Rodz. 2004;6(3):1213–5.
121. González CA, Agudo A. Carcinogenesis, prevention and early detection of gastric cancer: where we are and where we should go. Int J Cancer. 2012;130(4):745–53. https://doi.org/10.1002/ijc.26430.
122. Adani G, Filippini T, Wise LA, Halldorsson TI, Blaha L, Vinceti M. Dietary intake of acrylamide and risk of breast, endometrial, and ovarian cancers: a systematic review and dose–response meta-analysis. Cancer Epidemiol Biomark Prev. 2020;29(6):1095–106. https://doi.org/10.1158/1055-9965.EPI-19-1628.
123. Atabati H, Abouhamzeh B, Abdollahifar M-A, Sadat Javadinia S, Gharibian Bajestani S, Atamaleki A, Raoofi A, Fakhri Y, Oliveira CAF, Mousavi Khaneghah A. The association between high oral intake of acrylamide and risk of breast cancer: an updated systematic review and meta-analysis. Trends Food Sci Technol. 2020;100:155–63. https://doi.org/10.1016/j.tifs.2020.04.006.
124. Er R, Aydın B, Şekeroğlu V, Atlı Şekeroğlu Z. Protective effect of argan oil on mitochondrial function and oxidative stress against acrylamide-induced liver and kidney injury in rats. Biomarkers. 2020;25(6):458–67. https://doi.org/10.1080/1354750X.2020.1797877.

125. Çebi A. Acrylamide intake, its effects on tissues and cancer. In: Acrylamide in food. Elsevier; 2016. p. 63–91. https://doi.org/10.1016/B978-0-12-802832-2.00004-8.
126. Urban M, Kavvadias D, Riedel K, Scherer G, Tricker AR. Urinary mercapturic acids and a Hemoglobin adduct for the dosimetry of acrylamide exposure in smokers and nonsmokers. Inhal Toxicol. 2006;18(10):831–9. https://doi.org/10.1080/08958370600748430.
127. IARC. Monographs on the evaluation of carcinogenic risks to humans. Volume 60: some industrial chemicals. Lyon, International Agency for Research on Cancer; 1994.
128. U.S. Environmental Protection Agency (EPA). Integrated risk information system (IRIS) on acrylamide. Washington, DC: National Center for Environmental Assessment, Office of Research and Development; 1999.
129. Schulz M, Hoffmann K, Weikert C, Nöthlings U, Schulze MB, Boeing H. Identification of a dietary pattern characterized by high-fat food choices associated with increased risk of breast cancer: the European prospective investigation into cancer and nutrition (EPIC)-Potsdam study. Br J Nutr. 2008;100(5):942–6. https://doi.org/10.1017/S0007114508966149.
130. Fazel R, Krumholz HM, Wang Y, Ross JS, Chen J, Ting HH, Shah ND, Nasir K, Einstein AJ, Nallamothu BK. Exposure to low-dose ionizing radiation from medical imaging procedures. N Engl J Med. 2009;361(9):849–57. https://doi.org/10.1056/NEJMoa0901249.
131. Barbosa-Lorenzo R, Barros-Dios JM, Ruano-Ravina A. Radon and stomach cancer. Int J Epidemiol. 2017;46(2):767–8. https://doi.org/10.1093/ije/dyx011.
132. Lorenzo-González M, Torres-Durán M, Barbosa-Lorenzo R, Provencio-Pulla M, Barros-Dios JM, Ruano-Ravina A. Radon exposure: a major cause of lung cancer. Expert Rev Respir Med. 2019;13(9):839–50. https://doi.org/10.1080/17476348.2019.1645599.
133. Oeffinger KC, Hudson MM. Long-term complications following childhood and adolescent cancer: foundations for providing risk-based health care for survivors. CA Cancer J Clin. 2004;54(4):208–36. https://doi.org/10.3322/canjclin.54.4.208.
134. de Gruijl FR. Skin cancer and solar UV radiation. Eur J Cancer. 1999;35(14):2003–9. https://doi.org/10.1016/S0959-8049(99)00283-X.
135. Postrzech K, Welz K, Kopyra J, Reich A. Impact of ultraviolet B radiation on stratum corneum. Dermatol Rev. 2010;97:185–90.
136. Batycka-Baran A, Kuczborska I, Szepietowski J. Vitamin D and malignant melanoma—review of the literature. Clin Dermatol. 2012;14(1):37–41.
137. Stawczyk M, Łakis A, Ulatowska A, Szczerkowska-Dobosz A. Evaluation and comparison of the risk of sunbathing addiction among selected population of women. Dermatol Rev. 2011;98:305–11.
138. Halliwell B, Gutteridge J. Free radicals in biology and medicine. 4th ed. Oxford: Oxford University Press; 2007.
139. Valko M, Rhodes CJ, Moncol J, Izakovic M, Mazur M. Free radicals, metals and antioxidants in oxidative stress-induced cancer. Chem Biol Interact. 2006;160(1):1–40. https://doi.org/10.1016/j.cbi.2005.12.009.
140. Valko M, Leibfritz D, Moncol J, Cronin MTD, Mazur M, Telser J. Free radicals and antioxidants in normal physiological functions and human disease. Int J Biochem Cell Biol. 2007;39(1):44–84. https://doi.org/10.1016/j.biocel.2006.07.001.
141. Jen M, Murphy M, Grant-Kels JM. Childhood melanoma. Clin Dermatol. 2009;27(6):529–36. https://doi.org/10.1016/j.clindermatol.2008.09.011.
142. Lesiak A, Słowik-Rylska M, Kozłowski W, Sysa-Jędrzejowska A, Jochymski C, Rogowski-Tylman M, Narbutt J. Epidermal proliferation and intracellular adhesion impairment as one mode of action of ultraviolet-B radiation. Adv Dermatol Allergol. 2009;XXVI(4):180–5.
143. Burger M, Catto JWF, Dalbagni G, Grossman HB, Herr H, Karakiewicz P, Kassouf W, Kiemeney LA, La Vecchia C, Shariat S, Lotan Y. Epidemiology and risk factors of urothelial bladder cancer. Eur Urol. 2013;63(2):234–41. https://doi.org/10.1016/j.eururo.2012.07.033.
144. Juutilainen J. Do electromagnetic fields enhance the effects of environmental carcinogens? Radiat Prot Dosim. 2008;132(2):228–31. https://doi.org/10.1093/rpd/ncn258.

145. Johansen C. Electromagnetic fields and health effects—epidemiologic studies of cancer, diseases of the central nervous system and arrhythmia-related heart disease. Scand J Work Environ Health. 2004;30(Suppl 1):1–30.
146. Bray F, Jemal A, Grey N, Ferlay J, Forman D. Global cancer transitions according to the human development index (2008–2030): a population-based study. Lancet Oncol. 2012;13(8):790–801. https://doi.org/10.1016/S1470-2045(12)70211-5.
147. Boniol M, Autier P. Prevalence of main cancer lifestyle risk factors in Europe in 2000. Eur J Cancer. 2010;46(14):2534–44. https://doi.org/10.1016/j.ejca.2010.07.049.
148. Dart H, Wolin KY, Colditz GA. Commentary: eight ways to prevent cancer: a framework for effective prevention messages for the public. Cancer Causes Control. 2012;23(4):601–8. https://doi.org/10.1007/s10552-012-9924-y.
149. Khan N, Afaq F, Mukhtar H. Lifestyle as risk factor for cancer: evidence from human studies. Cancer Lett. 2010;293(2):133–43. https://doi.org/10.1016/j.canlet.2009.12.013.
150. Ligibel J. Lifestyle factors in cancer survivorship. J Clin Oncol. 2012;30(30):3697–704. https://doi.org/10.1200/JCO.2012.42.0638.
151. Parkin DM. Tobacco-attributable cancer burden in the UK in 2010. Br J Cancer. 2011;105(Suppl 2):S6–13.
152. Denissenko MF, Pao A, Tang M, Pfeifer GP. Preferential formation of benzo[*a*]pyrene adducts at lung cancer mutational hotspots in *P53*. Science. 1996;274(5286):430–2. https://doi.org/10.1126/science.274.5286.430.
153. Samet JM, Gupta PC, Ray CS. Tobacco smoking and smokeless tobacco use. In: Stewart BW, Wild CP, editors. World cancer report. IARC: Lyon Cedex, France; 2014. p. 88–95.
154. IARC Working Group on the Evaluation of Carcinogenic Risks to Humans. Tobacco smoke and involuntary smoking. IARC Monographs on the Evaluation of Carcinogenic Risks to Humans/World Health Organization, International Agency for Research on Cancer. 2004;83:1–1438.
155. Anto RJ. Cigarette smoke condensate activates nuclear transcription factor-kappaB through phosphorylation and degradation of IkappaBalpha: correlation with induction of cyclooxygenase-2. Carcinogenesis. 2002;23(9):1511–8. https://doi.org/10.1093/carcin/23.9.1511.
156. Shishodia S, Aggarwal BB. Cyclooxygenase (COX)-2 inhibitor celecoxib abrogates activation of cigarette smoke-induced nuclear factor (NF)-κB by suppressing activation of I-κB α kinase in human non-small cell lung carcinoma. Cancer Res. 2004;64(14):5004–12. https://doi.org/10.1158/0008-5472.CAN-04-0206.
157. Boffetta P, Hashibe M, La Vecchia C, et al. The burden of cancer attributable to alcohol drinking. Int J Cancer. 2006;119:884–7.
158. Bagnardi V, Rota M, Botteri E, Tramacere I, Islami F, Fedirko V, Scotti L, Jenab M, Turati F, Pasquali E, Pelucchi C, Galeone C, Bellocco R, Negri E, Corrao G, Boffetta P, La Vecchia C. Alcohol consumption and site-specific cancer risk: a comprehensive dose–response meta-analysis. Br J Cancer. 2015;112(3):580–93. https://doi.org/10.1038/bjc.2014.579.
159. Maier H, Sennewald E, Heller GFW-D, Weidauer H. Chronic alcohol consumption-the Key risk factor for pharyngeal cancer. Otolaryngol Head Neck Surg. 1994;110(2):168–73. https://doi.org/10.1177/019459989411000205.
160. Seitz HK, Stickel F, Homann N. Pathogenetic mechanisms of upper aerodigestive tract cancer in alcoholics. Int J Cancer. 2004;108(4):483–7. https://doi.org/10.1002/ijc.11600.
161. Tuyns AJ. Epidemiology of alcohol and cancer. Cancer Res. 1979;39:2840–3.
162. Longnecker MP, Newcomb PA, Mittendorf R, Greenberg ER, Clapp RW, Bogdan GF, Baron J, MacMahon B, Willett WC. Risk of breast cancer in relation to lifetime alcohol consumption. JNCI j Natl Cancer Inst. 1995;87(12):923–9. https://doi.org/10.1093/jnci/87.12.923.
163. Hamajima N, et al. Alcohol, tobacco and breast cancer—collaborative reanalysis of individual data from 53 epidemiological studies, including 58,515 women with breast cancer and 95,067 women without the disease. Br J Cancer. 2002;87:1234–45.
164. Seitz HK, Pöschl G, Simanowski UA. Alcohol and cancer. In: Recent developments in alcoholism. Springer; 1998. p. 67–95. https://doi.org/10.1007/0-306-47148-5_4.

165. Stickel F, Schuppan D, Hahn EG, Seitz HK. Cocarcinogenic effects of alcohol in hepatocarcinogenesis. Gut. 2002;51(1):132–9. https://doi.org/10.1136/gut.51.1.132.
166. Szabo G, Mandrekar P, Oak S, Mayerle J. Effect of ethanol on inflammatory responses. Pancreatology. 2007;7(2–3):115–23. https://doi.org/10.1159/000104236.
167. Kuratsune M, Kohchi S, Horie A. Carcinogenesis in the Esophagus. I. Penetration of Benzo(a) Pyrene and other hydrocarbons into the Esophageal mucosa. Gan. 1965;56:177–87.
168. Hashibe M, Brennan P, Chuang S, Boccia S, Castellsague X, Chen C, Curado MP, Dal Maso L, Daudt AW, Fabianova E, Fernandez L, Wünsch-Filho V, Franceschi S, Hayes RB, Herrero R, Kelsey K, Koifman S, La Vecchia C, Lazarus P, et al. Interaction between tobacco and alcohol use and the risk of head and neck cancer: pooled analysis in the international head and neck cancer epidemiology consortium. Cancer Epidemiol Biomark Prev. 2009;18(2):541–50. https://doi.org/10.1158/1055-9965.EPI-08-0347.
169. Bingham SA, Hughes R, Cross AJ. Effect of White versus red meat on endogenous N-Nitrosation in the human colon and Further evidence of a dose response. J Nutr. 2002;132(11):3522S–5S. https://doi.org/10.1093/jn/132.11.3522S.
170. Chao A. Meat consumption and risk of colorectal cancer. JAMA. 2005;293(2):172. https://doi.org/10.1001/jama.293.2.172.
171. Hogg N. Red meat and colon cancer: heme proteins and nitrite in the gut. A commentary on "diet-induced endogenous formation of nitroso compounds in the GI tract". Free Radic Biol Med. 2007;43(7):1037–9. https://doi.org/10.1016/j.freeradbiomed.2007.07.006.
172. Rodriguez C, McCullough ML, Mondul AM, Jacobs EJ, Chao A, Patel AV, Thun MJ, Calle EE. Meat consumption among black and white men and risk of prostate cancer in the cancer prevention study II nutrition cohort. Cancer Epidemiol Biomark Prev. 2006;15(2):211–6. https://doi.org/10.1158/1055-9965.EPI-05-0614.
173. García-Closas R, García-Closas M, Kogevinas M, Malats N, Silverman D, Serra C, Tardón A, Carrato A, Castaño-Vinyals G, Dosemeci M, Moore L, Rothman N, Sinha R. Food, nutrient and heterocyclic amine intake and the risk of bladder cancer. Eur J Cancer. 2007;43(11):1731–40. https://doi.org/10.1016/j.ejca.2007.05.007.
174. Tappel A. Heme of consumed red meat can act as a catalyst of oxidative damage and could initiate colon, breast and prostate cancers, heart disease and other diseases. Med Hypotheses. 2007;68(3):562–4. https://doi.org/10.1016/j.mehy.2006.08.025.
175. O'Hanlon LH. High meat consumption linked to gastric-cancer risk. Lancet Oncol. 2006;7(4):287. https://doi.org/10.1016/S1470-2045(06)70638-6.
176. Toporcov TN, Antunes JLF, Tavares MR. Fat food habitual intake and risk of oral cancer. Oral Oncol. 2004;40(9):925–31. https://doi.org/10.1016/j.oraloncology.2004.04.007.
177. Lauber SN, Gooderham NJ. The cooked meat–derived genotoxic carcinogen 2-amino-3-methylimidazo[4,5-*b*] pyridine has potent hormone-like activity: mechanistic support for a role in breast cancer. Cancer Res. 2007;67(19):9597–602. https://doi.org/10.1158/0008-5472.CAN-07-1661.
178. Sasaki YF, Kawaguchi S, Kamaya A, Ohshita M, Kabasawa K, Iwama K, Taniguchi K, Tsuda S. The comet assay with 8 mouse organs: results with 39 currently used food additives. Mutation Research/Genetic Toxicology and Environmental Mutagenesis. 2002;519(1–2):103–19. https://doi.org/10.1016/S1383-5718(02)00128-6.
179. Durando M, Kass L, Piva J, Sonnenschein C, Soto AM, Luque EH, Munoz-de-Toro M. Prenatal bisphenol A exposure induces preneoplastic lesions in the mammary gland in Wistar rats. Environ Health Perspect. 2007;115:80–6.
180. Ho SM, et al. Developmental exposure to estradiol and bisphenol A increases susceptibility to prostate carcinogenesis and epigenetically regulates phosphodiesterase type 4 variant. Cancer Res. 2006;66:5624–32.
181. Szymańska-Chabowska A, Antonowicz-Juchniewicz J, Andrzejak R. Some aspects of arsenic toxicity and carcinogenicity in living organism with special regard to its influence on cardiovascular system, blood and bone marrow. Int J Occup Med Environ Health. 2002;15(2):101–16.

182. Dhaka V, Gulia N, Ahlawat KS, Khatkar BS. Trans fats—sources, health risks and alternative approach—A review. J Food Sci Technol. 2011;48(5):534–41. https://doi.org/10.1007/s13197-010-0225-8.
183. Eyre H, Kahn R, Robertson RM, the ACS/ADA/AHA Collaborative Writing Committee, Clark NG, Doyle C, Gansler T, Glynn T, Hong Y, Smith RA, Taubert K, Thun MJ. Preventing cancer, cardiovascular disease, and diabetes: a common agenda for the American Cancer Society, the American Diabetes Association, and the American Heart Association. CA Cancer J Clin. 2004;54(4):190–207. https://doi.org/10.3322/canjclin.54.4.190.
184. Arnold M, Pandeya N, Byrnes G, Renehan AG, Stevens GA, Ezzati M, Ferlay J, Miranda JJ, Romieu I, Dikshit R, Forman D, Soerjomataram I. Global burden of cancer attributable to high body-mass index in 2012: a population-based study. Lancet Oncol. 2015;16(1):36–46. https://doi.org/10.1016/S1470-2045(14)71123-4.
185. Islami F, Goding Sauer A, Gapstur SM, Jemal A. Proportion of cancer cases attributable to excess body weight by US state, 2011-2015. JAMA Oncol. 2019;5(3):384. https://doi.org/10.1001/jamaoncol.2018.5639.
186. Division of Cancer Prevention and Control, Centers for Disease Control and Prevention (CDC). 2021. https://www.cdc.gov/cancer/obesity/index.htm.
187. Bhaskaran K, Douglas I, Forbes H, dos-Santos-Silva I, Leon DA, Smeeth L. Body-mass index and risk of 22 specific cancers: a population-based cohort study of 5·24 million UK adults. Lancet. 2014;384(9945):755–65. https://doi.org/10.1016/S0140-6736(14)60892-8.
188. Hursting D, Lashinger LM, Colbert LH, Rogers CJ, Wheatley KW, Nunez NP, Mahabir S, Barrett JC, Forman MR, Perkins SN. Energy balance and carcinogenesis: underlying pathways and targets for intervention. Curr Cancer Drug Targets. 2007;7:484–91.
189. Mc Tiernan A, Friedenreich CM, Katzmarzyk PT, Powell KE, Macko R, Buchner D, Pescatello LS, Bloodgood B, Tennant B, Vaux-Bjerke A, George SM, Troiano RP, Piercy KL. Physical activity in cancer prevention and survival: a systematic review. Med Sci Sports Exerc. 2019;51(6):1252–61. https://doi.org/10.1249/MSS.0000000000001937.
190. Patel AV, Friedenreich CM, Moore SC, Hayes SC, Silver JK, Campbell KL, Winters-Stone K, Gerber LH, George SM, Fulton JE, Denlinger C, Morris GS, Hue T, Schmitz KH, Matthews CE. American College of Sports Medicine roundtable report on physical activity, sedentary behavior, and cancer prevention and control. Med Sci Sports Exerc. 2019;51(11):2391–402. https://doi.org/10.1249/MSS.0000000000002117.
191. de Rezende LFM, de Sá TH, Markozannes G, Rey-López JP, Lee I-M, Tsilidis KK, Ioannidis JPA, Eluf-Neto J. Physical activity and cancer: an umbrella review of the literature including 22 major anatomical sites and 770 000 cancer cases. Br J Sports Med. 2018;52(13):826–33. https://doi.org/10.1136/bjsports-2017-098391.
192. Behrens G, Leitzmann MF. The association between physical activity and renal cancer: systematic review and meta-analysis. Br J Cancer. 2013;108(4):798–811. https://doi.org/10.1038/bjc.2013.37.
193. Clague J, Bernstein L. Physical activity and cancer. Curr Oncol Rep. 2012;14(6):550–8. https://doi.org/10.1007/s11912-012-0265-5.
194. Moore SC, Gierach GL, Schatzkin A, Matthews CE. Physical activity, sedentary behaviours, and the prevention of endometrial cancer. Br J Cancer. 2010;103(7):933–8. https://doi.org/10.1038/sj.bjc.6605902.
195. Singh S, Devanna S, Edakkanambeth Varayil J, Murad MH, Iyer PG. Physical activity is associated with reduced risk of esophageal cancer, particularly esophageal adenocarcinoma: a systematic review and meta-analysis. BMC Gastroenterol. 2014;14(1):101. https://doi.org/10.1186/1471-230X-14-101.
196. Behrens G, Jochem C, Keimling M, Ricci C, Schmid D, Leitzmann MF. The association between physical activity and gastroesophageal cancer: systematic review and meta-analysis. Eur J Epidemiol. 2014;29(3):151–70. https://doi.org/10.1007/s10654-014-9895-2.
197. Keimling M, Behrens G, Schmid D, Jochem C, Leitzmann MF. The association between physical activity and bladder cancer: systematic review and meta-analysis. Br J Cancer. 2014;110(7):1862–70. https://doi.org/10.1038/bjc.2014.77.

198. Byers T. Physical activity and gastric cancer: so what? An epidemiologist's confession. Cancer Prev Res. 2014;7(1):9–11. https://doi.org/10.1158/1940-6207.CAPR-13-0400.
199. McTiernan A. Mechanisms linking physical activity with cancer. Nat Rev Cancer. 2008;8(3):205–11. https://doi.org/10.1038/nrc2325.
200. Winzer BM, Whiteman DC, Reeves MM, Paratz JD. Physical activity and cancer prevention: a systematic review of clinical trials. Cancer Causes Control. 2011;22(6):811–26. https://doi.org/10.1007/s10552-011-9761-4.
201. de Martel C, Ferlay J, Franceschi S, Vignat J, Bray F, Forman D, Plummer M. Global burden of cancers attributable to infections in 2008: a review and synthetic analysis. Lancet Oncol. 2012;13(6):607–15. https://doi.org/10.1016/S1470-2045(12)70137-7.
202. Parkin DM. The global health burden of infection-associated cancers in the year 2002. Int J Cancer. 2006;118(12):3030–44. https://doi.org/10.1002/ijc.21731.
203. Pisani P, Parkin DM, Munoz N, Ferlay J. Cancer and infection: estimates of the attributable fraction in 1990. Cancer Epidemiol Biomark Prev. 1997;6:387–400.
204. Fung TT, Hu FB, Hankinson SE, Willett WC, Holmes MD. Low-carbohydrate diets, dietary approaches to Stop hypertension-style diets, and the risk of postmenopausal breast cancer. Am J Epidemiol. 2011;174(6):652–60. https://doi.org/10.1093/aje/kwr148.
205. O'Mahony M, Hegarty J. Help seeking for cancer symptoms: a review of the literature. Oncol Nurs Forum. 2009;36(4):E178–84. https://doi.org/10.1188/09.ONF.E178-E184.
206. Podlodowska J, Szumiło J, Podlodowski W, Starosławska E, Burdan F. Epidemiology and risk factors of the oral carcinoma. Polski Merkuriusz Lekarski: Organ Polskiego Towarzystwa Lekarskiego. 2012;32(188):135–7.
207. Venook AP, Papandreou C, Furuse J, Ladrón de Guevara L. The incidence and epidemiology of hepatocellular carcinoma: a global and regional perspective. Oncologist. 2010;15(S4):5–13. https://doi.org/10.1634/theoncologist.2010-S4-05.
208. Jemal A, Bray F, Center MM, Ferlay J, Ward E, Forman D. Global cancer statistics. CA Cancer J Clin. 2011;61(2):69–90. https://doi.org/10.3322/caac.20107.
209. Pierce ES. Could Mycobacterium avium subspecies paratuberculosis cause Crohn's disease, ulcerative colitis…and colorectal cancer? Infectious Agents and Cancer. 2018;13(1):1. https://doi.org/10.1186/s13027-017-0172-3.
210. Delbue S, Comar M, Ferrante P. Review on the role of the human polyomavirus JC in the development of tumors. Infect Agents Cancer. 2017;12(1):10. https://doi.org/10.1186/s13027-017-0122-0.
211. Maginnis MS, Atwood WJ. JC virus: an oncogenic virus in animals and humans? Semin Cancer Biol. 2009;19(4):261–9. https://doi.org/10.1016/j.semcancer.2009.02.013.
212. Cogliano VJ, Baan R, Straif K, Grosse Y, Lauby-Secretan B, El Ghissassi F, Bouvard V, Benbrahim-Tallaa L, Guha N, Freeman C, Galichet L, Wild CP. Preventable exposures associated with human cancers. JNCI J Natl Cancer Inst. 2011;103(24):1827–39. https://doi.org/10.1093/jnci/djr483.
213. Guan Y-S, He Q, Wang M-Q, Li P. Nuclear factor kappa B and hepatitis viruses. Expert Opin Ther Targets. 2008;12(3):265–80. https://doi.org/10.1517/14728222.12.3.265.
214. Takayama S, Takahashi H, Matsuo Y, Okada Y, Manabe T. Effects of helicobacter pylori infection on human pancreatic cancer cell line. Hepato-Gastroenterology. 2007;54(80): 2387–91.
215. Aricò M, Schrappe M, Hunger SP, Carroll WL, Conter V, Galimberti S, Manabe A, Saha V, Baruchel A, Vettenranta K, Horibe K, Benoit Y, Pieters R, Escherich G, Silverman LB, Pui C-H, Valsecchi MG. Clinical Outcome of children with newly diagnosed Philadelphia chromosome–positive acute lymphoblastic leukemia treated between 1995 and 2005. J Clin Oncol. 2010;28(31):4755–61. https://doi.org/10.1200/JCO.2010.30.1325.
216. Chen Y-C, Hunter DJ. Molecular epidemiology of cancer. CA Cancer J Clin. 2005;55(1):45–54. https://doi.org/10.3322/canjclin.55.1.45.
217. de Martel C, Georges D, Bray F, Ferlay J, Clifford GM. Global burden of cancer attributable to infections in 2018: a worldwide incidence analysis. Lancet Glob Health. 2020;8(2):e180–90. https://doi.org/10.1016/S2214-109X(19)30488-7.

218. Ferlay J, Ervik M, Lam F, Colombet M, Mery L, Piñeros M, et al. Global cancer observatory: cancer today. Lyon: International Agency for Research on Cancer; 2020. https://gco.iarc.fr/today. Accessed Feb 2021.
219. Monsalve J, Kapur J, Malkin D, Babyn PS. Imaging of cancer predisposition syndromes in children. Radiographics. 2011;31(1):263–80. https://doi.org/10.1148/rg.311105099.
220. Weiderpass E. Lifestyle and cancer risk. J Prev Med Public Health. 2010;43(6):459. https://doi.org/10.3961/jpmph.2010.43.6.459.
221. Ferlay J, Colombet M, Soerjomataram I, Dyba T, Randi G, Bettio M, Gavin A, Visser O, Bray F. Cancer incidence and mortality patterns in Europe: estimates for 40 countries and 25 major cancers in 2018. Eur J Cancer. 2018;103:356–87. https://doi.org/10.1016/j.ejca.2018.07.005.
222. McNeill KA. Epidemiology of brain Tumors. Neurol Clin. 2016;34(4):981–98. https://doi.org/10.1016/j.ncl.2016.06.014.
223. The American Association of Neurological Surgeons (AANS). Brain tumors. 2022. https://www.aans.org/Patients/Neurosurgical-Conditions-and-Treatments/Brain-Tumors.
224. Karlsson P, Holmberg E, Lundell M, Mattsson A, Holm L-E, Wallgren A. Intracranial Tumors after exposure to ionizing radiation during infancy: a pooled analysis of two Swedish cohorts of 28,008 infants with skin Hemangioma. Radiat Res. 1998;150(3):357. https://doi.org/10.2307/3579984.
225. Ron E, Modan B, Boice JD, Alfandary E, Stovall M, Chetrit A, Katz L. Tumors of the brain and nervous system after radiotherapy in childhood. N Engl J Med. 1988;319(16):1033–9. https://doi.org/10.1056/NEJM198810203191601.
226. Shore RE, Albert RE, Pasternack BS. Follow-up study of patients treated by X-ray epilation for tinea capitis: resurvey of post-treatment illness and mortality experience. Arch Environ Health Int J. 1976;31(1):21–8. https://doi.org/10.1080/00039896.1976.10667184.
227. Mathews JD, Forsythe AV, Brady Z, Butler MW, Goergen SK, Byrnes GB, Giles GG, Wallace AB, Anderson PR, Guiver TA, McGale P, Cain TM, Dowty JG, Bickerstaffe AC, Darby SC. Cancer risk in 680 000 people exposed to computed tomography scans in childhood or adolescence: data linkage study of 11 million Australians. BMJ. 2013;346(may21 1):f2360. https://doi.org/10.1136/bmj.f2360.
228. Pearce MS, Salotti JA, Little MP, McHugh K, Lee C, Kim KP, Howe NL, Ronckers CM, Rajaraman P, Craft AW, Parker L, Berrington de González A. Radiation exposure from CT scans in childhood and subsequent risk of leukaemia and brain tumours: a retrospective cohort study. Lancet. 2012;380(9840):499–505. https://doi.org/10.1016/S0140-6736(12)60815-0.
229. Shintani T, Hayakawa N, Hoshi M, Sumida M, Kurisu K, Oki S, Kodama Y, Kajikawa H, Inai K, Kamada N. High incidence of meningioma among Hiroshima atomic bomb survivors. J Radiat Res. 1999;40(1):49–57. https://doi.org/10.1269/jrr.40.49.
230. Vienne-Jumeau A, Tafani C, Ricard D. Environmental risk factors of primary brain tumors: a review. Rev Neurol. 2019;175(10):664–78. https://doi.org/10.1016/j.neurol.2019.08.004.
231. Kyrtopoulos SA. N-nitroso compound formation in human gastric juice. Cancer Surv. 1989;8(2):423–42.
232. Dietrich M, Block G, Pogoda JM, Buffler P, Hecht S, Martin SP. A review: dietary and endogenously formed N-nitroso compounds and risk of childhood brain tumors. Cancer Causes Control. 2005;16(6):619–35. https://doi.org/10.1007/s10552-005-0168-y.
233. Carles C, Bouvier G, Esquirol Y, Piel C, Migault L, Pouchieu C, Fabbro-Peray P, Lebailly P, Baldi I. Residential proximity to agricultural land and risk of brain tumor in the general population. Environ Res. 2017;159:321–30. https://doi.org/10.1016/j.envres.2017.08.025.
234. Fallahi P, Foddis R, Cristaudo A, Antonelli A. High risk of brain tumors in farmers: a mini-review of the literature, and report of the results of a case control study. Clin Ter. 2017;168(5):e290–2. https://doi.org/10.7417/T.2017.2022.
235. Piel C, Pouchieu C, Migault L, Béziat B, Boulanger M, Bureau M, Carles C, Grüber A, Lecluse Y, Rondeau V, Schwall X, Tual S, Lebailly P, Baldi I, The AGRICAN group, Arveux P, Bara S, Bouvier AM, Busquet T, et al. Increased risk of central nervous system tumours with carbamate insecticide use in the prospective cohort AGRICAN. Int J Epidemiol. 2019;48(2):512–26. https://doi.org/10.1093/ije/dyy246.

236. Provost D, Cantagrel A, Lebailly P, Jaffre A, Loyant V, Loiseau H, Vital A, Brochard P, Baldi I. Brain tumours and exposure to pesticides: a case-control study in southwestern France. Occup Environ Med. 2007;64(8):509–14. https://doi.org/10.1136/oem.2006.028100.
237. Samanic CM, De Roos AJ, Stewart PA, Rajaraman P, Waters MA, Inskip PD. Occupational exposure to pesticides and risk of adult brain tumors. Am J Epidemiol. 2008;167(8):976–85. https://doi.org/10.1093/aje/kwm401.
238. IARC. Monographs on the evaluation of carcinogenic risks to humans—IARC. https://monographs.iarc.fr/iarcmonographs-on-the-evaluation-of-carcinogenic-risks-tohumans-14/. Accessed 27 May 2019.
239. Jennifer M, Connelly MD, Mark GM, MD, FRCPC. Environmental risk factors for brain tumors. Curr Neurol Neurosci Rep. 2007;7:208–14.
240. Schüz J. Exposure to extremely low-frequency magnetic fields and the risk of childhood cancer: update of the epidemiological evidence. Prog Biophys Mol Biol. 2011;107(3):339–42. https://doi.org/10.1016/j.pbiomolbio.2011.09.008.
241. Kleinerman RA. Self-reported electrical appliance use and risk of adult brain Tumors. Am J Epidemiol. 2005;161(2):136–46. https://doi.org/10.1093/aje/kwi013.
242. American Cancer Society (ACS). Key statistics for bladder cancer. 2022. https://www.cancer.org/cancer/types/bladder-cancer/about/key-statistics.html#how-common-is-bladder-cancer?
243. Kuper H, Boffetta P, Adami H-O. Tobacco use and cancer causation: association by tumour type. J Intern Med. 2002;252(3):206–24. https://doi.org/10.1046/j.1365-2796.2002.01022.x.
244. Volanis D, Kadiyska T, Galanis A, Delakas D, Logotheti S, Zoumpourlis V. Environmental factors and genetic susceptibility promote urinary bladder cancer. Toxicol Lett. 2010;193(2):131–7. https://doi.org/10.1016/j.toxlet.2009.12.018.
245. Brennan P, Bogillot O, Cordier S, Greiser E, Schill W, Vineis P, Lopez-Abente G, Tzonou A, Chang-Claude J, Bolm-Audorff U, Jöckel KH, Donato F, Serra C, Wahrendorf J, Hours M, T'Mannetje A, Kogevinas M, Boffetta P. Cigarette smoking and bladder cancer in men: a pooled analysis of 11 case-control studies. Int J Cancer. 2000;86(2):289–94. https://doi.org/10.1002/(sici)1097-0215(20000415)86:2<289::aid-ijc21>3.0.co;2-m.
246. Brennan P, Bogillot O, Greiser E, Chang-Claude J, Wahrendorf J, Cordier S, Jöckel K-H, Lopez-Abente G, Tzonou A, Vineis P, Donato F, Hours M, Serra C, Bolm-Audorff U, Schill W, Kogevinas M, Boffetta P. No title found. Cancer Causes Control. 2001;12(5):411–7. https://doi.org/10.1023/A:1011214222810.
247. Zeegers MPA, Kellen E, Buntinx F, van den Brandt PA. The association between smoking, beverage consumption, diet and bladder cancer: a systematic literature review. World J Urol. 2004;21(6):392–401. https://doi.org/10.1007/s00345-003-0382-8.
248. Stewart SL, Cardinez CJ, Richardson LC, Norman L, Kaufmann R, Pechacek TF, Thompson TD, Weir HK, Sabatino SA. Surveillance for cancers associated with tobacco use–United States, 1999-2004. MMWR Surveill Summ. 2008;57:1–33.
249. Iscovich J, Castelletto R, Estè J, Muñoz N, Colanzi R, Coronel A, Deamezola I, Tassi V, Arslan A. Tobacco smoking, occupational exposure and bladder cancer in Argentina. Int J Cancer. 1987;40(6):734–40. https://doi.org/10.1002/ijc.2910400604.
250. Meliker JR, Nriagu JO. Arsenic in drinking water and bladder cancer: review of epidemiological evidence. In: Trace metals and other contaminants in the environment, vol. 9. Elsevier; 2007. p. 551–84. https://doi.org/10.1016/S1875-1121(06)09021-3.
251. Silvera SAN, Rohan TE. Trace elements and cancer risk: a review of the epidemiologic evidence. Cancer Causes Control. 2007;18(1):7–27. https://doi.org/10.1007/s10552-006-0057-z.
252. Mink PJ, Alexander DD, Barraj LM, Kelsh MA, Tsuji JS. Low-level arsenic exposure in drinking water and bladder cancer: a review and meta-analysis. Regul Toxicol Pharmacol. 2008;52(3):299–310. https://doi.org/10.1016/j.yrtph.2008.08.010.
253. Tapio S, Grosche B. Arsenic in the aetiology of cancer. Mutation Research/Reviews in Mutation Research. 2006;612(3):215–46. https://doi.org/10.1016/j.mrrev.2006.02.001.
254. Anetor JI, Wanibuchi H, Fukushima S. Arsenic exposure and its health effects and risk of cancer in developing countries: micronutrients as host defence. Asian Pacific J Cancer Prevent APJCP. 2007;8(1):13–23.

255. Moore LE. P53 alterations in bladder tumors from arsenic and tobacco exposed patients. Carcinogenesis. 2003;24(11):1785–91. https://doi.org/10.1093/carcin/bgg136.
256. Andrew AS, Mason RA, Kelsey KT, Schned AR, Marsit CJ, Nelson HH, Karagas MR. DNA repair genotype interacts with arsenic exposure to increase bladder cancer risk☆. Toxicol Lett. 2009;187(1):10–4. https://doi.org/10.1016/j.toxlet.2009.01.013.
257. Villanueva CM. Meta-analysis of studies on individual consumption of chlorinated drinking water and bladder cancer. J Epidemiol Community Health. 2003;57(3):166–73. https://doi.org/10.1136/jech.57.3.166.
258. Chen H-I, Liou S-H, Loh C-H, Uang S-N, Yu Y-C, Shih T-S. Bladder cancer screening and monitoring of 4,4′-Methylenebis(2-chloroaniline) exposure among workers in Taiwan. Urology. 2005;66(2):305–10. https://doi.org/10.1016/j.urology.2005.02.031.
259. García-Pérez J, Pollán M, Boldo E, Pérez-Gómez B, Aragonés N, Lope V, Ramis R, Vidal E, López-Abente G. Mortality due to lung, laryngeal and bladder cancer in towns lying in the vicinity of combustion installations. Sci Total Environ. 2009;407(8):2593–602. https://doi.org/10.1016/j.scitotenv.2008.12.062.
260. Snyderwine EG, Sinha R, Felton JS, Ferguson LR. Highlights of the eighth international conference on carcinogenic/mutagenic N-substituted aryl compounds. Mutation Research/Fundamental and Molecular Mechanisms of Mutagenesis. 2002;506–507:1–8. https://doi.org/10.1016/S0027-5107(02)00146-X.
261. Yu MC, Skipper PL, Tannenbaum SR, Chan KK, Ross RK. Arylamine exposures and bladder cancer risk. Mutat Res/Fundamental Mol Mech Mutagenesis. 2002;506–507:21–8. https://doi.org/10.1016/S0027-5107(02)00148-3.
262. Case RAM, Hosker ME. Tumour of the urinary bladder as an occupational disease in the rubber industry in England and Wales. J Epidemiol Community Health. 1954;8(2):39–50. https://doi.org/10.1136/jech.8.2.39.
263. Golka K, Wiese A, Assennato G, Bolt HM. Occupational exposure and urological cancer. World J Urol. 2004;21(6):382–91. https://doi.org/10.1007/s00345-003-0377-5.
264. Centers for Disease Control and Prevention (CDC). Special occupational hazard review for benzidine-based dyes. 2014. https://www.cdc.gov/niosh/docs/80-109/default.html
265. Ma Q, Lin G, Qin Y, Lu D, Golka K, Geller F, Chen J, Shen J. GSTP1 A1578G (Ile105Val) polymorphism in benzidine-exposed workers: an association with cytological grading of exfoliated urothelial cells. Pharmacogenetics. 2003;13(7):409–15. https://doi.org/10.1097/00008571-200307000-00006.
266. Lokeshwar S, Klaassen Z, Terris M. A contemporary review of risk factors for bladder. Clinics in Oncol. 2016;1:1–3.
267. Ohkuma Y, Hiraku Y, Oikawa S, Yamashita N, Murata M, Kawanishi S. Distinct mechanisms of oxidative DNA damage by two metabolites of carcinogenic o-toluidine. Arch Biochem Biophys. 1999;372(1):97–106. https://doi.org/10.1006/abbi.1999.1461.
268. Watanabe C, Egami T, Midorikawa K, Hiraku Y, Oikawa S, Kawanishi S, Murata M. DNA damage and estrogenic activity induced by the environmental pollutant 2-nitrotoluene and its metabolite. Environ Health Prev Med. 2010;15(5):319–26. https://doi.org/10.1007/s12199-010-0146-1.
269. American Society of Clinical Oncology (ASCO). Bladder cancer. 2021. https://www.cancer.net/cancertypes/bladdercancer/introduction.
270. Steliarova-Foucher E, Colombet M, Ries LAG, Moreno F, Dolya A, Bray F, Hesseling P, Shin HY, Stiller CA, IICC-3 contributors. International incidence of childhood cancer, 2001-10: a population-based registry study. Lancet Oncol. 2017;18(6):719–31. https://doi.org/10.1016/S1470-2045(17)30186-9.
271. Namayandeh SM, Khazaei Z, Lari Najafi M, Goodarzi E, Moslem A. GLOBAL Leukemia in children 0-14 statistics 2018, incidence and mortality and human development index (HDI): GLOBOCAN sources and methods. Asian Pac J Cancer Prev. 2020;21(5):1487–94. https://doi.org/10.31557/APJCP.2020.21.5.1487.
272. Pereira FAC, Mirra AP, de Oliveira Latorre MRD, Vicente de Assunção J. Environmental risk factors and acute lymphoblastic leukaemia in childhood. Rev Cienc Salud. 2016;15(1):129–44.

273. Kutanzi K, Lumen A, Koturbash I, Miousse I. Pediatric exposures to ionizing radiation: carcinogenic considerations. Int J Environ Res Public Health. 2016;13(11):1057. https://doi.org/10.3390/ijerph13111057.
274. Murray R, Heckel P, Hempelmann LH. Leukemia in children exposed to ionizing radiation. N Engl J Med. 1959;261(12):585–9. https://doi.org/10.1056/NEJM195909172611203.
275. Richardson D, Sugiyama H, Nishi N, Sakata R, Shimizu Y, Grant EJ, Soda M, Hsu W-L, Suyama A, Kodama K, Kasagi F. Ionizing radiation and Leukemia mortality among Japanese atomic bomb survivors, 1950–2000. Radiat Res. 2009;172(3):368–82. https://doi.org/10.1667/RR1801.1.
276. Simpson CL, Hempelmann LH, Fuller LM. Neoplasia in children treated with X-rays in infancy for thymic enlargement. Radiology. 1955;64(6):840–5. https://doi.org/10.1148/64.6.840.
277. UNSCEAR. Nations scientific committee on the effects of atomic radiation 2013 report, sources, effects and risks of ionizing radiation, vol. II annex B, effects of radiation exposure of children. United Nations: New York, NY; 2013.
278. Schmidt J-A, Hornhardt S, Erdmann F, Sánchez-García I, Fischer U, Schüz J, Ziegelberger G. Risk factors for childhood leukemia: radiation and beyond. Front Public Health. 2021;9:805757. https://doi.org/10.3389/fpubh.2021.805757.
279. Folley JH, Borges W, Yamawaki T. Incidence of leukemia in survivors of the atomic bomb in Hiroshima and Nagasaki, Japan. Am J Med. 1952;13(3):311–21. https://doi.org/10.1016/0002-9343(52)90285-4.
280. Kendall GM, Little MP, Wakeford R. A review of studies of childhood cancer and natural background radiation. Int J Radiat Biol. 2021;97(6):769–81. https://doi.org/10.1080/09553002.2020.1867926.
281. Ahlbom A, Day N, Feychting M, Roman E, Skinner J, Dockerty J, Linet M, McBride M, Michaelis J, Olsen JH, Tynes T, Verkasalo PK. A pooled analysis of magnetic fields and childhood leukaemia. Br J Cancer. 2000;83(5):692–8. https://doi.org/10.1054/bjoc.2000.1376.
282. Greenland S, Sheppard AR, Kaune WT, Poole C, Kelsh MA. A pooled analysis of magnetic fields, wire codes, and childhood leukemia. Epidemiology. 2000;11(6):624–34. https://doi.org/10.1097/00001648-200011000-00003.
283. Wünsch-Filho V, Pelissari DM, Barbieri FE, Sant'Anna L, de Oliveira CT, de Mata JF, Tone LG, de Lee MLM, de Andréa MLM, Bruniera P, Epelman S, Filho VO, Kheifets L. Exposure to magnetic fields and childhood acute lymphocytic leukemia in São Paulo, Brazil. Cancer Epidemiol. 2011;35(6):534–9. https://doi.org/10.1016/j.canep.2011.05.008.
284. Daniels JL, Olshan AF, Savitz DA. Pesticides and childhood cancers. Environ Health Perspect. 1997;105(10):1068–77. https://doi.org/10.1289/ehp.971051068.
285. Menegaux F. Household exposure to pesticides and risk of childhood acute leukaemia. Occup Environ Med. 2006;63(2):131–4. https://doi.org/10.1136/oem.2005.023036.
286. Reynolds P, Behrens JV, Günter R, Goldberg DE, Hertz A. Agricultural pesticides and lymphoproliferative childhood cancer in California. Scand J Work Environ Health. 2005;31(Suppl1):46–54.
287. Zahm SH, Ward MH. Pesticides and childhood cancer. Environ Health Perspect. 1998;106(suppl 3):893–908. https://doi.org/10.1289/ehp.98106893.
288. Rudant J, Baccaïni B, Ripert M, Goubin A, Bellec S, Hémon D, Clavel J. Population-mixing at the place of residence at the time of birth and incidence of childhood leukaemia in France. Eur J Cancer. 2006;42(7):927–33. https://doi.org/10.1016/j.ejca.2005.12.015.
289. Kinlen L. Evidence for an infective cause of childhood leukaemia: comparison of a Scottish new town with nuclear reprocessing sites in Britain. Lancet. 1988;332(8624):1323–7. https://doi.org/10.1016/S0140-6736(88)90867-7.
290. Greaves MF. Speculations on the cause of childhood acute lymphoblastic leukemia. Leukemia. 1988;2(2):120–5.
291. Hauer J, Fischer U, Borkhardt A. Toward prevention of childhood ALL by early-life immune training. Blood. 2021;138(16):1412–28.
292. Lichtenstein P, Holm NV, Verkasalo PK, Iliadou A, Kaprio J, Koskenvuo M, Pukkala E, Skytthe A, Hemminki K. Environmental and heritable factors in the causation of cancer—analyses

of cohorts of twins from Sweden, Denmark, and Finland. N Engl J Med. 2000;343(2):78–85. https://doi.org/10.1056/NEJM200007133430201.
293. Russo J, Hu Y-F, Yang X, Russo IH. Chapter 1: developmental, cellular, and molecular basis of human breast cancer. JNCI Monographs. 2000;2000(27):17–37. https://doi.org/10.1093/oxfordjournals.jncimonographs.a024241.
294. Strumylaitė L, Mechonošina K, Tamašauskas Š. Environmental factors and breast cancer. Medicina. 2010;46(12):867. https://doi.org/10.3390/medicina46120121.
295. Evans JS, Wennberg JE, McNeil BJ. The influence of diagnostic radiography on the incidence of breast cancer and leukemia. N Engl J Med. 1986;315(13):810–5. https://doi.org/10.1056/NEJM198609253151306.
296. American Cancer Society. Types of breast cancer. 2021. https://www.cancer.org/cancer/breast-cancer/about/types-of-breast-cancer.html.
297. McElroy JA, Shafer MM, Trentham-Dietz A, Hampton JM, Newcomb PA. Cadmium exposure and breast cancer risk. JNCI: J Natl Cancer Inst. 2006;98(12):869–73. https://doi.org/10.1093/jnci/djj233.
298. Ionescu JG, Novotny J, Stejskal V, Lätsch A, Blaurock-Busch E, Eisenmann-Klein M. Increased levels of transition metals in breast cancer tissue. Neuro Endocrinol Lett. 2006;27(Suppl 1):36–9.
299. Strumylaitė L, Boguševičius A, Ryselis S, Pranys D, Poškienė L, Kregždytė R, Abdrachmanovas O, Asadauskaitė R. Association between cadmium and breast cancer. Medicina. 2008;44(6):415. https://doi.org/10.3390/medicina44060054.
300. Antila E, Mussalo-Rauhamaa H, Kantola M, Atroshi F, Westermarck T. Association of cadmium with human breast cancer. Sci Total Environ. 1996;186(3):251–6. https://doi.org/10.1016/0048-9697(96)05119-4.
301. Garcia-Morales P, Saceda M, Kenney N, Kim N, Salomon DS, Gottardis MM, Solomon HB, Sholler PF, Jordan VC, Martin MB. Effect of cadmium on estrogen receptor levels and estrogen-induced responses in human breast cancer cells. J Biol Chem. 1994;269(24):16896–901. https://doi.org/10.1016/S0021-9258(19)89474-7.
302. Stoica A, Katzenellenbogen BS, Martin MB. Activation of Estrogen receptor-α by the heavy metal cadmium. Mol Endocrinol. 2000;14(4):545–53. https://doi.org/10.1210/mend.14.4.0441.
303. Johnson MD, Kenney N, Stoica A, Hilakivi-Clarke L, Singh B, Chepko G, Clarke R, Sholler PF, Lirio AA, Foss C, Reiter R, Trock B, Paik S, Martin MB. Cadmium mimics the in vivo effects of estrogen in the uterus and mammary gland. Nat Med. 2003;9(8):1081–4. https://doi.org/10.1038/nm902.
304. Ronckers CM, Erdmann CA, Land CE. Radiation and breast cancer: a review of current evidence. Breast Cancer Res. 2005;7:21–32.
305. Preston DL, Ron E, Tokuoka S, Funamoto S, Nishi N, Soda M, Mabuchi K, Kodama K. Solid cancer incidence in atomic bomb survivors: 1958–1998. Radiat Res. 2007;168(1):1–64. https://doi.org/10.1667/RR0763.1.
306. Ronckers CM, Doody MM, Lonstein JE, Stovall M, Land CE. Multiple diagnostic x-rays for spine deformities and risk of breast cancer. Cancer Epidemiol Biomark Prev. 2008;17:605–13.
307. Travis LB, Hill D, Dores GM, Gospodarowicz M, van Leeuwen FE, Holowaty E, Glimelius B, Andersson M, Pukkala E, Lynch CF, Pee D, Smith SA, Van't Veer MB, Joensuu T, Storm H, Stovall M, Boice JD, Gilbert E, Gail MH. Cumulative absolute breast cancer risk for Young women treated for Hodgkin lymphoma. JNCI: J Natl Cancer Inst. 2005;97(19):1428–37. https://doi.org/10.1093/jnci/dji290.
308. Preston DL, Mattsson A, Holmberg E, Shore R, Hildreth NG, Boice JD. Radiation effects on breast cancer risk: a pooled analysis of eight cohorts. Radiat Res. 2002;158:220–35.
309. van Leeuwen FE, Klokman WJ, Stovall M, Dahler EC, van't Veer MB, Noordijk EM, Crommelin MA, Aleman BMP, Broeks A, Gospodarowicz M, Travis LB, Russell NS. Roles of radiation dose, chemotherapy, and hormonal factors in breast cancer following Hodgkin's disease. JNCI J Natl Cancer Institute. 2003;95(13):971–80. https://doi.org/10.1093/jnci/95.13.971.

310. De Bruin ML, Sparidans J, van't Veer MB, Noordijk EM, Louwman MWJ, Zijlstra JM, van den Berg H, Russell NS, Broeks A, Baaijens MHA, Aleman BMP, van Leeuwen FE. Breast cancer risk in female survivors of Hodgkin's lymphoma: lower risk after smaller radiation volumes. J Clin Oncol. 2009;27(26):4239–46. https://doi.org/10.1200/JCO.2008.19.9174.
311. Carmichael A, Sami AS, Dixon JM. Breast cancer risk among the survivors of atomic bomb and patients exposed to therapeutic ionising radiation. Eur J Surg Oncol (EJSO). 2003;29(5):475–9. https://doi.org/10.1016/S0748-7983(03)00010-6.
312. Stevens RG. Electric power use and breast cancer: a hypothesis. Am J Epidemiol. 1987;125(4):556–61. https://doi.org/10.1093/oxfordjournals.aje.a114569.
313. Boscoe FP, Schymura MJ. Solar ultraviolet-B exposure and cancer incidence and mortality in the United States, 1993–2002. BMC Cancer. 2006;6(1):264. https://doi.org/10.1186/1471-2407-6-264.
314. McKay JD, McCullough ML, Ziegler RG, Kraft P, Saltzman BS, Riboli E, Barricarte A, Berg CD, Bergland G, Bingham S, Brustad M, Bueno-de-Mesquita HB, Burdette L, Buring J, Calle EE, Chanock SJ, Clavel-Chapelon F, Cox DG, Dossus L, et al. Vitamin D receptor polymorphisms and breast cancer risk: results from the National Cancer Institute breast and prostate cancer cohort consortium. Cancer Epidemiol Biomark Prev. 2009;18(1):297–305. https://doi.org/10.1158/1055-9965.EPI-08-0539.
315. Ooi LL, Zhou H, Kalak R, Zheng Y, Conigrave AD, Seibel MJ, Dunstan CR. Vitamin D deficiency promotes human breast cancer growth in a murine model of bone metastasis. Cancer Res. 2010;70(5):1835–44. https://doi.org/10.1158/0008-5472.CAN-09-3194.
316. Rodgers KM, Udesky JO, Rudel RA, Brody JG. Environmental chemicals and breast cancer: an updated review of epidemiological literature informed by biological mechanisms. Environ Res. 2018;160:152–82. https://doi.org/10.1016/j.envres.2017.08.045.
317. Lee H-R, Hwang K-A, Nam K-H, Kim H-C, Choi K-C. Progression of breast cancer cells was enhanced by endocrine-disrupting chemicals, triclosan and Octylphenol, via an Estrogen receptor-dependent Signaling pathway in cellular and mouse xenograft models. Chem Res Toxicol. 2014;27(5):834–42. https://doi.org/10.1021/tx5000156.
318. Rudel R, Ackerman J, Attfield K, Dodson R, Brody J. Exposure biomarkers as tools for breast cancer epidemiology, biomonitoring, and prevention. ISEE Conference Abstracts. 2014;2014(1):2726. https://doi.org/10.1289/isee.2014.P3-721.
319. NTP. Report on carcinogens. 14th ed. Research Triangle Park, NC: United States National Toxicology Program; 2016. http://ntp.niehs.nih.gov/go/roc14. Accessed 08 Mar 2017.
320. Murray IA, Patterson AD, Perdew GH. Aryl hydrocarbon receptor ligands in cancer: friend and foe. Nat Rev Cancer. 2014;14(12):801–14. https://doi.org/10.1038/nrc3846.
321. Hahn ME, Allan LL, Sherr DH. Regulation of constitutive and inducible AHR signaling: complex interactions involving the AHR repressor. Biochem Pharmacol. 2009;77(4):485–97. https://doi.org/10.1016/j.bcp.2008.09.016.
322. Vogel CFA, Chang WLW, Kado S, McCulloh K, Vogel H, Wu D, Haarmann-Stemmann T, Yang G, Leung PSC, Matsumura F, Gershwin ME. Transgenic overexpression of aryl hydrocarbon receptor repressor (AhRR) and AhR-mediated induction of CYP1A1, cytokines, and acute toxicity. Environ Health Perspect. 2016;124(7):1071–83. https://doi.org/10.1289/ehp.1510194.
323. Hoover RN, Hyer M, Pfeiffer RM, Adam E, Bond B, Cheville AL, Colton T, Hartge P, Hatch EE, Herbst AL, Karlan BY, Kaufman R, Noller KL, Palmer JR, Robboy SJ, Saal RC, Strohsnitter W, Titus-Ernstoff L, Troisi R. Adverse health outcomes in women exposed in utero to diethylstilbestrol. N Engl J Med. 2011;365(14):1304–14. https://doi.org/10.1056/NEJMoa1013961.
324. Palmer JR, Hatch EE, Rosenberg CL, Hartge P, Kaufman RH, Titus-Ernstoff L, Noller KL, Herbst AL, Rao RS, Troisi R, Colton T, Hoover RN. Risk of breast cancer in women exposed to diethylstilbestrol in utero: prelimiinary results (United States). Cancer Causes Control: CCC. 2002;13(8):753–8. https://doi.org/10.1023/a:1020254711222.
325. Palmer JR, Wise LA, Hatch EE, Troisi R, Titus-Ernstoff L, Strohsnitter W, Kaufman R, Herbst AL, Noller KL, Hyer M, Hoover RN. Prenatal Diethylstilbestrol exposure and

risk of breast cancer. Cancer Epidemiol Biomark Prev. 2006;15(8):1509–14. https://doi.org/10.1158/1055-9965.EPI-06-0109.
326. Calle EE, Mervis CA, Thun MJ, Rodriguez C, Wingo PA, Heath CW. Diethylstilbestrol and risk of fatal breast cancer in a prospective cohort of US women. Am J Epidemiol. 1996;144(7):645–52. https://doi.org/10.1093/oxfordjournals.aje.a008976.
327. Colton T. Breast cancer in mothers prescribed Diethylstilbestrol in pregnancy: further follow-up. JAMA. 1993;269(16):2096. https://doi.org/10.1001/jama.1993.03500160066033.
328. Greenberg ER, Barnes AB, Resseguie L, Barrett JA, Burnside S, Lanza LL, Neff RK, Stevens M, Young RH, Colton T. Breast cancer in mothers given Diethylstilbestrol in pregnancy. N Engl J Med. 1984;311(22):1393–8. https://doi.org/10.1056/NEJM198411293112201.
329. Titus-Ernstoff L, Hatch EE, Hoover RN, Palmer J, Greenberg ER, Ricker W, Kaufman R, Noller K, Herbst AL, Colton T, Hartge P. Long-term cancer risk in women given diethylstilbestrol (DES) during pregnancy. Br J Cancer. 2001;84(1):126–33. https://doi.org/10.1054/bjoc.2000.1521.
330. Fielden M. Normal mammary gland morphology in pubertal female mice following in utero and lactational exposure to genistein at levels comparable to human dietary exposure. Toxicol Lett. 2002;133(2–3):181–91. https://doi.org/10.1016/S0378-4274(02)00154-6.
331. Hovey RC, Asai-Sato M, Warri A, Terry-Koroma B, Colyn N, Ginsburg E, Vonderhaar BK. Effects of neonatal exposure to diethylstilbestrol, tamoxifen, and toremifene on the BALB/c mouse mammary Gland1. Biol Reprod. 2005;72(2):423–35. https://doi.org/10.1095/biolreprod.104.029769.
332. Rudel RA, Fenton SE, Ackerman JM, Euling SY, Makris SL. Environmental exposures and mammary gland development: state of the science, public health implications, and research recommendations. Environ Health Perspect. 2011;119(8):1053–61. https://doi.org/10.1289/ehp.1002864.
333. Meyer JE, Narang T, Schnoll-Sussman FH, Pochapin MB, Christos PJ, Sherr DL. Increasing incidence of rectal cancer in patients aged younger than 40 years: an analysis of the surveillance, epidemiology, and end results database. Cancer. 2010;116(18):4354–9. https://doi.org/10.1002/cncr.25432.
334. Kimmie N, Folasade PM, Deborah S. US preventive services task force recommendations for colorectal cancer screening. JAMA. 2021;325(19):1943.
335. Rattray NJW, Charkoftaki G, Rattray Z, Hansen JE, Vasiliou V, Johnson CH. Environmental influences in the etiology of colorectal cancer: the premise of metabolomics. Curr Pharmacol Rep. 2017;3(3):114–25. https://doi.org/10.1007/s40495-017-0088-z.
336. Murphy N, Moreno V, Hughes DJ, Vodicka L, Vodicka P, Aglago EK, Gunter MJ, Jenab M. Lifestyle and dietary environmental factors in colorectal cancer susceptibility. Mol Asp Med. 2019;69:2–9. https://doi.org/10.1016/j.mam.2019.06.005.
337. Bouvard V, Loomis D, Guyton KZ, Grosse Y, Ghissassi FE, Benbrahim-Tallaa L, Guha N, Mattock H, Straif K. Carcinogenicity of consumption of red and processed meat. Lancet Oncol. 2015;16(16):1599–600. https://doi.org/10.1016/S1470-2045(15)00444-1.
338. Vieira AR, Abar L, Chan DSM, Vingeliene S, Polemiti E, Stevens C, Greenwood D, Norat T. Foods and beverages and colorectal cancer risk: a systematic review and meta-analysis of cohort studies, an update of the evidence of the WCRF-AICR continuous update project. Ann Oncol. 2017;28(8):1788–802. https://doi.org/10.1093/annonc/mdx171.
339. Cascella M, Bimonte S, Barbieri A, Del Vecchio V, Caliendo D, Schiavone V, Fusco R, Granata V, Arra C, Cuomo A. Dissecting the mechanisms and molecules underlying the potential carcinogenicity of red and processed meat in colorectal cancer (CRC): an overview on the current state of knowledge. Infect Agents Cancer. 2018;13(1):3. https://doi.org/10.1186/s13027-018-0174-9.
340. Ashmore JH, Rogers CJ, Kelleher SL, Lesko SM, Hartman TJ. Dietary iron and colorectal cancer risk: a review of human population studies. Crit Rev Food Sci Nutr. 2016;56(6):1012–20. https://doi.org/10.1080/10408398.2012.749208.

341. Májer F, Sharma R, Mullins C, Keogh L, Phipps S, Duggan S, Kelleher D, Keely S, Long A, Radics G, Wang J, Gilmer JF. New highly toxic bile acids derived from deoxycholic acid, chenodeoxycholic acid and lithocholic acid. Bioorg Med Chem. 2014;22(1):256–68. https://doi.org/10.1016/j.bmc.2013.11.029.
342. Lauby-Secretan B, Scoccianti C, Loomis D, Grosse Y, Bianchini F, Straif K. Body fatness and cancer—viewpoint of the IARC working group. N Engl J Med. 2016;375(8):794–8. https://doi.org/10.1056/NEJMsr1606602.
343. Ning Y, Wang L, Giovannucci EL. A quantitative analysis of body mass index and colorectal cancer: findings from 56 observational studies. Obes Rev. 2010;11(1):19–30. https://doi.org/10.1111/j.1467-789X.2009.00613.x.
344. Botteri E, Iodice S, Raimondi S, Maisonneuve P, Lowenfels AB. Cigarette smoking and adenomatous polyps: a meta-analysis. Gastroenterology. 2008;134(2):388–395.e3. https://doi.org/10.1053/j.gastro.2007.11.007.
345. Figueiredo JC, Crockett SD, Snover DC, Morris CB, McKeown-Eyssen G, Sandler RS, Ahnen DJ, Robertson DJ, Burke CA, Bresalier RS, Church JM, Church TR, Baron JA. Smoking-associated risks of conventional adenomas and serrated polyps in the colorectum. Cancer Causes Control. 2015;26(3):377–86. https://doi.org/10.1007/s10552-014-0513-0.
346. Carr PR, Alwers E, Bienert S, Weberpals J, Kloor M, Brenner H, Hoffmeister M. Lifestyle factors and risk of sporadic colorectal cancer by microsatellite instability status: a systematic review and meta-analyses. Ann Oncol. 2018;29(4):825–34. https://doi.org/10.1093/annonc/mdy059.
347. Liang PS, Chen T-Y, Giovannucci E. Cigarette smoking and colorectal cancer incidence and mortality: systematic review and meta-analysis. Int J Cancer. 2009;124(10):2406–15. https://doi.org/10.1002/ijc.24191.
348. Seitz HK. Alcohol and cancer—individual risk factors. Addiction. 2017;112(2):232–3.
349. Moskal A, Norat T, Ferrari P, Riboli E. Alcohol intake and colorectal cancer risk: a dose–response meta-analysis of published cohort studies. Int J Cancer. 2007;120(3):664–71. https://doi.org/10.1002/ijc.22299.
350. World Health Organization. Global status report on alcohol and health. 2014.
351. IARC Working Group. Alcoholic beverage consumption and ethyl carbamate (urethane). International Agency for Research on Cancer, 2010. IARC monographs on the evaluation of carcinogenic risks to humans No. 96.
352. Wu S, Rhee K-J, Albesiano E, Rabizadeh S, Wu X, Yen H-R, Huso DL, Brancati FL, Wick E, McAllister F, Housseau F, Pardoll DM, Sears CL. A human colonic commensal promotes colon tumorigenesis via activation of T helper type 17 T cell responses. Nat Med. 2009;15(9):1016–22. https://doi.org/10.1038/nm.2015.
353. Kostic AD, Chun E, Robertson L, Glickman JN, Gallini CA, Michaud M, et al. Aug 14. Fusobacterium nucleatum potentiates intestinal tumorigenesis and modulates the tumor-immune microenvironment. Cell Host Microbe. 2013;14(2):207–15.
354. Mima K, Cao Y, Chan AT, Qian ZR, Nowak JA, Masugi Y, Shi Y, Song M, da Silva A, Gu M, Li W, Hamada T, Kosumi K, Hanyuda A, Liu L, Kostic AD, Giannakis M, Bullman S, Brennan CA, et al. Fusobacterium nucleatum in colorectal carcinoma tissue according to tumor location. Clin Transl Gastroenterol. 2016;7(11):e200. https://doi.org/10.1038/ctg.2016.53.
355. Arthur JC, Perez-Chanona E, Mühlbauer M, Tomkovich S, Uronis JM, Fan T-J, Campbell BJ, Abujamel T, Dogan B, Rogers AB, Rhodes JM, Stintzi A, Simpson KW, Hansen JJ, Keku TO, Fodor AA, Jobin C. Intestinal inflammation targets cancer-inducing activity of the microbiota. Science. 2012;338(6103):120–3. https://doi.org/10.1126/science.1224820.
356. Abed J, Emgård JEM, Zamir G, Faroja M, Almogy G, Grenov A, Sol A, Naor R, Pikarsky E, Atlan KA, Mellul A, Chaushu S, Manson AL, Earl AM, Ou N, Brennan CA, Garrett WS, Bachrach G. Fap2 mediates fusobacterium nucleatum colorectal adenocarcinoma enrichment by binding to tumor-expressed gal-GalNAc. Cell Host Microbe. 2016;20(2):215–25. https://doi.org/10.1016/j.chom.2016.07.006.

357. Mima K, Sukawa Y, Nishihara R, Qian ZR, Yamauchi M, Inamura K, Kim SA, Masuda A, Nowak JA, Nosho K, Kostic AD, Giannakis M, Watanabe H, Bullman S, Milner DA, Harris CC, Giovannucci E, Garraway LA, Freeman GJ, et al. *Fusobacterium nucleatum* and T cells in colorectal carcinoma. JAMA Oncol. 2015;1(5):653. https://doi.org/10.1001/jamaoncol.2015.1377.
358. Rubinstein MR, Wang X, Liu W, Hao Y, Cai G, Han YW. Fusobacterium nucleatum promotes colorectal carcinogenesis by modulating E-cadherin/β-catenin Signaling via its FadA adhesin. Cell Host Microbe. 2013;14(2):195–206. https://doi.org/10.1016/j.chom.2013.07.012.
359. Scott AJ, Merrifield CA, Alexander JL, Marchesi JR, Kinross JM. Highlights from the inaugural international cancer microbiome consortium meeting (ICMC), 5–6 September 2017, London, UK. Ecancermedicalscience. 2017;11:791. https://doi.org/10.3332/ecancer.2017.791.
360. Safiri S, Kolahi AA, Mansournia MA, et al. The burden of kidney cancer and its attributable risk factors in 195 countries and territories, 1990–2017. Sci Rep. 2020;10:13862. https://doi.org/10.1038/s41598-020-70840-2.
361. Global Burden of Disease Cancer Collaboration, Fitzmaurice C, Akinyemiju TF, Al Lami FH, Alam T, Alizadeh-Navaei R, Allen C, Alsharif U, Alvis-Guzman N, Amini E, Anderson BO, Aremu O, Artaman A, Asgedom SW, Assadi R, Atey TM, Avila-Burgos L, Awasthi A, Ba Saleem HO, et al. Global, regional, and national cancer incidence, mortality, years of life lost, years lived with disability, and disability-adjusted life-years for 29 cancer groups, 1990 to 2016: a systematic analysis for the global burden of disease study. JAMA Oncol. 2018;4(11):1553. https://doi.org/10.1001/jamaoncol.2018.2706.
362. Chow W-H, Dong LM, Devesa SS. Epidemiology and risk factors for kidney cancer. Nat Rev Urol. 2010;7(5):245–57. https://doi.org/10.1038/nrurol.2010.46.
363. Scelo G, Larose TL. Epidemiology and risk factors for kidney cancer. J Clin Oncol. 2018;36(36):3574–81. https://doi.org/10.1200/JCO.2018.79.1905.
364. American Cancer Society. Cancer facts & figures 2022. Atlanta, GA: American Cancer Society; 2022a.
365. American Cancer Society. Key statistics about kidney cancer. 2022b. https://www.cancer.org/cancer/kidney-cancer/about/key-statistics.html.
366. Cancer Treatment Centers of America. Types of kidney cancer. 2022a. https://www.cancercenter.com/cancer-types/kidney-cancer/types.
367. Cancer Treatment Centers of America. Liver cancer types. 2022b. https://www.cancercenter.com/cancer-types/liver-cancer/types.
368. U.S. Department of Health and Human Services (HHS). The health consequences of smoking: a report of the surgeon general. Atlanta, GA: Centers for Disease Control and Prevention, National Center for Chronic Disease Prevention and Health Promotion, Office on Smoking and Health; 2004.
369. Theis RP, Dolwick Grieb SM, Burr D, Siddiqui T, Asal NR. Smoking, environmental tobacco smoke, and risk of renal cell cancer: a population-based case-control study. BMC Cancer. 2008;8(1):387. https://doi.org/10.1186/1471-2407-8-387.
370. Hunt JD, van der Hel OL, McMillan GP, Boffetta P, Brennan P. Renal cell carcinoma in relation to cigarette smoking: meta-analysis of 24 studies. Int J Cancer. 2005;114(1):101–8. https://doi.org/10.1002/ijc.20618.
371. Sharifi N, Farrar WL. Perturbations in hypoxia detection: a shared link between hereditary and sporadic tumor formation? Med Hypotheses. 2006;66(4):732–5. https://doi.org/10.1016/j.mehy.2005.11.003.
372. Clague J, Shao L, Lin J, Chang S, Zhu Y, Wang W, Wood CG, Wu X. Sensitivity to NNKOAc is associated with renal cancer risk. Carcinogenesis. 2009;30(4):706–10. https://doi.org/10.1093/carcin/bgp045.
373. Zhu Y, Horikawa Y, Yang H, Wood CG, Habuchi T, Wu X. BPDE induced lymphocytic chromosome 3p deletions may predict renal cell carcinoma risk. J Urol. 2008;179(6):2416–21. https://doi.org/10.1016/j.juro.2008.01.092.

374. Adams KF, Leitzmann MF, Albanes D, Kipnis V, Moore SC, Schatzkin A, Chow W-H. Body size and renal cell cancer incidence in a large US cohort study. Am J Epidemiol. 2008;168(3):268–77. https://doi.org/10.1093/aje/kwn122.
375. Oh SW, Yoon YS, Shin S-A. Effects of excess weight on cancer incidences depending on cancer sites and histologic findings among men: Korea national health insurance corporation study. J Clin Oncol. 2005;23(21):4742–54. https://doi.org/10.1200/JCO.2005.11.726.
376. Pischon T, Lahmann PH, Boeing H, Tjønneland A, Halkjaer J, Overvad K, Klipstein-Grobusch K, Linseisen J, Becker N, Trichopoulou A, Benetou V, Trichopoulos D, Sieri S, Palli D, Tumino R, Vineis P, Panico S, Monninkhof E, Peeters PHM, et al. Body size and risk of renal cell carcinoma in the European prospective investigation into cancer and nutrition (EPIC). Int J Cancer. 2006;118(3):728–38. https://doi.org/10.1002/ijc.21398.
377. Reeves GK, Pirie K, Beral V, Green J, Spencer E, Bull D. Cancer incidence and mortality in relation to body mass index in the million women study: cohort study. BMJ. 2007;335(7630):1134. https://doi.org/10.1136/bmj.39367.495995.AE.
378. Renehan AG, et al. Body-mass index and incidence of cancer: a systematic review and meta-analysis of prospective observational studies. Lancet. 2008;371:569–78.
379. Klinghoffer Z, Yang B, Kapoor A, Pinthus JH. Obesity and renal cell carcinoma: epidemiology, underlying mechanisms and management considerations. Expert Rev Anticancer Ther. 2009;9(7):975–87. https://doi.org/10.1586/era.09.51.
380. Pialoux V, Brown AD, Leigh R, Friedenreich CM, Poulin MJ. Effect of cardiorespiratory fitness on vascular regulation and oxidative stress in postmenopausal women. Hypertension. 2009;54(5):1014–20. https://doi.org/10.1161/HYPERTENSIONAHA.109.138917.
381. Richardson CR, Newton TL, Abraham JJ, Sen A, Jimbo M, Swartz AM. A meta-analysis of pedometer-based walking interventions and weight loss. Ann Family Med. 2008;6(1):69–77. https://doi.org/10.1370/afm.761.
382. Solomon TP, Haus JM, Kelly KR, Cook MD, Riccardi M, Rocco M, Kashyap SR, Barkoukis H, Kirwan JP. Randomized trial on the effects of a 7-d low-glycemic diet and exercise intervention on insulin resistance in older obese humans. Am J Clin Nutr. 2009;90(5):1222–9. https://doi.org/10.3945/ajcn.2009.28293.
383. Törnqvist M. Acrylamide in food: the discovery and its implications. In: Friedman M, Mottram D, editors. Chemistry and safety of acrylamide in food, vol. 561. Springer; 2005. p. 1–19. https://doi.org/10.1007/0-387-24980-X_1.
384. Lee JE, Männistö S, Spiegelman D, Hunter DJ, Bernstein L, van den Brandt PA, Buring JE, Cho E, English DR, Flood A, Freudenheim JL, Giles GG, Giovannucci E, Håkansson N, Horn-Ross PL, Jacobs EJ, Leitzmann MF, Marshall JR, McCullough ML, et al. Intakes of fruit, vegetables, and carotenoids and renal cell cancer risk: a pooled analysis of 13 prospective studies. Cancer Epidemiol Biomark Prev. 2009;18(6):1730–9. https://doi.org/10.1158/1055-9965.EPI-09-0045.
385. Lew JQ, Chow W-H, Hollenbeck AR, Schatzkin A, Park Y. Alcohol consumption and risk of renal cell cancer: the NIH-AARP diet and health study. Br J Cancer. 2011;104(3):537–41. https://doi.org/10.1038/sj.bjc.6606089.
386. Chiu WA, Caldwell JC, Keshava N, Scott CS. Key scientific issues in the health risk assessment of trichloroethylene. Environ Health Perspect. 2006;114(9):1445–9. https://doi.org/10.1289/ehp.8690.
387. Scott CS, Chiu WA. Trichloroethylene cancer epidemiology: a consideration of select issues. Environ Health Perspect. 2006;114(9):1471–8. https://doi.org/10.1289/ehp.8949.
388. Henschler D, Vamvakas S, Lammert M, Dekant W, Kraus B, Thomas B, Ulm K. Increased incidence of renal cell tumors in a cohort of cardboard workers exposed to trichloroethene. Arch Toxicol. 1995;69(5):291–9. https://doi.org/10.1007/s002040050173.
389. Buzio L, Tondel M, De Palma G, Buzio C, Franchini I, Mutti A, Axelson O. Occupational risk factors for renal cell cancer. An Italian case-control study. Med Lav. 2002;93(4):303–9.
390. Chow W-H, Devesa SS. Contemporary epidemiology of renal cell cancer. Cancer J. 2008;14(5):288–301. https://doi.org/10.1097/PPO.0b013e3181867628.

391. Hu J. Renal cell carcinoma and occupational exposure to chemicals in Canada. Occup Med. 2002;52(3):157–64. https://doi.org/10.1093/occmed/52.3.157.
392. Mattioli S, Truffelli D, Baldasseroni A, Risi A, Marchesini B, Giacomini C, Bacchini P, Violante FS, Buiatti E. Occupational risk factors for renal cell cancer: a case-control study in northern Italy. J Occup Environ Med. 2002;44(11):1028–36. https://doi.org/10.1097/00043764-200211000-00009.
393. Parent ME, Hua Y, Siemiatycki J. Occupational risk factors for renal cell carcinoma in Montreal. Am J Ind Med. 2000;38(6):609–18. https://doi.org/10.1002/1097-0274(200012)38:6<609::aid-ajim1>3.0.co;2-4.
394. Pesch B, Haerting J, Ranft U, Klimpel A, Oelschlägel B, Schill W. Occupational risk factors for renal cell carcinoma: agent-specific results from a case-control study in Germany. Int J Epidemiol. 2000;29(6):1014–24. https://doi.org/10.1093/ije/29.6.1014.
395. Ryerson AB, Eheman CR, Altekruse SF, Ward JW, Jemal A, Sherman RL, Henley SJ, Holtzman D, Lake A, Noone A-M, Anderson RN, Ma J, Ly KN, Cronin KA, Penberthy L, Kohler BA. Annual report to the nation on the status of cancer, 1975-2012, featuring the increasing incidence of liver cancer: report on status of cancer, 1975-2012. Cancer. 2016;122(9):1312–37. https://doi.org/10.1002/cncr.29936.
396. Tapper EB, Parikh ND. Mortality due to cirrhosis and liver cancer in the United States, 1999-2016: observational study. BMJ. 2018;k2817 https://doi.org/10.1136/bmj.k2817.
397. Wong MCS, Jiang JY, Goggins WB, Liang M, Fang Y, Fung FDH, Leung C, Wang HHX, Wong GLH, Wong VWS, Chan HLY. International incidence and mortality trends of liver cancer: a global profile. Sci Rep. 2017;7(1):45846. https://doi.org/10.1038/srep45846.
398. El-Serag HB, Rudolph KL. Hepatocellular carcinoma: epidemiology and molecular carcinogenesis. Gastroenterology. 2007;132(7):2557–76.
399. Watson J, Hydon K, Lodge P. Primary and secondary liver tumours. InnovAiT: Education and Inspiration forGeneral Practice. 2016;9(8):477–82. https://doi.org/10.1177/1755738016653419.
400. Ledda C, Loreto C, Zammit C, Marconi A, Fago L, Matera S, Costanzo V, Sanzà GF, Palmucci S, Ferrante M, Costa C, Fenga C, Biondi A, Pomara C, Rapisarda V. Non-infective occupational risk factors for hepatocellular carcinoma: a review. Mol Med Rep. 2017;15(2):511–33. https://doi.org/10.3892/mmr.2016.6046.
401. Melaram R. Environmental risk factors implicated in liver disease: a mini-review. Front Public Health. 2021;9:683719.
402. Jepsen P, Ott P, Andersen PK, Sørensen HT, Vilstrup H. Risk for hepatocellular carcinoma in patients with alcoholic cirrhosis: a Danish nationwide cohort study. Ann Intern Med. 2012;156(12):841. https://doi.org/10.7326/0003-4819-156-12-201206190-00004.
403. McGlynn KA, London WT. Epidemiology and natural history of hepatocellular carcinoma. Best Pract Res Clin Gastroenterol. 2005;19(1):3–23. https://doi.org/10.1016/j.bpg.2004.10.004.
404. Vieira VM, Hoffman K, Shin H-M, Weinberg JM, Webster TF, Fletcher T. Perfluorooctanoic acid exposure and cancer outcomes in a contaminated community: a geographic analysis. Environ Health Perspect. 2013;121(3):318–23. https://doi.org/10.1289/ehp.1205829.
405. Dich J, Zahm SH, Hanberg A, Adami H-O, Adami H-O. No title found. Cancer Causes Control. 1997;8(3):420–43. https://doi.org/10.1023/A:1018413522959.
406. Williams JH, Phillips TD, Jolly PE, Stiles JK, Jolly CM, Aggarwal D. Human aflatoxicosis in developing countries: a review of toxicology, exposure, potential health consequences, and interventions. Am J Clin Nutr. 2004;80(5):1106–22. https://doi.org/10.1093/ajcn/80.5.1106.
407. Gomez-Quiroz LE, Roman S. Influence of genetic and environmental risk factors in the development of hepatocellular carcinoma in Mexico. Ann Hepatol. 2022;27:100649. https://doi.org/10.1016/j.aohep.2021.100649.
408. Forrester LM, Neal GE, Judah DJ, Glancey MJ, Wolf CR. Evidence for involvement of multiple forms of cytochrome P-450 in aflatoxin B1 metabolism in human liver. Proc Natl Acad Sci. 1990;87(21):8306–10. https://doi.org/10.1073/pnas.87.21.8306.

409. IARC Working Group on the Evaluation of Carcinogenic Risks to Humans. DDT, lindane, and 2,4-D. IARC monographs on the evaluation of carcinogenic risks to humans. Lyon (FR); 2018.
410. VoPham T, Bertrand KA, Hart JE, Laden F, Brooks MM, Yuan J-M, Talbott EO, Ruddell D, Chang C-CH, Weissfeld JL. Pesticide exposure and liver cancer: a review. Cancer Causes Control. 2017;28(3):177–90. https://doi.org/10.1007/s10552-017-0854-6.
411. IARC Monographs on the Evaluation of Carcinogenic Risks to Humans, No. 109. IARC working group on the evaluation of carcinogenic risks to humans. Lyon: International Agency for Research on Cancer; 2016.
412. International Agency for Research on Cancer. Outdoor air pollution. In: IARC monographs on the evaluation of carcinogenic risks to humans. Geneva: WHO Press; 2016.
413. Pan W-C, Wu C-D, Chen M-J, Huang Y-T, Chen C-J, Su H-J, Yang H-I. Fine particle pollution, alanine transaminase, and liver cancer: a Taiwanese prospective cohort study (REVEAL-HBV). JNCI J Natl Cancer Inst. 2016;108(3) https://doi.org/10.1093/jnci/djv341.
414. Pasetto R, Ranzi A, De Togni A, Ferretti S, Pasetti P, Angelini P, Comba P. Cohort study of residents of a district with soil and groundwater industrial waste contamination. Annali Dell'Istituto Superiore Di Sanita. 2013;49(4):354–7. https://doi.org/10.4415/ANN_13_04_06.
415. Pedersen M, Andersen ZJ, Stafoggia M, Weinmayr G, Galassi C, Sørensen M, Eriksen KT, Tjønneland A, Loft S, Jaensch A, Nagel G, Concin H, Tsai M-Y, Grioni S, Marcon A, Krogh V, Ricceri F, Sacerdote C, Ranzi A, et al. Ambient air pollution and primary liver cancer incidence in four European cohorts within the ESCAPE project. Environ Res. 2017;154:226–33. https://doi.org/10.1016/j.envres.2017.01.006.
416. Su Y, Zhao B, Guo F, Bin Z, Yang Y, Liu S, Han Y, Niu J, Ke X, Wang N, Geng X, Jin C, Dai Y, Lin Y. Interaction of benzo[a]pyrene with other risk factors in hepatocellular carcinoma: a case-control study in Xiamen, China. Ann Epidemiol. 2014;24(2):98–103. https://doi.org/10.1016/j.annepidem.2013.10.019.
417. Tian M, Zhao B, Zhang J, Martin FL, Huang Q, Liu L, Shen H. Association of environmental benzo[a]pyrene exposure and DNA methylation alterations in hepatocellular carcinoma: a Chinese case–control study. Sci Total Environ. 2016;541:1243–52. https://doi.org/10.1016/j.scitotenv.2015.10.003.
418. Chen J, Xie P, Li L, Xu J. First identification of the hepatotoxic microcystins in the serum of a chronically exposed human population together with indication of hepatocellular damage. Toxicol Sci. 2009;108(1):81–9. https://doi.org/10.1093/toxsci/kfp009.
419. Shi L, Du X, Liu H, Chen X, Ma Y, Wang R, Tian Z, Zhang S, Guo H, Zhang H. Update on the adverse effects of microcystins on the liver. Environ Res. 2021;195:110890. https://doi.org/10.1016/j.envres.2021.110890.
420. Siegel RL, Kimberly DM, Ahmedin J. Cancer statistics, 2019. CA Cancer J Clin. 2019;69(1):7–34.
421. American Cancer Society. Key statistics about kidney cancer. About pancreatic cancer. 2019. https://www.cancer.org/cancer/pancreatic-cancer/about/what-is-pancreatic-cancer.
422. Sarantis P, Koustas E, Papadimitropoulou A, Papavassiliou AG, Karamouzis MV. Pancreatic ductal adenocarcinoma: treatment hurdles, tumor microenvironment and immunotherapy. World J Gastrointest Oncol. 2020;12(2):173–81. https://doi.org/10.4251/wjgo.v12.i2.173.
423. Johns Hopkins Medicine. Pancreatic cancer types. 2022. https://www.hopkinsmedicine.org/health/conditions-and-diseases/pancreatic-cancer/pancreatic-cancer-types.
424. Brown HA, Dotto J, Robert M, Salem RR. Squamous cell carcinoma of the pancreas. J Clin Gastroenterol. 2005;39(10):915–9. https://doi.org/10.1097/01.mcg.0000180636.74387.e6.
425. Wahab A, Gonzalez JJ, Devarkonda V, Saint-Phard T, Singh T, Adekolujo OS. Squamous cell carcinoma—a rare pancreatic exocrine malignancy. Cancer Biol Ther. 2019;20(5):593–6. https://doi.org/10.1080/15384047.2018.1539291.
426. Madura JA. Adenosquamous carcinoma of the pancreas. Arch Surg. 1999;134(6):599. https://doi.org/10.1001/archsurg.134.6.599.

427. Adsay NV, Pierson C, Sarkar F, Abrams J, Weaver D, Conlon KC, Brennan MF, Klimstra DS. Colloid (mucinous noncystic) carcinoma of the pancreas. Am J Surg Pathol. 2001;25(1):26–42. https://doi.org/10.1097/00000478-200101000-00003.
428. American Cancer Society. Family cancer syndromes. 2020a. https://www.cancer.org/healthy/cancer-causes/genetics/family-cancer-syndromes.html.
429. American Cancer Society. About pancreatic neuroendocrine tumors. 2020b. https://www.cancer.org/cancer/pancreatic-neuroendocrine-tumor/about/what-is-pnet.html.
430. Lucenteforte E, La Vecchia C, Silverman D, Petersen GM, Bracci PM, Ji BT, Bosetti C, Li D, Gallinger S, Miller AB, et al. Alcohol consumption and pancreatic cancer: a pooled analysis in the international pancreatic cancer Case-control consortium (PanC4). Ann Oncol. 2012;23:374–82.
431. Raimondi S, Maisonneuve P, Lowenfels AB. Epidemiology of pancreatic cancer: an overview. Nat Rev Gastroenterol Hepatol. 2009;6(12):699–708. https://doi.org/10.1038/nrgastro.2009.177.
432. Lugo A, Peveri G, Bosetti C, Bagnardi V, Crippa A, Orsini N, Rota M, Gallus S. Strong excess risk of pancreatic cancer for low frequency and duration of cigarette smoking: a comprehensive review and meta-analysis. Eur J Cancer. 2018;104:117–26. https://doi.org/10.1016/j.ejca.2018.09.007.
433. Bosetti C, Lucenteforte E, Silverman DT, Petersen G, Bracci PM, Ji BT, Negri E, Li D, Risch HA, Olson SH, Gallinger S, Miller AB, Bueno-de-Mesquita HB, Talamini R, Polesel J, Ghadirian P, Baghurst PA, Zatonski W, Fontham E, et al. Cigarette smoking and pancreatic cancer: an analysis from the international pancreatic cancer Case-control consortium (Panc4). Ann Oncol. 2012;23(7):1880–8. https://doi.org/10.1093/annonc/mdr541.
434. Lynch SM, Vrieling A, Lubin JH, Kraft P, Mendelsohn JB, Hartge P, Canzian F, Steplowski E, Arslan AA, Gross M, Helzlsouer K, Jacobs EJ, LaCroix A, Petersen G, Zheng W, Albanes D, Amundadottir L, Bingham SA, Boffetta P, et al. Cigarette smoking and pancreatic cancer: a pooled analysis from the pancreatic cancer cohort consortium. Am J Epidemiol. 2009;170(4):403–13. https://doi.org/10.1093/aje/kwp134.
435. Ding Y, Yu C, Han Z, Xu S, Li D, Meng X, Chen D. Environmental tobacco smoke and pancreatic cancer: a case-control study. Int J Clin Exp Med. 2015;8(9):16729–32.
436. Blackford A, Parmigiani G, Kensler TW, Wolfgang C, Jones S, Zhang X, Parsons DW, Lin JC-H, Leary RJ, Eshleman JR, Goggins M, Jaffee EM, Iacobuzio-Donahue CA, Maitra A, Klein A, Cameron JL, Olino K, Schulick R, Winter J, et al. Genetic mutations associated with cigarette smoking in pancreatic cancer. Cancer Res. 2009;69(8):3681–8. https://doi.org/10.1158/0008-5472.CAN-09-0015.
437. Porta M, Crous-Bou M, Wark PA, Vineis P, Real FX, Malats N, Kampman E. Cigarette smoking and K-ras mutations in pancreas, lung and colorectal adenocarcinomas: etiopathogenic similarities, differences and paradoxes. Mutat Res Rev Mutat Res. 2009;682(2–3):83–93. https://doi.org/10.1016/j.mrrev.2009.07.003.
438. Momi N, Ponnusamy MP, Kaur S, Rachagani S, Kunigal SS, Chellappan S, Ouellette MM, Batra SK. Nicotine/cigarette smoke promotes metastasis of pancreatic cancer through α7nAChR-mediated MUC4 upregulation. Oncogene. 2013;32(11):1384–95. https://doi.org/10.1038/onc.2012.163.
439. Askari MDF, Tsao M-S, Schuller HM. The tobacco-specific carcinogen, 4-(methylnitrosamino)-1-(3-pyridyl)-1-butanone stimulates proliferation of immortalized human pancreatic duct epithelia through β-adrenergic transactivation of EGF receptors. J Cancer Res Clin Oncol. 2005;131(10):639–48. https://doi.org/10.1007/s00432-005-0002-7.
440. Weddle DL. Beta-adrenergic growth regulation of human cancer cell lines derived from pancreatic ductal carcinomas. Carcinogenesis. 2001;22(3):473–9. https://doi.org/10.1093/carcin/22.3.473.
441. Park C-H, Lee I-S, Grippo P, Pandol SJ, Gukovskaya AS, Edderkaoui M. Akt kinase mediates the prosurvival effect of smoking compounds in pancreatic ductal cells. Pancreas. 2013;42(4):655–62. https://doi.org/10.1097/MPA.0b013e3182762928.
442. Edderkaoui M, Thrower E. Smoking and pancreatic disease. J Cancer Ther. 2013;4:34–40.

443. Genkinger JM, Spiegelman D, Anderson KE, Bergkvist L, Bernstein L, van den Brandt PA, English DR, Freudenheim JL, Fuchs CS, Giles GG, Giovannucci E, Hankinson SE, Horn-Ross PL, Leitzmann M, Männistö S, Marshall JR, McCullough ML, Miller AB, Reding DJ, et al. Alcohol intake and pancreatic cancer risk: a pooled analysis of fourteen cohort studies. Cancer Epidemiol Biomark Prev. 2009;18(3):765–76. https://doi.org/10.1158/1055-9965.EPI-08-0880.
444. Tuma DJ, Casey CA. Dangerous byproducts of alcohol breakdown—focus on adducts. Alcohol Res Health. 2003;27(4):285–90.
445. Yu H-S, Oyama T, Isse T, Kitagawa K, Pham T-T-P, Tanaka M, Kawamoto T. Formation of acetaldehyde-derived DNA adducts due to alcohol exposure. Chem Biol Interact. 2010;188(3):367–75. https://doi.org/10.1016/j.cbi.2010.08.005.
446. Tsai H-J, Chang JS. Environmental risk factors of pancreatic cancer. J Clin Med. 2019;8(9):1427. https://doi.org/10.3390/jcm8091427.
447. Chang JS, Hsiao J-R, Chen C-H. ALDH2 polymorphism and alcohol-related cancers in Asians: a public health perspective. J Biomed Sci. 2017;24(1):19. https://doi.org/10.1186/s12929-017-0327-y.
448. Paluszkiewicz P, Smolińska K, Dębińska I, Turski WA. Main dietary compounds and pancreatic cancer risk. The quantitative analysis of case–control and cohort studies. Cancer Epidemiol. 2012;36(1):60–7. https://doi.org/10.1016/j.canep.2011.05.004.
449. Zheng W, Lee S-A. Well-done meat intake, heterocyclic amine exposure, and cancer risk. Nutr Cancer. 2009;61(4):437–46. https://doi.org/10.1080/01635580802710741.
450. Anderson KE, Mongin SJ, Sinha R, Stolzenberg-Solomon R, Gross MD, Ziegler RG, Mabie JE, Risch A, Kazin SS, Church TR. Pancreatic cancer risk: associations with meat-derived carcinogen intake in the prostate, lung, colorectal, and ovarian cancer screening trial (PLCO) cohort. Mol Carcinog. 2012;51(1):128–37. https://doi.org/10.1002/mc.20794.
451. Chan JM, Gong Z, Holly EA, Bracci PM. Dietary patterns and risk of pancreatic cancer in a large population-based case-control study in the San Francisco Bay Area. Nutr Cancer. 2013;65(1):157–64. https://doi.org/10.1080/01635581.2012.725502.
452. Jansen RJ, Robinson DP, Stolzenberg-Solomon RZ, Bamlet WR, de Andrade M, Oberg AL, Rabe KG, Anderson KE, Olson JE, Sinha R, Petersen GM. Nutrients from fruit and vegetable consumption reduce the risk of pancreatic cancer. J Gastrointest Cancer. 2013;44(2):152–61. https://doi.org/10.1007/s12029-012-9441-y.
453. Taunk P, Hecht E, Stolzenberg-Solomon R. Are meat and heme iron intake associated with pancreatic cancer? Results from the NIH-AARP diet and health cohort: meat and Heme iron intake and pancreatic cancer. Int J Cancer. 2016;138(9):2172–89. https://doi.org/10.1002/ijc.29964.
454. Wu Q-J, Wu L, Zheng L-Q, Xu X, Ji C, Gong T-T. Consumption of fruit and vegetables reduces risk of pancreatic cancer: evidence from epidemiological studies. Eur J Cancer Prev. 2016;25(3):196–205. https://doi.org/10.1097/CEJ.0000000000000171.
455. Casari I, Falasca M. Diet and pancreatic cancer prevention. Cancers. 2015;7(4):2309–17. https://doi.org/10.3390/cancers7040892.
456. Falasca M, Casari I, Maffucci T. Cancer chemoprevention with nuts. JNCI J Natl Cancer Inst. 2014;106(9):dju238–dju238. https://doi.org/10.1093/jnci/dju238.
457. Mao Q-Q, Lin Y-W, Chen H, Qin J, Zheng X-Y, Xu X, Xie L-P. Dietary fiber intake is inversely associated with risk of pancreatic cancer: a meta-analysis. Asia Pac J Clin Nutr. 2017;26(1):89–96. https://doi.org/10.6133/apjcn.102015.03.
458. Rawla P, Thandra KC, Sunkara T. Pancreatic cancer and obesity: epidemiology, mechanism, and preventive strategies. Clin J Gastroenterol. 2019;12(4):285–91.
459. Gerlovin H, Michaud DS, Cozier YC, Palmer JR. Oral health in relation to pancreatic cancer risk in African American women. Cancer Epidemiol Biomark Prev. 2019;28(4):675–9. https://doi.org/10.1158/1055-9965.EPI-18-1053.
460. Maisonneuve P, Amar S, Lowenfels AB. Periodontal disease, edentulism, and pancreatic cancer: a meta-analysis. Ann Oncol. 2017;28(5):985–95. https://doi.org/10.1093/annonc/mdx019.

461. Stolzenberg-Solomon RZ, Dodd KW, Blaser MJ, Virtamo J, Taylor PR, Albanes D. Tooth loss, pancreatic cancer, and helicobacter pylori. Am J Clin Nutr. 2003;78:176–81.
462. Fan X, Alekseyenko AV, Wu J, Peters BA, Jacobs EJ, Gapstur SM, Purdue MP, Abnet CC, Stolzenberg-Solomon R, Miller G, Ravel J, Hayes RB, Ahn J. Human oral microbiome and prospective risk for pancreatic cancer: a population-based nested case-control study. Gut. 2018;67(1):120–7. https://doi.org/10.1136/gutjnl-2016-312580.
463. Michaud DS, Izard J, Wilhelm-Benartzi CS, You D-H, Grote VA, Tjønneland A, Dahm CC, Overvad K, Jenab M, Fedirko V, Boutron-Ruault MC, Clavel-Chapelon F, Racine A, Kaaks R, Boeing H, Foerster J, Trichopoulou A, Lagiou P, Trichopoulos D, et al. Plasma antibodies to oral bacteria and risk of pancreatic cancer in a large European prospective cohort study. Gut. 2013;62(12):1764–70. https://doi.org/10.1136/gutjnl-2012-303006.
464. Coussens LM, Werb Z. Inflammation and cancer. Nature. 2002;420(6917):860–7. https://doi.org/10.1038/nature01322.
465. Hayashi C, Gudino CV, Gibson FC III, Genco CA. REVIEW: pathogen-induced inflammation at sites distant from oral infection: bacterial persistence and induction of cell-specific innate immune inflammatory pathways: pathogen-induced inflammation. Mol Oral Microbiol. 2010;25(5):305–16. https://doi.org/10.1111/j.2041-1014.2010.00582.x.
466. Andreotti G, Silverman DT. Occupational risk factors and pancreatic cancer: a review of recent findings. Mol Carcinog. 2012;51(1):98–108. https://doi.org/10.1002/mc.20779.
467. US Department of Health and Human Services. Cadmium and cadmium compounds. Report on carcinogens. 12th ed; 2011. p. 80.
468. American Cancer Society. Cancer facts & figures 2013. Atlanta: American Cancer Society; 2013.
469. Moffitt Cancer Center. Types of stomach cancer 2022. https://moffitt.org/cancers/stomach-gastric-cancer/diagnosis/types.
470. Bray F, Ferlay J, Soerjomataram I, Siegel RL, Torre LA, Jemal A. Global cancer statistics 2018: GLOBOCAN estimates of incidence and mortality worldwide for 36 cancers in 185 countries. CA Cancer J Clin. 2018;68(6):394–424. https://doi.org/10.3322/caac.21492.
471. Parkin DM, Stjernswärd J, Muir CS. Estimates of the worldwide frequency of twelve major cancers. Bull World Health Organ. 1984;62(2):163–82.
472. Yin J, Wu X, Li S, Li C, Guo Z. Impact of environmental factors on gastric cancer: a review of the scientific evidence, human prevention and adaptation. J Environ Sci. 2020;89:65–79. https://doi.org/10.1016/j.jes.2019.09.025.
473. Malnasi G, Jakab S, Incze A, Apostol A, Csap JM, Szab E, et al. An assay for selecting high risk population for gastric cancer by studying environmental factors. Neoplasma. 1976;23:333e341.
474. Zhang XH, Zhao WY, Zhang ZG, Wang JL, Yan X, Xie TX, et al. Comparative studies on drinking water between high and relatively low risk areas of gastric cancer in Zanhuang county. Chin J Publ Heal. 1999;18:172e173.
475. Tromp SW, Diehl JC. A statistical study on the possible influence of soil and other geological conditions on cancer. Experientia. 1954;10(12):510–8. https://doi.org/10.1007/BF02166190.
476. Chen BY. Bio-geochemical characteristics of high and lowincidence area of stomach cancer in the coastal area of Fujian Province. Geol Fujian. 2008;27:29e36.
477. Ren H, Wan X, Yang F, Shi X, Xu J, Zhuang D, Yang G. Association between changing mortality of digestive tract cancers and water pollution: a case study in the Huai River basin, China. Int J Environ Res Public Health. 2014;12(1):214–26. https://doi.org/10.3390/ijerph120100214.
478. Wang ZQ, Chen Y, Lin YC, Wang ZH, Su XB, He J, et al. Relationship between quality of dringking water and gastric cancer mortality from 11 counties in Fujian Province. Chin JPubl Heal. 1997;16:79e80.
479. Yuan W, Yang N, Li X. Advances in understanding how heavy metal pollution triggers gastric cancer. Biomed Res Int. 2016;2016:1–10. https://doi.org/10.1155/2016/7825432.
480. Wang M, Song H, Chen W-Q, Lu C, Hu Q, Ren Z, Yang Y, Xu Y, Zhong A, Ling W. Cancer mortality in a Chinese population surrounding a multi-metal sulphide mine in

Guangdong province: an ecologic study. BMC Public Health. 2011a;11(1):319. https://doi.org/10.1186/1471-2458-11-319.
481. Wang M, Xu Y, Pan S, Zhang J, Zhong A, Song H, Ling W. Long-term heavy metal pollution and mortality in a Chinese population: an ecologic study. Biol Trace Elem Res. 2011b;142(3):362–79. https://doi.org/10.1007/s12011-010-8802-2.
482. Gunduz O, Bakar C, Simsek C, Baba A, Elci A, Gurleyuk H, Mutlu M, Cakir A. Statistical analysis of causes of death (2005–2010) in villages of Simav plain, Turkey, with high arsenic levels in drinking water supplies. Arch Environ Occup Health. 2015;70(1):35–46. https://doi.org/10.1080/19338244.2013.872076.
483. Beaumont J, Sedman R, Reynolds S, Sherman C, Li L-H, Howd R, Sandy M, Zeise L, Alexeeff G. Analysis of cancer mortality data from five villages in China with hexavalent chromium-contaminated drinking water. Am J Epidemiol. 2006;163(suppl_11):S115. https://doi.org/10.1093/aje/163.suppl_11.S115-c.
484. Chiu H-F, Kuo C-H, Tsai S-S, Chen C-C, Wu D-C, Wu T-N, Yang C-Y. Effect modification by drinking water hardness of the association between nitrate levels and gastric cancer: evidence from an ecological study. J Toxic Environ Health A. 2012;75(12):684–93. https://doi.org/10.1080/15287394.2012.688486.
485. Jensen OM. Nitrate in drinking water and cancer in northern Jutland, Denmark, with special reference to stomach cancer. Ecotoxicol Environ Saf. 1982;6(3):258–67. https://doi.org/10.1016/0147-6513(82)90016-1.
486. Sandor J, Kiss I, Farkas O, Ember I. No title found. Eur J Epidemiol. 2001;17(5):443–7. https://doi.org/10.1023/A:1013765016742.
487. Taneja P, Labhasetwar P, Nagarnaik P, Ensink JHJ. The risk of cancer as a result of elevated levels of nitrate in drinking water and vegetables in Central India. J Water Health. 2017;15(4):602–14. https://doi.org/10.2166/wh.2017.283.
488. Yang C-Y, Cheng M-F, Tsai S-S, Hsieh Y-L. Calcium, magnesium, and nitrate in drinking water and gastric cancer mortality. Jpn J Cancer Res. 1998;89(2):124–30. https://doi.org/10.1111/j.1349-7006.1998.tb00539.x.
489. Blondell JM. The anticarcinogenic effect of magnesium. Med Hypotheses. 1980;6(8):863–71. https://doi.org/10.1016/0306-9877(80)90010-9.
490. Lipkin M, Newmark H. Effect of added dietary calcium on colonic epithelial-cell proliferation in subjects at high risk for familial colonic cancer. N Engl J Med. 1985;313(22):1381–4. https://doi.org/10.1056/NEJM198511283132203.
491. Legon CD. A note on geographical variations in cancer mortality, with special reference to gastric cancer in Wales. Br J Cancer. 1951;5(2):175–9. https://doi.org/10.1038/bjc.1951.19.
492. Legon CD. Aetiological significance of geographical variations in cancer mortality. BMJ. 1952;2(4786):700–2. https://doi.org/10.1136/bmj.2.4786.700.
493. Jackson ML, Zhang JZ, Li CS, Martin DF. The geochemical availability of soil Zn and Mo in relation to stomach and esophageal cancer in the People's Republic of China and U.S.A. Appl Geochem. 1986;1:487e492.
494. Nagel G, Stafoggia M, Pedersen M, Andersen ZJ, Galassi C, Munkenast J, Jaensch A, Sommar J, Forsberg B, Olsson D, Oftedal B, Krog NH, Aamodt G, Pyko A, Pershagen G, Korek M, De Faire U, Pedersen NL, Östenson C-G, et al. Air pollution and incidence of cancers of the stomach and the upper aerodigestive tract in the European study of cohorts for air pollution effects (ESCAPE). Int J Cancer. 2018;143(7):1632–43. https://doi.org/10.1002/ijc.31564.
495. Weinmayr G, Pedersen M, Stafoggia M, Andersen ZJ, Galassi C, Munkenast J, Jaensch A, Oftedal B, Krog NH, Aamodt G, Pyko A, Pershagen G, Korek M, De Faire U, Pedersen NL, Östenson C-G, Rizzuto D, Sørensen M, Tjønneland A, et al. Particulate matter air pollution components and incidence of cancers of the stomach and the upper aerodigestive tract in the European study of cohorts of air pollution effects (ESCAPE). Environ Int. 2018;120:163–71. https://doi.org/10.1016/j.envint.2018.07.030.
496. Lee YY, Derakhshan MH. Environmental and lifestyle risk factors of gastric cancer. Arch Iran Med. 2013;16(6):358–65.

497. Kreuzer M, Straif K, Marsh JW, Dufey F, Grosche B, Nosske D, Sogl M. Occupational dust and radiation exposure and mortality from stomach cancer among German uranium miners, 1946–2003. Occup Environ Med. 2012;69(3):217–23. https://doi.org/10.1136/oemed-2011-100051.
498. Wilkinson GS. Gastric cancer in New Mexico counties with significant deposits of uranium. Arch Environ Health Int J. 1985;40(6):307–12. https://doi.org/10.1080/00039896.1985.10545938.
499. Amani F, Ahari SS, Barzegari S, Hassanlouei B, Sadrkabir M, Farzaneh E. Analysis of relationships between altitude and distance from volcano with stomach cancer incidence using a geographic information system. Asian Pac J Cancer Prev. 2015;16(16):6889–94. https://doi.org/10.7314/APJCP.2015.16.16.6889.
500. Goto H, Watanabe T, Miyao M, Fukuda H, Sato Y, Oshida Y. Cancer mortality among atomic bomb survivors exposed as children. Environ Health Prev Med. 2012;17(3):228–34. https://doi.org/10.1007/s12199-011-0246-6.
501. Sauvaget C, Lagarde F, Nagano J, Soda M, Koyama K, Kodama K. Lifestyle factors, radiation and gastric cancer in atomic-bomb survivors (Japan). Cancer Causes Control. 2005;16(7):773–80. https://doi.org/10.1007/s10552-005-5385-x.
502. Wiseman M. The second World Cancer Research Fund/American Institute for Cancer Research expert report. Food, nutrition, physical activity, and the prevention of cancer: a global perspective: nutrition society and BAPEN medical symposium on 'nutrition support in cancer therapy'. Proc Nutr Soc. 2008;67(3):253–6. https://doi.org/10.1017/S002966510800712X.
503. World Health Organization. Diet, nutrition and the prevention of chronic diseases, vol. 916.:i-viii, 1–149, backcover. World Health Organ Tech Rep Ser; 2003.
504. D'Elia L, Rossi G, Ippolito R, Cappuccio FP, Strazzullo P. Habitual salt intake and risk of gastric cancer: a meta-analysis of prospective studies. Clin Nutr. 2012;31(4):489–98. https://doi.org/10.1016/j.clnu.2012.01.003.
505. Fox JG, Dangler CA, Taylor NS, King A, Koh TJ, Wang TC. High-salt diet induces gastric epithelial hyperplasia and parietal cell loss, and enhances helicobacter pylori colonization in C57BL/6 mice. Cancer Res. 1999;59(19):4823–8.
506. Kato S, Tsukamoto T, Mizoshita T, Tanaka H, Kumagai T, Ota H, Katsuyama T, Asaka M, Tatematsu M. High salt diets dose-dependently promote gastric chemical carcinogenesis inHelicobacter pylori-infected Mongolian gerbils associated with a shift in mucin production from glandular to surface mucous cells. Int J Cancer. 2006;119(7):1558–66. https://doi.org/10.1002/ijc.21810.
507. Loh JT, Torres VJ, Cover TL. Regulation of *Helicobacter pylori cagA* expression in response to salt. Cancer Res. 2007;67(10):4709–15. https://doi.org/10.1158/0008-5472.CAN-06-4746.
508. Takahashi M, Nishikawa A, Furukawa F, Enami T, Hasegawa T, Hayashi Y. Dose-dependent promoting effects of sodium chloride (NaCI) on rat glandular stomach carcinogenesis initiated with *Nmethyl-N'*-nitro- *N*-nitrosoguanidine. Carcinogenesis. 1994;15(7):1429–32. https://doi.org/10.1093/carcin/15.7.1429.
509. Tatematsu M, Takahashi M, Fukushima S, Hananouchi M, Shirai T. Effects in rats of sodium chloride on experimental gastric cancers induced by N-methyl-N′-nitro-N-nitrosoguanidine or 4-nitroquinoline-1-oxide2. JNCI: J Natl Cancer Inst. 1975;55(1):101–6. https://doi.org/10.1093/jnci/55.1.101.
510. Koizumi Y, Tsubono Y, Nakaya N, Kuriyama S, Shibuya D, Matsuoka H, Tsuji I. Cigarette smoking and the risk of gastric cancer: a pooled analysis of two prospective studies in Japan. Int J Cancer. 2004;112(6):1049–55. https://doi.org/10.1002/ijc.20518.
511. Ladeiras-Lopes R, Pereira AK, Nogueira A, Pinheiro-Torres T, Pinto I, Santos-Pereira R, Lunet N. Smoking and gastric cancer: systematic review and meta-analysis of cohort studies. Cancer Causes Control. 2008;19(7):689–701. https://doi.org/10.1007/s10552-008-9132-y.
512. Nishino Y, Inoue M, Tsuji I, Wakai K, Nagata C, Mizoue T, Tanaka K, Tsugane S. Tobacco smoking and gastric cancer risk: an evaluation based on a systematic review of epidemiologic evidence among the Japanese population. Jpn J Clin Oncol. 2006;36(12):800–7. https://doi.org/10.1093/jjco/hyl112.

513. Nomura AMY, Wilkens LR, Henderson BE, Epplein M, Kolonel LN. The association of cigarette smoking with gastric cancer: the multiethnic cohort study. Cancer Causes Control. 2012;23(1):51–8. https://doi.org/10.1007/s10552-011-9854-0.
514. Li LF, Chan RLY, Lu L, Shen J, Zhang L, Wu WKK, Wang L, Hu T, Li MX, Cho CH. Cigarette smoking and gastrointestinal diseases: the causal relationship and underlying molecular mechanisms (review). Int J Mol Med. 2014;34(2):372–80. https://doi.org/10.3892/ijmm.2014.1786.
515. Shin VY, Cho C-H. Nicotine and gastric cancer. Alcohol. 2005;35(3):259–64. https://doi.org/10.1016/j.alcohol.2005.04.007.
516. Duell EJ, Travier N, Lujan-Barroso L, Clavel-Chapelon F, Boutron-Ruault M-C, Morois S, Palli D, Krogh V, Panico S, Tumino R, Sacerdote C, Quirós JR, Sánchez-Cantalejo E, Navarro C, Gurrea AB, Dorronsoro M, Khaw K-T, Allen NE, Key TJ, et al. Alcohol consumption and gastric cancer risk in the European prospective investigation into cancer and nutrition (EPIC) cohort. Am J Clin Nutr. 2011;94(5):1266–75. https://doi.org/10.3945/ajcn.111.012351.
517. Rota M, Pelucchi C, Bertuccio P, Matsuo K, Zhang Z-F, Ito H, Hu J, Johnson KC, Palli D, Ferraroni M, Yu G-P, Muscat J, Lunet N, Peleteiro B, Ye W, Song H, Zaridze D, Maximovitch D, Guevara M, et al. Alcohol consumption and gastric cancer risk-A pooled analysis within the StoP project consortium: alcohol consumption and gastric cancer risk. Int J Cancer. 2017;141(10):1950–62. https://doi.org/10.1002/ijc.30891.
518. Tramacere I, Negri E, Pelucchi C, Bagnardi V, Rota M, Scotti L, Islami F, Corrao G, La Vecchia C, Boffetta P. A meta-analysis on alcohol drinking and gastric cancer risk. Ann Oncol. 2012;23(1):28–36. https://doi.org/10.1093/annonc/mdr135.
519. Wang P-L, Xiao F-T, Gong B-C, Liu F-N. Alcohol drinking and gastric cancer risk: a meta-analysis of observational studies. Oncotarget. 2017;8(58):99013–23. https://doi.org/10.18632/oncotarget.20918.
520. Travis WD, Brambilla E, Nicholson AG, Yatabe Y, Austin JHM, Beasley MB, Chirieac Lucian R, Dacic S, Duhig E, Flieder DB, Geisinger K, Hirsch FR, Ishikawa Y, Kerr KM, Noguchi M, Pelosi G, Powell CA, Tsao MS, Wistuba I. The 2015 World Health Organization classification of lung tumors. J Thorac Oncol. 2015;10(9):1243–60. https://doi.org/10.1097/JTO.0000000000000630.
521. Zappa C, Mousa SA. Non-small cell lung cancer: current treatment and future advances. Transl Lung Cancer Res. 2016;5(3):288–300. https://doi.org/10.21037/tlcr.2016.06.07.
522. GLOBOCAN. International Agency for Research on Cancer. WHO; 2018. http://gco.iarc.fr/.
523. U.S. Department of Health and Human Services. The health consequences of smoking-50 years of progress: a report of the surgeon general. Atlanta: U.S. Department of Health and Human Services, Centers for Disease Control and Prevention, National Center for Chronic Disease Prevention and Health Promotion, Office on Smoking and Health; 2014.
524. Division of Cancer Prevention and Control, Centers for Disease Control and Prevention. What are the risk factors for lung cancer? 2022. https://www.cdc.gov/cancer/lung/basic_info/risk_factors.htm.
525. World Health Organization. Global health risks: mortality and burden of disease attributable to selected major risks. Geneva: World Health Organization; 2009. http://www.who.int/iris/handle/10665/44203.
526. IARC. Household use of solid fuels and high temperature frying. IARC monographs on the evaluation of carcinogenic risks to humans. WHO; 2010.
527. IARC. Environment and Lifestyle Epidemiology Branch (ENV). 2022. https://www.iarc.who.int/branches-env/.
528. Öberg M, Woodward A, Jaakkola MS, et al. Global estimate of the burden of disease from second-hand smoke. Geneva: WHO; 2010. p. 18.
529. Avino P, Scungio M, Stabile L, Cortellessa G, Buonanno G, Manigrasso M. Second-hand aerosol from tobacco and electronic cigarettes: evaluation of the smoker emission rates and doses and lung cancer risk of passive smokers and vapers. Sci Total Environ. 2018;642:137–47. https://doi.org/10.1016/j.scitotenv.2018.06.059.

530. IARC. Agents classified by the IARC monographs, Volumes 1–123 [Online]. WHO. https://monographs.iarc.fr/agents-classified-by-the-iarc/.
531. Das S, Kundu M, Jena BC, Mandal M. Causes of cancer: physical, chemical, biological carcinogens, and viruses. In: Biomaterials for 3D tumor modeling. Elsevier; 2020. p. 607–41. https://doi.org/10.1016/B978-0-12-818128-7.00025-3.
532. Herceg Z, Vaissière T. Epigenetic mechanisms and cancer: an interface between the environment and the genome. Epigenetics. 2011;6(7):804–19. https://doi.org/10.4161/epi.6.7.16262.

Air Pollution and Cancer

2

Ethan Burns and Eric H. Bernicker

Sources of Air Pollution

While human activity has affected the environment for millennia, it is mostly since the industrial revolution that human commercial activity has led to the production of gases and particulate matter that have adverse effects on human health and the environment. Air pollution covers a wide number of constituents that are produced by varied human activities; some, such as methane, do not have significant immediate health effects but can have significant impacts on anthropogenic climate change. In modern societies, air pollution arises from the burning of wood (whether intentional or from wildfires), vegetation, coal, oil, gasoline, liquid biofuels and waste [1]. The two primary types of air pollution responsible for urban pollution are smog from the burning of coal, and photochemical smog from the reaction of emitted hydrocarbons and nitrogen oxides in the presence of sunlight [2].

The common picture of air pollution often is the rise of Dickensian air pollution in industrial London, although there is data from 2000 years ago from Greenland ice cores that the Greeks and Romans' smelting of lead caused a significant rise in lead pollution (and the Romans love of lead might have had something to do with the decline of the Empire [3]. Still, the transformation of much of the economy of the West to a capitalist economy based on the burning of fossil fuels has led to the unchecked release of carbon and particles into the environment. Because of the slow pace of health consequences of exposure to airborne particles, the adverse effects of pollution on health has very slowly come to the attention of health care workers. In addition, other societal changes in how humans live and structure their environments have maximized the conditions whereby air pollution can be damaging to

E. Burns · E. H. Bernicker (✉)
Neal Cancer Center, Houston Methodist Hospital, Houston, TX, USA
e-mail: EricBernicker@Centura.org

E. H. Bernicker (ed.), *Environmental Oncology*,
https://doi.org/10.1007/978-3-031-33750-5_2

human health: the migration of people from the country to mega -cities and slums, especially in low- and middle-income countries (LMIC) have increased the density of human lives in the shadow of industrial plans, increasing the number of individuals living in areas of poor air quality [4].

Particle emissions can be natural (volcanoes, biomass fires) or anthropogenic (fossil fuels, industrial emissions, vehicles, construction) [5]. Gases that contribute to air pollution consist of carbon monoxide, radon, sulfur dioxide, ozone and others; aerosol particle components for example are black carbon, organic matter, soil dust, sea spray, and soil dust. The main culprit when it comes to air pollutions effect on human health is particulate matter less than 2.5 M, referred to as $PM_{2.5}$.

This chapter will review the data behind air pollution and human health with a focus on cancer impact, although we will briefly discuss other adverse health consequences of air pollution.

Biology of Interaction Between Particles and the Respiratory Epithelium

The air we breathe is a mixture of fundamental components comprised of nitrogen (78%) and oxygen (21%), along with a mixture of carbon dioxide (~1%) and other gasses [6]. Environmental pollution is the product of both natural phenomena and man-made industry, which ultimately culminates in aerosolized particulate that is potentially harmful to the human respiratory tract. The harmful impact of outdoor air pollution is not a new concept; in the 1960s, when coal combustion was the primary culprit contributing to air pollution, a study discovered an association with the rising incidence of bronchitis and diminished respiratory function [7]. Fast forward to the 1980s, and ambient pollution was reported as a cause of respiratory disease and implicated as the second most common cause driving respiratory mortality in patients with pre-existing chronic obstructive pulmonary disease (COPD) [8]. Either via direct or indirect mechanisms, chronic exposure to pollution in children, adolescents, and adults have long term deleterious impacts on respiratory function, including reduction in forced vital capacity (FVC), forced expiratory volume in 1 s (FEV [1]), and higher COPD prevalence [9, 10]. Unfortunately, our growing knowledge has not been a harbinger of change, and the number of subjects with chronic respiratory diseases continues to rise. However, a mechanistic understanding that describes how chronic inhalation of air pollution damages our airway is an important first step to address a discovery made over one-half century ago.

What dilutes the air we breathe and how does it alter the anatomic and physiologic workings of our respiratory tract? Are there specific ambient compounds that independently increase the risk of respiratory disease and cancer? While the answers to these questions remains a work in progress, it is clear that air pollution has direct and indirect impacts on our airways. The following section aims to summarize what is known about these alterations.

Respiratory Epithelium

The pulmonary epithelium is the first line of defense against inhaled toxins and pathogens. It is comprised of a continuous layer of pseudostratified epithelium made up of approximately 10 [10] cells covering a surface area of 2500 cm [2] [11]. The airway epithelium is made up of 4 major cell types including ciliated, secretory, undifferentiated intermediate, and basal cells, which line a continuous basement membrane [12]. As a unit, the respiratory epithelium functions as a physical and immunologic barrier with multiple functions including mucociliary clearance, ion secretions, production of anti-inflammatory and pro-inflammatory substances, and regulating the adaptive and innate immune response. The respiratory epithelium has key roles in both inflammation and remodeling in respiratory disease [13–15]. Many air pollutants impose damage to the respiratory epithelium by causing barrier dysfunction, imposing oxidative stress, generating pro-inflammatory cytokines, and impairing mucociliary clearance [14].

Barrier Dysfunction

Multiple studies have demonstrated the association between air pollutants and airway epithelial barrier dysfunction [16–20]. $PM_{2.5}$, also known as fine PM, can be deposited at the level of the alveoli. The smallest fraction of $PM_{2.5}$ contain nanoparticles (known as ultrafine PM [UFP]) which is defined as particles ≤ 0.1 μm in aerodynamic diameter [21]. It is hypothesized that UFPs may contribute to the majority of toxicological effects in the respiratory epithelium [22]. $PM_{2.5}$ and other atmospheric agents including ozone O_3, ammonia (NH_3), and sulfur dioxide (SO_2) can cause epithelial barrier dysfunction [14, 18, 19, 23–25]. $PM_{2.5}$ disrupts and/or reduces the expression of tight junctions, E-cadherin, ZO-1, transepithelial electrical resistance (TEER), increases in paracellular permeability, and stimulates the production of reactive oxygen species (ROS) [19, 20, 26, 27].

ROS has a central role in epithelial barrier disruption. The generation of ROS may result from direct suppression of lung antioxidant mechanisms, which increases the concentration of harmful ROS [14]. PM may result in the reduction of transcriptional responses including activation of superoxide dismutase, nuclear factor ER-related factor 2, and antioxidant responsive element [27, 28]. Individual components that make up PM, including quinones, polyaromatic hydrocarbons (PAH), and other inorganic materials such as transition metals, nitrogen oxide (NOx), and O_3 increase free radical burden [29]. PAH also drives mitochondrial dysfunction, which causes dysregulated ROS generation and oxidative phosphorylation [14, 30–33]. Elevation in ROS induces upregulation of epidermal growth factor and downstream extracellular signal-related kinase (ERK), which impairs both tight and adherens junctions and results in epithelial barrier disruption [34]. Other studies have shown that PM-induced EGFR activation in airway epithelial cells (AECs) induces barrier disruption [35, 36].

Mucociliary Dysfunction

Ciliated cells are the dominant cell type of the airway. The apical surface of ciliated cells contains numerous microvilli which have important roles in transepithelial movement of fluids and electrolytes [37]. Ciliated cells are connected to one another by tight junctions, which are structures that regulate homeostatic movement of solutes and ions across the respiratory epithelium, and E-cadherin based junctions, which promote cell-cell adhesion [11, 15]. The primary role of ciliated cells is to maintain the mucociliary escalator, a highly coordinated in-plane beating of cilia to facilitate movement of the mucus gel layer across the epithelial surface [38]. The job of these cells is contingent on the number, length, and beat frequency of the cilia.

A dysfunctional escalator diminishes the epithelial defense system, resulting in lung disease and infection. There are both inherited and acquired causes for this. Cigarette smoking is a well-recognized factor that suppresses mucociliary clearance immediately after smoking, and chronically in those that eventually develop bronchitis [39, 40]. Smoking marijuana and cocaine can also impact the escalator by resultant loss of cilia and goblet cell hyperplasia, which may correlate with increased mucus secretion and dysfunctional mucociliary clearance [41, 42].

The effect of air pollutants on ciliary cells is well described. Nasal biopsies from individuals residing in Mexico City exposed to high levels of air pollution showed patches of short cilia and regions of cilia loss [43]. Diesel exhaust may reduce ciliary beat frequency amongst bronchial epithelial cells [44, 45]. Exposure of bronchial epithelial cells to NO_2 also impacts ciliary function [44]. Air pollutant exposure may also have a time dependent effect. While 24-hour exposure with higher concentrations of $PM_{2.5}$ (6 and 12 μg·mm [2]) reduced ciliary beat in cultured nasal epithelial cells, 12-hour exposure with lower concentrations of $PM_{2.5}$ (1.5 μg·mm − 2) promoted ciliary function, suggesting an adaptive response in ciliary beat upon exposure to lower concentrations and shorter PM exposures [46]. Loss of cilia has been observed in mice exposed to $PM_{2.5}$ for 28 days [47]. Chronic exposure of rats to airborne PM for 7 months induced COPD-like phenotypes in the airways with an increase in mucin production and mucus metaplasia [48]. Other pollutants have a notable dose-dependent exposure effect. Nasal samples of nonsmokers exposed to 0.4 ppm of O_3 or 0.75 ppm SO_2 at 0.75 ppm failed to show alterations to ciliary structure, and only animal models exposed to higher levels of ozone (4 ppm) had disruption of trachea cilia structure, with blebbing and vesiculation of ciliary membranes [49].

Other compounds frequently seen with indoor air pollution, such as formaldehyde, acrolein, and ammonia, appear to have detrimental effects on the frequency of ciliary beating and mucus flow [11, 50]. Workplace exposures such as cadmium and nickel reduces ciliary beat frequency; nickel also causes epithelial cell damage and uncoordinated cilia beats. Hairspray may reduce mucociliary clearance in hairdressers, and wood dust exposure decreases mucociliary mucus clearance and loss of ciliated epithelium [11, 50].

Immune-Mediated Alterations in Airway Epithelial Barrier Function upon Exposure to Air Pollutants

An inflammatory response to a noxious stimulus is a balancing act that eliminates the offending pathogen without destroying the normal tissue. Both the innate and adaptive arms of the immune system have crucial roles in protecting the airway, and both may be dysregulated by air pollution. Following exposure to an inhaled stimulus, respiratory epithelial cells increase production of interleukins (ILs), including IL-1α, IL-1β, IL-6, IL-8, and IL-18, and tumor necrosis factor-α (TNF -α) [51]. ILs facilitate cell-cell communication and promote an inflammatory response in the respiratory epithelium [6]. Continuous exposure to a noxious antigen results in the dysregulation and over-production of pro-inflammatory agents, which alters cellular transport mechanisms in AECs and generates ROS [52, 53]. In an absence of negative feedback, ROS has a fundamental role in tissue damage, mucous production, squamous metaplasia, disruption of cellular repair mechanisms, and ultimately, pulmonary fibrosis [53]. Thus, immune dysregulation at any level by ambient air pollution may predispose to airway epithelial barrier disruption [14].

Chronic exposure to pollutants appears to increase the risk of an inflammatory phenotype. For example, rats exposed to $PM_{2.5}$ from biomass fuels increased neutrophil recruitment and precipitated epithelial barrier disruption [48]. Accumulation of neutrophils leads to the secretion of the neutrophil-derived cytokine oncostatin M and serine proteases, which disrupts cell-cell adhesion in AECs [14, 54, 55]. $PM_{2.5}$ and other air pollutants also promotes IL-1β secretion which can reduce TEER through upregulation of human epidermal growth factor receptor 2 (HER2) [56–58]. Diesel exhaust upregulates IL-17, a proinflammatory cytokine secreted by TH17 regulatory cells [59]. IL-17 activates NF-κB subunit p65, leading to hypersecretion of MUC5B and MUC5AC and thus enhances mucus production, suppression of E-Cadherin expression, and precipitates mitochondrial dysfunction which augments ROS production [59–61]. In contrast, IL-33, a cytokine that increases in response to O_3, diesel exhaust and $PM_{2.5}$ exposure, has a protective effect on AECs, and in vivo has demonstrated the restoration of epithelial barrier effects including E-cadherin, ZO-1 and claudin-4 proteins [25, 62].

Chronic overactivation of the adaptive immune response, including cytokines produced by T-helper (Th)1, Th2, and Th17 cells precipitates AEC damage in patients with pre-existing asthma and COPD [14, 34, 63]. The adaptive arm of the immune system is augmented from exposure to air pollutants and overproduction of pro-inflammatory cytokines including IFN-γ, TNF-α, IL-13, IL-17A and IL-22 interact with their receptors expressed on AEC, which culminates in barrier dysfunction [59, 64, 65]. IFN-γ and TNF-α produced by Th1 cells and stimulated by airway exposure to pollution disrupts tight junctions through EGFR mediated activation [48, 66]. Th2 phenotype, like what is seen in patients with allergic asthma, is observed in $PM_{2.5}$ exposure and increases circulating Th2 cytokines IL-4 and IL-13. Mice models with chronic exposure to $PM_{2.5}$ had E-cadherin and claudin-1 expression which disrupted tight junctions through Janus-associated kinase (JAK) [56, 63, 64, 67].

Alveolar Macrophages

Alveolar macrophages (AMs) reside on the respiratory epithelial surface and are a pivotal early line of lung defense and barrier immunity. Through phagocytosis of invading pathogens, they prevent tissue damage and introduce essential innate and adaptive immune responses [68]. The inhalation of polluted particles induces an immune response initiated by AMs and AECs in the lungs. Chronic exposure to air pollution impedes the ability of macrophages to phagocytose pathogens [69]. Furthermore, O_3 exposure decreases macrophage efferocytosis, a process that prevents secondary necrosis and the production of proinflammatory cytokines [70–72]. Autophagy, a self-regulatory process that allows removal of damaged organelles, cellular debris, and intracellular pathogens for homeostatic balance, is inhibited by generation of ROS and other factors from PM inhalation [73]. In addition to reducing macrophage function, PM promotes apoptosis of AMs. This is precipitated by PM binding of scavenger receptor class A, diminished expression of Bcl-2, increased expression of Bax, and dysregulated oxidative stress from augmented ROS production [74–76].

Air Pollution and Cancer Development

Until recently, the association between ambient air pollution exposure and cancer risk was grossly underestimated. Why do we say this? First, there was a six-decade gap between Doll and Hill's pivotal study suggesting an association between pollution (vehicle exhaust, tar from road paving, coal combustion, and tobacco smoke) and cancer risk, and when air pollution was finally classified as a Group 1 human carcinogenic risk by the International Agency for Research on Cancer in 2013 (IARC) [77, 78]. Second, it is possible that the risk of air pollution was considered diminutive when studied in the shadow of tobacco smoking. Furthermore, many of the epidemiologic studies assessing the association with pollution and cancer risk were likely diluted with confounders including tobacco smoking and other factors that were associated with tumorigenesis [21]. A renewed interest in recent decades is shedding light on airway exposures and carcinogenic risk. For instance, secondhand tobacco smoke, ambient air pollution, radon, asbestos exposure, household burning of fuels, and occupational exposures including rubber manufacturing, paving, roofing, painting, and chimney sweeping, are now well-known risk factors for lung cancer [21, 79, 80].

There is a growing number of epidemiologic studies implicating ambient air pollution as a risk factor for cancer development. In fact, air pollution exposure in low-middle and low socioeconomic regions may now pose a greater risk than cigarette smoking [81]. $PM_{2.5}$ is associated with a greater risk of lung-cancer related mortality in several studies, and in 2017, the global proportion of lung cancer deaths attributable to ambient $PM_{2.5}$ was second only to tobacco smoking (14.1% vs 63.2%) [82, 83]. In a recent study, the hazard ratio (HR) of lung cancer mortality was 1.13 (1.07–1.20), with similar risks across North America, Europe, and Asia [84].

Worldwide, ambient $PM_{2.5}$ is estimated to have contributed to 265,267 lung cancer deaths (95% uncertainty interval [UI], 182,903–350,835 lung cancer deaths) in 2017, or 14.1% (95% UI, 9.8%–18.7%) of all lung cancer deaths [83]. In 2014, a large meta-analysis in predominately North America and Europe reported a 9% (95% confidence interval [CI] 4%, 14%), increase in lung cancer incidence or mortality for each 10 μg/m3 increase of $PM_{2.5}$ exposure [85].

Other inhalational toxins have been studied. In 2015, an epidemiological study assessing the relationship of exposures of NOx from traffic and air pollution to lung cancer found a meta-estimate of 3% (95% CI: 1%, 5%) for a 10-μg/m3 increase in exposure to NO_2 [86]. Another study found a higher risk of lung cancer mortality (relative risk [RR] 1.05 [95% CI: 1.02, 1.08] [87]. While other studies noted that NO_2 exposure increases the RR of death from all cancers (1.021; 95% CI: 1.017, 1.025),several Danish studies did not find a meaningful association between NO_2, O_3, and lung cancer mortality [88–90].

While this risk is now well-established, a definitive mechanistic explanation continues to be elusive; however, there are a growing number of studies attempting to unlock these answers. Like the multifactorial impacts on AECs, air pollution likely precipitates dysplasia, the first step in carcinogenesis via several different mechanisms. Air pollution is theorized to follow a multistep process that includes initiation, promotion, and progression [21, 91]. Studies have shown that the number of mutations occurring in patients with lung cancer attributed to air pollution is 3 times higher than in patients from low-exposed regions [92]. It is theorized that individual genetics, time-dependent doses of pollutants, and inhalation of specific carcinogens are all likely to play a role in carcinogenesis. Specific carcinogens interfere with various molecular processes, through direct mechanism such as DNA binding, transcriptional or epigenetic modification, oncogene activation, TP53 activation, chromosomal instability, induction of cellular proliferation, gene mutations, or indirect mechanisms such as generation of ROS [21]. The presence and retention of known mutagens and carcinogens likely also play a role. Following inhalation, retained particles and gas can generate both local and systemic inflammatory effects and oxidative stress [93]. PAH (benzo[a]pyrene and polar compounds) are known to result in covalent bonds with DNA, also termed DNA-adduct, which is a well-documented risk factor for human carcinogenesis [94–98]. Dioxin (2,3,7,8-Tetrachlorodibenzo-p-dioxin) has been associated with a number of malignancies across a range of species and is postulated to be through its interaction with the Ah receptor, a transcriptional enhancer that interacts with a number of downstream regulatory cascades including heat shock proteins, kinases, and DNA binding species [99]. In mice treated with inhaled SO2, there were dose-dependent increases DNA damage in multiple organs, suggesting that inhalation can cause carcinogenesis [91, 100–107].

Transcriptional changes brought on by $PM_{2.5}$ in bronchial cells may precipitate neoplastic transformation [108]. For instance, down regulation of miRNA-802 leads to dysregulation of the actin cytoskeleton and resultant cellular dysplasia [47]. In another study of respiratory cell lines treated with $PM_{2.5}$, miRNA was suppressed and aberrant expression of mRNA was noted, along with increased oncogenic expression of SLC30A1, SERPINB2 and AKR1C1, which may serve as an explanation for carcinogenesis [108]. $PM_{2.5}$ causes transcriptional changes in hundreds of genes, influencing inflammatory responses, immune regulation, generation of ROS, and DNA damage repair mechanisms, all of which may increase the risk for carcinogenesis [109].

Epigenetic modification may also augment carcinogenesis. Repeated exposure to $PM_{2.5}$ may induce epigenetic silencing of TP53 in human alveolar cells [110]. TP53 is a tumor suppressor gene that regulates cellular proliferation, apoptosis, and damage repair. When mutated, TP53 has been implicated in a wide range of malignancies. Outdoor air pollution has also been linked to post-translational modification of histones, hydroxymethylation, and methylation; hypermethylation can result in gene silencing and hypomethylating can result in chromosomal instability [111–114]. Other studies have demonstrated that human epithelial cells exposed to $PM_{2.5}$ are susceptible to hypomethylation and transcriptional activation of several genes and miRNA, which can substantially modify oncogenic signaling pathways [21, 115].

In lung cancer models, there may be a relationship between pollution exposure and driver mutations. In a recent report presented at the European Society of Medical Oncology (ESMO), an analysis of the association of PM exposure in mice models in cancer promotion was assessed. Notably, increasing PM levels were associated with increased risk of cancer, including EGFR mutated NSCLC in England, South Korea, and Taiwan, along with an increased risk of mesothelioma (HR = 1.19), lung (HR = 1.16), anal (HR = 1.23), small intestine (HR = 1.30), glioblastoma multiforme (HR = 1.19), lip, oral cavity and pharynx (HR: 1.15) and laryngeal carcinomas (HR = 1.26) for each 1ug/m3 $PM_{2.5}$ increment. Mechanistically, PM promoted a macrophage response and a progenitor-like state in lung epithelium harboring mutant EGFR, and increased tumor burden in three EGFR or KRAS driven lung cancer models in a dose dependent manner. PM also upregulated the IL1B inflammatory axis, with anti-IL1B therapy preventing PM-induced mouse tumor formation further suggestive of a mechanistic driver [116].

As we get closer to fully illuminating the causality between air pollution and carcinogenesis, one thing is becoming apparent: there are potentially multiple mechanisms and carcinogens in the air we breathe.

Air Pollution as Complicating Factor in Establish Cancer patient's Care

$PM_{2.5}$ is a known contributor to the global burden of disease and mortality rates [117, 118]. As previously described, there is accumulating data associating chronic air pollution exposure to the risk of cancer and cancer specific mortality and growing data hypothesizing various mechanisms of carcinogenesis. There is also data amassing on the implications of air pollution on patients with known cancers. A deeper understanding of the long-term impacts of air pollution is necessary to determine strategies to reduce exposures in this patient population.

Multiple pollutants have been linked with detrimental outcomes in patients with cancer. In a population-based study of 352,053 patients with newly diagnosed lung cancer in California between 1988–2009, HR for all-cause mortality were 1.13 (95% CI 1.12, 1.13) for NO_2, 1.02 (95% CI 1.02, 1.03) for O_3, 1.11 (95% CI 1.11, 1.12) for PM_{10}, and 1.16 (95% CI 1.16, 1.17) for $PM_{2.5}$, which were all similar for lung cancer-specific mortality. Also notable in this study, the HR associated with these pollutants were largest in early-stage non-small cell cancers, particularly adenocarcinoma, and lower in small cell lung cancers [119]. In the Cancer Prevention II (CPS-II) study, $PM_{2.5}$ was associated with higher mortality with kidney cancers (HR per 4.4 μg/m3 = 1.14 [95% CI 1.03, 1.27]), bladder cancer (HR = 1.13 [95% CI 1.03, 1.23]), and NO_2 was associated with colorectal cancer mortality (HR per 6.5 ppb = 1.06 [95% CI 1.02, 1.10]). In patients with breast cancer in California, NOx (per 50 ppb) and NO_2 (per 20 ppb), $PM_{2.5}$ (per 10 μg/m3), and PM_{10} (per 10 μg/m3) were associated with higher all-cause mortality (HR range = 1.13–1.25), breast cancer (HR range = 1.19–1.45), and cardiovascular disease mortality (HR range = 1.37–1.60) [120]. In a study of older individuals in Hong Kong, $PM_{2.5}$ exposure (HR per 10 μg/m [3]) was associated with increased risk of mortality for all "natural causes" of death (HR 1.13 [95% CI 1.08, 1.19]), all causes of cancer (HR, 1.22 [95% CI 1.11, 1.34]), cancers of all digestive organs (HR 122 [95% CI 1.05, 1.42]), breast (HR 1.80 [95% CI 1.26, 2.55]) and genital tract (HR 1.73 [95% CI 1.17, 2.54]) in females; and lung (HR 1.36 [95% CI 1.05, 1.77]) in males [121]. In a recent analysis of the Surveillance, Epidemiology, and End Results (SEER) Program data linked with county-level estimates of long-term average concentrations of $PM_{2.5}$ from 2000 to 2016, $PM_{2.5}$ had higher all-cause mortality (HR 1.01 [95% CI 1.00, 1.03]) per 10 μg/m3 increase in $PM_{2.5}$. In addition, $PM_{2.5}$ was associated with cardiovascular (HR 1.32 [95% CI = 1.26, 1.39]), COPD (HR 1.10 [95% CI 1.01, 1.20]), influenza/pneumonia (HR 1.55 [95% CI 1.33, 1.80]), and cardiopulmonary mortality (HR 1.25 [95% CI 1.21, 1.30]). $PM_{2.5}$-cardiopulmonary mortality was higher for cancer patients who received chemotherapy or radiation treatments, and statistically significant adverse $PM_{2.5}$-mortality associations for cancer mortality were observed for oral and oropharyngeal, rectal, skin, breast, and ill-defined cancer types. Although no associations were observed between exposure to $PM_{2.5}$ and all-cancer survival, there were associations for cancers with relatively high survival rates (HR 1.05 [95% CI 1.01, 1.10]) and for those individuals treated with chemotherapy or radiation (HR 1.33 [95% CI 1.25, 1.42]). A total of 26% of cancer patients died of noncancer causes, primarily from cardiopulmonary disease [122]. Chemotherapy or radiation may have adverse effects on the cardiovascular and respiratory systems, which may contribute to susceptibility from air pollution exposure.

Aside from an increased risk of cancer specific death, these studies showed a higher risk of all-cause mortality, death from infections, cardiovascular and pulmonary diseases, and higher mortality rates in patients who received chemotherapy and radiation. These findings may be confounded by factors including patient-level demographics, therapy utilized in cancer treatment, and socioeconomic status, to name a few. However, these epidemiologic studies demonstrate on a global scale the detrimental impact of air pollution in individuals with cancer.

A keen awareness of the potential health influences of air pollutions in patients with pre-existing comorbidities can improve cancer patient care. Cancer is frequently a disease of older patients, and many individuals will have other medical comorbidities that may compromise the oncologist's ability to provide standard of care therapies. Cancer therapies may have deleterious effects on the immune, respiratory, and cardiovascular systems, all of which may be further impacted by exposure to air pollution. For example, the Multi-Ethnic Study of Atherosclerosis and Air Pollution (MESA) study found that increased long-term exposure to $PM_{2.5}$, NOx, and O_3 was associated with progression of coronary atherosclerosis [123]. Air pollution has also been associated with elevations in blood pressure and heightened risk of cardiovascular mortality from myocardial ischemia, arrhythmias, and heart failure [124, 125]. Multiple cancer therapies including chemotherapeutics, monoclonal antibodies, and newer generation targeted therapies have been implicated in heart failure, worsening hypertension, and generating arrhythmias and whether the combination of air pollution and approved systemic therapies have a higher combined effect on patient cardiovascular comorbidities needs to be evaluated.

The association of air pollution and morbidity from COPD and other respiratory diseases is well-established [126, 127]. Progression of air-way disease or frequent hospitalizations arising from complications of airway disease may cause delays in provision of lifesaving/prolonging oncologic therapies or prevent clinicians from using therapies known to cause pulmonary toxicities. For instance, Bleomycin, a chemotherapy backbone in testicular cancers, can impart a significant risk of pulmonary fibrosis and fatal pulmonary disease. Multiple other therapies including immunecheckpoint inhibitor therapies, antibody drug conjugates, and targeted therapies can increase the risk of pneumonitis, which may be fatal in patients with pre-existing pulmonary diseases.

Long-term exposure to $PM_{2.5}$, PM_{10}, and NO_2 has been associated with an increased risk of the development of diabetes, a risk that is more pronounced in overweight and obese individuals [128, 129]. Certain oncologic therapies may also augment this risk, primarily in patients with hematologic malignancies such as multiple myeloma or non-Hodgkin lymphoma, where regular high dose steroids are a recommended therapeutic strategy. Other treatments such as immune checkpoint inhibitor therapies can also rarely predispose to immune mediated diabetes. The complication of diabetes, including peripheral neuropathy or diabetic ketoacidosis may also limit therapies, such as the platinum-based chemotherapies known to precipitate or worsen pre-existing neuropathies.

The genitourinary system has also been implicated as a bystander in air pollution mediated damage. Exposure to PM_{10}, carbon monoxide, and NO_2 has been associated with progression of chronic kidney disease, end stage renal disease, and reduction in glomerular filtration rate [130]. Many cancer therapies such as platinum chemotherapies, methotrexate, pemetrexed, and cytarabine require either dose modifications or are contraindicated depending on the severity of renal dysfunction, due to risk of worsening kidney function or treatment toxicities.

The pro-inflammatory state enhanced by chronic air pollution exposure may also impact patients with rheumatologic diseases. A Canadian study demonstrated that

$PM_{2.5}$ exposure may be associated with an increased risk of systemic autoimmune disease [131]. Traffic pollution and O_3 may be associated with a higher incidence of rheumatoid arthritis [132]. While rheumatologic diseases have many independent long-term health implications, they can also impact oncologic care. For instance, immune checkpoint inhibitor therapy may exacerbate pre-existing autoimmune disease. Furthermore, the immune suppressive therapy often utilized for these diseases, in combination with immunosuppressive effects of cancer and cytotoxic chemotherapies predispose patients to infections, a common culprit in morbidity and mortality in this patient population.

PM may impact efficacy of chemotherapeutic agents. For instance, cellular models exposed to $PM_{2.5}$ decreased cytotoxic effects of doxorubicin by preventing intracellular accumulation [133]. Chronic $PM_{2.5}$ can also modify the extracellular matrix of cells and upregulate the NOTCH signaling pathway which gives cells stem-cell like qualities, which may impact chemotherapy efficacy [134].

Air Pollution and Other Non-cancerous Illnesses

It is not difficult to recognize that the biological processes triggered by exposure to $PM_{2.5}$ would not be solely confined to carcinogenesis. Recent research has solidified our understanding of respiratory illnesses caused by polluted air, most of which are intuitively obvious, such as the link between air quality and asthma exacerbations in children growing up in polluted urban environments [135]. However increasing data suggests that exposure to particulate matter can affect everything from pregnancy outcomes, childhood development, coronary disease and dementia and cognitive decline. While it is beyond the scope of this chapter to review all of the adverse health outcomes from pollution that harm human health, we will highlight some recent data that concerns the health of future generations.

Data continues to accrue that exposure to air pollution above a certain level confers an increased risk of dementia. Calderon-Garciduenas et al. examined 134 autopsies on long time citizens of Mexico City under the age of 30, a sprawling congested and polluted mega-city [136]. Notably, 99% of the examined brains had pathologic evidence of changes similar to those seen in Alzheimer's disease—absent in patients who had not lived in highly polluted cities. While there was some criticism of the findings being more suggestive than causative, further studies have strengthened the link between breathing in $PM_{2.5}$ particles and neurodegeneration over the life span. Chandra et al. performed a systematic review that looked at 84 studies that described associations between exposure to air pollutants and an increased risk of lower cognitive function among children and adolescents [137]. The authors also found a relationship between cognitive impairment seen in adults and dementia among older adults, coupled with supportive evidence of neuroimaging and inflammatory biomarkers. The data is suggestive to the point that the Lancet Commission in 2020 recognized air pollution as a risk factor for the development of dementia [138]. In many cases, the primary source of particles involved in urban smog comes from vehicular sources, so action towards improving public transportation and improving fuel standards could make a substantial impact on the cognitive health of future generations.

Vohra et al. used a chemical transport model GEOS-Chem to estimate global exposure to fossil fuel generated $PM_{2.5}$ in 2012; they estimated a global total of 10.2 million premature deaths annually, with the greatest mortality impacts being in China, India and the eastern USA [139]. Further projects sampled observations of air pollution from space-based instruments over 46 fast-growing tropical cities between 2005–2018 [140]. They discovered significant increases in N02, reactive VOCs, and AOD in megacities 2–3 times faster than seen in national and regional trends. Alongside these increases, the authors calculated a significant increase in 13,000 premature deaths per year as the population of these cities face increasing pollution.

Teasing out the relationship between exposure to air pollution and health outcomes is difficult, especially as there are often other factors that impact health outcomes…tobacco use, obesity, co-morbid conditions—that are often mixed in the toxic brew of rural communities which are often health care deserts. Nevertheless, there is data that air pollution can adversely affect outcomes of pregnancy. Tran et al. looked at pregnancy outcome data in mother's related to their proximity to oil and gas production (OGD) [141]. The authors performed a retrospective cohort study of 2,918,089 births to mothers living within 10 km of at least one production well between January 1, 2006 and December 31, 2015. They examined associations between overall and trimester specific OGD exposures and term birth weight (tBW), low birth weight (LBW), preterm birth (PTB), and small for gestational age birth (SGA). They also attempted to assess effect modification by urban/rural community type. They found that in rural areas, increasing production volume was associated with increasing the probability of adverse outcomes such as low birth weight.

Further work has looked at exposure to air pollution during fetal life and subsequent cognitive performance. Guxens et al. performed a study to ascertain whether air pollution exposure during fetal life alters brain morphology and whether the imaging changes were associated between maternal air pollution exposure during pregnancy and the child's cognitive function in school [142]. They found that children exposed to higher particulate matter levels during fetal life had thinner cortex in several brain regions of both hemispheres. Specific studies have looked at the relationship between certain pollutants and cognitive outcomes. Choi et al. analyzed the associations between prenatal sulfur dioxide exposure and genome wide DNA methylation levels at ages 2 and 6 [143]. They found that changes in DNA methylation levels associated with prenatal sulfur dioxide exposure during early childhood were associated with increases in attention deficit hyperactivity disorder (ADHD) in later childhood. Thus, as data continues to be confirmed that air pollution significantly impacts the pulmonary and neurocognitive health of children and that there are significant health disparities in who is exposed to these particles, it is imperative that health care workers and legislators be aware of this data and work to mitigate air pollution.

Summary and Call to Action

Air pollution remains a significant cause of morbidity and mortality across the world. In addition to the obvious concerns of fossil fuel production adding to the increasing amount of greenhouse gases in the atmosphere driving anthropogenic global warming, there are immediate and longer-term adverse health outcomes to people exposed to particulate matter. In a planetary system where noncommunicable diseases will undoubtedly continue to cause human suffering as well as strain the delivery of health care, the increased rates of cancer, coronary disease and cognitive decline will pose a significant challenge to improving human health. Thus, attempts to educate the public about the harms of pollution and the various remedies that can have significant effects are an essential job for health care workers to perform as we work to improve human flourishing.

References

1. Jacobson MZ. Air pollution and global warming: history, science and solutions. Cambridge University Press; 2012. p. 101.
2. Jacobson MZ. Air pollution and global warming: history, science and solutions. Cambridge University Press; 2012. p. 73.
3. Hong S, Candelone J, Patterson C, Boutron C. History of ancient copper smelting pollution during Roman and medieval times recorded in Greenland ice. Science. 1996;272:246–9.
4. Davis M. Planet of slums, vol. 23. London: Verso; 2006. p. 6–11.
5. Jacobson MZ. Air pollution and global warming: history, science and solutions. Cambridge University Press; 2012. p. 103.
6. Albano GD, Montalbano AM, Gagliardo R, Anzalone G, Profita M. Impact of air pollution in airway diseases: role of the epithelial cells (cell models and biomarkers). Int J Mol Sci. 2022;23(5):2799.
7. Holland WW, Reid DD. The urban factor in chronic bronchitis. Lancet. 1965;1:445–8.
8. Schwartz J. Lung function and chronic exposure to air pollution: a cross-sectional analysis of NHANES II. Environ Res. 1989;50:309–21.
9. Doiron D, de Hoogh K, Probst-Hensch N, Fortier I, Cai Y, De Matteis S, et al. Air pollution, lung function and COPD: results from the population-based UK biobank study. Eur Respir J. 2019;54(1):1802140.
10. Gauderman WJ, Avol E, Gilliland F, Vora H, Thomas D, Berhane K, et al. The effect of air pollution on lung development from 10 to 18 years of age. N Engl J Med. 2004;351:1057–67.
11. Tilley AE, Walters MS, Shaykhiev R, Crystal RG. Cilia dysfunction in lung disease. Annu Rev Physiol. 2015;77:379–406.
12. Crystal RG, Randell SH, Engelhardt JF, Voynow J, Sunday ME. Airway epithelial cells: current concepts and challenges. Proc Am Thorac Soc. 2008;5:772–7.
13. Tam A, Wadsworth S, Dorscheid D, Man SF, Sin DD. The airway epithelium: more than just a structural barrier. Ther Adv Respir Dis. 2011;5:255–73.
14. Aghapour M, Ubags ND, Bruder D, Hiemstra PS, Sidhaye V, Rezaee F, et al. Role of air pollutants in airway epithelial barrier dysfunction in asthma and COPD. Eur Respir Rev. 2022;31(163):210112.
15. Knight DA, Holgate ST. The airway epithelium: structural and functional properties in health and disease. Respirology. 2003;8:432–46.
16. Rynning I, Neca J, Vrbova K, Libalova H, Rossner P Jr, Holme JA, et al. In vitro transformation of human bronchial epithelial cells by diesel exhaust particles: gene expression profiling and early toxic responses. Toxicol Sci. 2018;166(1):51–64.

17. Fukuoka A, Matsushita K, Morikawa T, Takano H, Yoshimoto T. Diesel exhaust particles exacerbate allergic rhinitis in mice by disrupting the nasal epithelial barrier. Clin Exp Allergy. 2016;46(1):142–52.
18. Byun J, Song B, Lee K, Kim B, Hwang HW, Ok MR, et al. Identification of urban particulate matter-induced disruption of human respiratory mucosa integrity using whole transcriptome analysis and organ-on-a chip. J Biol Eng. 2019;13:88.
19. Zhao C, Wang Y, Su Z, Pu W, Niu M, Song S, et al. Respiratory exposure to PM2.5 soluble extract disrupts mucosal barrier function and promotes the development of experimental asthma. Sci Total Environ. 2020;730:139145.
20. Xian M, Ma S, Wang K, Lou H, Wang Y, Zhang L, et al. Particulate matter 2.5 causes deficiency in barrier integrity in human nasal epithelial cells. Allergy asthma. Immunol Res. 2020;12(1):56–71.
21. Turner MC, Andersen ZJ, Baccarelli A, Diver WR, Gapstur SM, Pope CA 3rd, et al. Outdoor air pollution and cancer: an overview of the current evidence and public health recommendations. CA Cancer J Clin. 2020;70:460–79. https://doi.org/10.3322/caac.21632.
22. Leikauf GD, Kim SH, Jang AS. Mechanisms of ultrafine particle-induced respiratory health effects. Exp Mol Med. 2020;52:329–37.
23. Wang X, Wang M, Chen S, Wei B, Gao Y, Huang L, Liu C, Huang T, Yu M, Zhao SH, Li X. Ammonia exposure causes lung injuries and disturbs pulmonary circadian clock gene network in a pig study. Ecotoxicol Environ Saf. 2020;205:111050.
24. Joelsson JP, Kricker JA, Arason AJ, Sigurdsson S, Valdimarsdottir B, Gardarsson FR, Page CP, Lehmann F, Gudjonsson T, Ingthorsson S. Azithromycin ameliorates sulfur dioxide-induced airway epithelial damage and inflammatory responses. Respir Res. 2020;21(1):233.
25. Michaudel C, Mackowiak C, Maillet I, Fauconnier L, Akdis CA, Sokolowska M, et al. Ozone exposure induces respiratory barrier biphasic injury and inflammation controlled by IL-33. J Allergy Clin Immunol. 2018;142(3):942–58.
26. Sidhaye VK, Chau E, Breysse PN, King LS. Septin-2 mediates airway epithelial barrier function in physiologic and pathologic conditions. Am J Respir Cell Mol Biol. 2011;45(1):120–6.
27. Zarcone MC, Duistermaat E, van Schadewijk A, Jedynska A, Hiemstra PS, Kooter IM. Cellular response of mucociliary differentiated primary bronchial epithelial cells to diesel exhaust. Am J Physiol Lung Cell Mol Physiol. 2016;311(1):L111–23.
28. Pardo M, Qiu X, Zimmermann R, Rudich Y. Particulate matter toxicity is Nrf2 and mitochondria dependent: the roles of metals and polycyclic aromatic hydrocarbons. Chem Res Toxicol. 2020;33(5):1110–20.
29. Lodovici M, Bigagli E. Oxidative stress and air pollution exposure. J Toxicol. 2011;2011:487074.
30. Aghapour M, Remels AHV, Pouwels SD, Bruder D, Hiemstra PS, Cloonan SM, et al. Mitochondria: at the crossroads of regulating lung epithelial cell function in chronic obstructive pulmonary disease. Am J Physiol Lung Cell Mol Physiol. 2020;318(1):L149–64.
31. Ferecatu I, Borot MC, Bossard C, Leroux M, Boggetto N, Marano F, et al. Polycyclic aromatic hydrocarbon components contribute to the mitochondria-antiapoptotic effect of fine particulate matter on human bronchial epithelial cells via the aryl hydrocarbon receptor. Part Fibre Toxicol. 2010;7:18.
32. Gurbani D, Bharti SK, Kumar A, Pandey AK, Ana GR, Verma A, et al. Polycyclic aromatic hydrocarbons and their quinones modulate the metabolic profile and induce DNA damage in human alveolar and bronchiolar cells. Int J Hyg Environ Health. 2013;216(5):553–65.
33. Leclercq B, Kluza J, Antherieu S, Sotty J, Alleman LY, Perdrix E, et al. Air pollution-derived PM2.5 impairs mitochondrial function in healthy and chronic obstructive pulmonary diseased human bronchial epithelial cells. Environ Pollut. 2018;243(Pt B):1434–49.
34. Aghapour M, Raee P, Moghaddam SJ, Hiemstra PS, Heijink IH. Airway epithelial barrier dysfunction in chronic obstructive pulmonary disease: role of cigarette smoke exposure. Am J Respir Cell Mol Biol. 2018;58(2):157–69.
35. Pourazar J, Blomberg A, Kelly FJ, Davies DE, Wilson SJ, Holgate ST, et al. Diesel exhaust increases EGFR and phosphorylated C-terminal Tyr 1173 in the bronchial epithelium. Part Fibre Toxicol. 2008;5:8.

36. Blanchet S, Ramgolam K, Baulig A, Marano F, Baeza-Squiban A. Fine particulate matter induces amphiregulin secretion by bronchial epithelial cells. Am J Respir Cell Mol Biol. 2004;30(4):421–7.
37. Chang MMJ, Shih L, Wu R. Pulmonary epithelium: cell types and functions. In: Proud D, editor. The pulmonary epithelium in health and disease. John Wiley & Songs, Ltd: Chichester; 2008. p. 1–16.
38. Knowles MR, Boucher RC. Mucus clearance as a primary innate defense mechanism for mammalian airways. J Clin Invest. 2002;109:571–7.
39. Albert RE, Lippmann M, Briscoe W. The characteristics of bronchial clearance in humans and the effects of cigarette smoking. Arch Environ Health. 1969;18:738–55.
40. Foster WM, Langenback EG, Bergofsky EH. Disassociation in the mucociliary function of central and peripheral airways of asymptomatic smokers. Am Rev Respir Dis. 1985;132:633–9.
41. Fligiel SE, Roth MD, Kleerup EC, Barsky SH, Simmons MS, Tashkin DP. Tracheobronchial histopathology in habitual smokers of cocaine, marijuana, and/or tobacco. Chest. 1997;112:319–26.
42. Roth MD, Arora A, Barsky SH, Kleerup EC, Simmons M, Tashkin DP. Airway inflammation in young marijuana and tobacco smokers. Am J Respir Crit Care Med. 1998;157:928–37.
43. Calderon-Garciduenas L, Rodriguez-Alcaraz A, Villarreal-Calderon A, Lyght O, Janszen D, Morgan KT. Nasal epithelium as a sentinel for airborne environmental pollution. Toxicol Sci. 1998;46:352–64.
44. Riechelmann H, Kienast K, Schellenberg J, Mann WJ. An in vitro model to study effects of airborne pollutants on human ciliary activity. Rhinology. 1994;32:105–8.
45. Bayram H, Devalia JL, Sapsford RJ, Ohtoshi T, Miyabara Y, Sagai M, et al. The effect of diesel exhaust particles on cell function and release of inflammatory mediators from human bronchial epithelial cells in vitro. Am J Respir Cell Mol Biol. 1998;18(3):441–8. Xian M, Ma S, Wang K, Lou H, Wang Y, Zhang L, Wang C, Akdis CA. Particulate Matter 2.5 Causes Deficiency in Barrier Integrity in Human Nasal Epithelial Cells. Allergy Asthma Immunol Res. 2020 Jan;12(1):56–71. doi: 10.4168/aair.2020.12.1.56. PMID: 31743964; PMCID: PMC6875480
46. Jia J, Xia J, Zhang R, Bai Y, Liu S, Dan M, et al. Investigation of the impact of PM2.5 on the ciliary motion of human nasal epithelial cells. Chemosphere. 2019;233:309–18.
47. Li X, Lv Y, Gao N, Sun H, Lu R, Yang H, et al. microRNA-802/Rnd3 pathway imposes on carcinogenesis and metastasis of fine particulate matter exposure. Oncotarget. 2016;7(23):35026–43.
48. He F, Liao B, Pu J, Li C, Zheng M, Huang L, et al. Exposure to ambient particulate matter induced COPD in a rat model and a description of the underlying mechanism. Sci Rep. 2017;7:45666.
49. Carson JL, Collier AM, Fernald GW, Hu SC. Microtubular discontinuities as acquired ciliary defects in airway epithelium of patients with chronic respiratory diseases. Ultrastruct Pathol. 1994;18:327–32.
50. Pedersen M. Ciliary activity and pollution. Lung. 1990;168(Suppl):368–76.
51. Chow AW, Liang JF, Wong JS, Fu Y, Tang NL, Ko WH. Polarized secretion of interleukin (IL)-6 and IL-8 by human airway epithelia 16HBE14o- cells in response to cationic polypeptide chal-lenge. PLoS One. 2010;5:e12091.
52. Gao W, Li L, Wang Y, Zhang S, Adcock IM, Barnes PJ, et al. Bronchial epithelial cells: the key effector cells in the pathogenesis of chronic obstructive pulmonary disease? Respirology. 2015;20:722–9.
53. Beale J, Jayaraman A, Jackson DJ, Macintyre JD, Edwards MR, Walton RP, et al. Rhinovirus-induced IL-25 in asthma exacerbation drives type 2 immunity and allergic pulmonary inflammation. Sci Transl Med. 2014;6:256ra134.
54. Kao SS, Ramezanpour M, Bassiouni A, Wormald PJ, Psaltis AJ, Vreugde S. The effect of neutrophil serine proteases on human nasal epithelial cell barrier function. Int Forum Allergy Rhinol. 2019;9(10):1220–6.

55. Pothoven KL, Norton JE, Suh LA, Carter RG, Harris KE, Biyasheva A, et al. Neutrophils are a major source of the epithelial barrier disrupting cytokine oncostatin M in patients with mucosal airways disease. J Allergy Clin Immunol. 2017;139(6):1966–1978.e9.
56. Ramanathan M Jr, London NR Jr, Tharakan A, Surya N, Sussan TE, Rao X, et al. Airborne particulate matter induces nonallergic eosinophilic Sinonasal inflammation in mice. Am J Respir Cell Mol Biol. 2017;57(1):59–65.
57. Finigan JH, Faress JA, Wilkinson E, Mishra RS, Nethery DE, Wyler D, et al. Neuregulin-1-human epidermal receptor-2 signaling is a central regulator of pulmonary epithelial permeability and acute lung injury. J Biol Chem. 2011;286(12):10660–70.
58. Ma X, Yu X, Zhou Q. The IL1β-HER2-CLDN18/CLDN4 axis mediates lung barrier damage in ARDS. Aging. 2020;12:3249–65.
59. Weng CM, Lee MJ, He JR, Chao MW, Wang CH, Kuo HP. Diesel exhaust particles up-regulate interleukin-17A expression via ROS/NF-κB in airway epithelium. Biochem Pharmacol. 2018;151:1–8.
60. Fujisawa T, Chang MM, Velichko S, Thai P, Hung LY, Huang F, et al. NF-κB mediates IL-1β- and IL-17A-induced MUC5B expression in airway epithelial cells. Am J Respir Cell Mol Biol. 2011;45(2):246–52.
61. Cong LH, Li T, Wang H, Wu YN, Wang SP, Zhao YY, et al. IL-17A-producing T cells exacerbate fine particulate matter-induced lung inflammation and fibrosis by inhibiting PI3K/Akt/mTOR-mediated autophagy. J Cell Mol Med. 2020;24(15):8532–44.
62. Gabryelska A, Kuna P, Antczak A, Białasiewicz P, Panek M. IL-33 mediated inflammation in chronic respiratory diseases-understanding the role of the member of IL-1 superfamily. Front Immunol. 2019;10:692.
63. Georas SN, Rezaee F. Epithelial barrier function: at the front line of asthma immunology and allergic airway inflammation. J Allergy Clin Immunol. 2014;134:509–20.
64. Shadie AM, Herbert C, Kumar RK. Ambient particulate matter induces an exacerbation of airway inflammation in experimental asthma: role of interleukin-33. Clin Exp Immunol. 2014;177:491–9.
65. Pourazar J, Frew AJ, Blomberg A, Helleday R, Kelly FJ, Wilson S, et al. Diesel exhaust exposure enhances the expression of IL-13 in the bronchial epithelium of healthy subjects. Respir Med. 2004;98(9):821–5.
66. Petecchia L, Sabatini F, Usai C, Caci E, Varesio L, Rossi GA. Cytokines induce tight junction disassembly in airway cells via an EGFR-dependent MAPK/ERK1/2-pathway. Lab Investig. 2012;92(8):1140–8.
67. Saatian B, Rezaee F, Desando S, Emo J, Chapman T, Knowlden S, et al. Interleukin-4 and interleukin-13 cause barrier dysfunction in human airway epithelial cells. Tissue Barriers. 2013;1(2):e24333.
68. Epelman S, Lavine KJ, Randolph GJ. Origin and functions of tissue macrophages. Immunity. 2014;41:21–35.
69. Li CH, Tsai ML, Chiou HC, Lin YC, Liao WT, Hung CH. Role of macrophages in air pollution exposure related asthma. Int J Mol Sci. 2022;23(20):12337.
70. Doran AC, Yurdagul A, Tabas I. Efferocytosis in health and disease. Nat Rev Immunol. 2020;20:254–67. https://doi.org/10.1038/s41577-019-0240-6.
71. de Souza Xavier Costa N, Ribeiro Júnior G, Dos Santos Alemany AA, Belotti L, Schalch AS, Cavalcante MF, et al. Air pollution impairs recovery and tissue remodeling in a murine model of acute lung injury. Sci Rep. 2020;10:15314.
72. Hodge MX, Reece SW, Madenspacher JH, Gowdy KM. In vivo assessment of alveolar macrophage Efferocytosis following ozone exposure. J Vis Exp. 2019;152:e60109.
73. Li Y, Yong YL, Yang M, Wang W, Qu X, Dang X, et al. Fine particulate matter inhibits phagocytosis of macrophages by disturbing autophagy. FASEB J. 2020;34:16716–35.
74. Rahmani H, Sadeghi S, Taghipour N, Roshani M, Amani D, Ghazanfari T, et al. The effects of particulate matter on C57BL/6 peritoneal and alveolar macrophages. Iran J Allergy Asthma Immunol. 2020;19:647–59.

75. Obot CJ, Morandi MT, Beebe TP, Hamilton RF, Holian A. Surface components of airborne particulate matter induce macrophage apoptosis through scavenger receptors. Toxicol Appl Pharmacol. 2002;184:98–106.
76. Xiong Q, Ru Q, Chen L, Yue K, Tian X, Ma B, et al. Combined effects of fine particulate matter and lipopolysaccharide on apoptotic responses in NR8383 macrophages. J Toxicol Environ Health A. 2015;78:443–52.
77. Doll R, Hill AB. Smoking and carcinoma of the lung. Br Med J. 1950;2:739–48.
78. Loomis D, Grosse Y, Lauby-Secretan B, El Ghissassi F, Bouvard V, Benbrahim-Tallaa L, Guha N, et al. International Agency for Research on Cancer monograph working group IARC. The carcinogenicity of outdoor air pollution. Lancet Oncol. 2013;14(13):1262–3.
79. Bray F, Ferlay J, Soerjomataram I, Siegel RL, Torre LA, Jemal A. Global cancer statistics 2018: GLOBOCAN estimates of incidence and mortality worldwide for 36 cancers in 185 countries. CA Cancer J Clin. 2018;68:394–424.
80. Loomis D, Guha N, Hall AL, Straif K. Identifying occupational carcinogens: an update from the IARC monographs. Occup Environ Med. 2018;75:593–603.
81. Deng YJ, Zhao P, Zhou LH, Xiang D, Hu JJ, Liu Y, et al. Epidemiological trends of tracheal, bronchus, and lung cancer at the global, regional, and national levels: a population-based study. J Hematol Oncol. 2020;13:98.
82. Turner MC, Krewski D, Pope CA III, Chen Y, Gapstur SM, Thun MJ. Long-term ambient fine particulate matter air pollution and lung cancer in a large cohort of never smokers. Am J Respir Crit Care Med. 2011;184:1374–81.
83. Global Burden of Disease (GBD) 2017 Risk Factor Collaborators. Global, regional, and national comparative risk assessment of 84 behavioural, environmental and occupational, and metabolic risks or clusters of risks for 195 countries and territories, 1990–2017: a systematic analysis for the Global Burden of Disease Study 2017. Lancet. 2018;392:1923–94.
84. Pope CA 3rd, Coleman N, Pond ZA, Burnett RT. Fine particulate air pollution and human mortality: 25+ years of cohort studies. Environ Res. 2020;183:108924. Erratum in: Environ Res. 2020 Dec;191:109974.
85. Hamra GB, Guha N, Cohen A, Laden F, Raaschou-Nielsen O, Samet JM, et al. Outdoor particulate matter exposure and lung cancer: a systematic review and meta-analysis. Environ Health Perspect. 2014;122(9):906–11. Erratum in: Environ Health Perspect. 2014 Sep;122(9):A236
86. Hamra GB, Laden F, Cohen AJ, Raaschou-Nielsen O, Brauer M, Loomis D. Lung cancer and exposure to nitrogen dioxide and traffic: a systematic review and meta-analysis. Environ Health Perspect. 2015;123(11):1107–12.
87. Atkinson RW, Butland BK, Anderson HR, Maynard RL. Long-term concentrations of nitrogen dioxide and mortality: a meta-analysis of cohort studies. Epidemiology. 2018;29(4):460–72.
88. Eum KD, Kazemiparkouhi F, Wang B, Manjourides J, Pun V, Pavlu V, et al. Long-term NO2 exposures and cause-specific mortality in American older adults. Environ Int. 2019;124:10–5.
89. Hvidtfeldt UA, Geels C, Sørensen M, Ketzel M, Khan J, Tjønneland A, et al. Long-term residential exposure to PM2.5 constituents and mortality in a Danish cohort. Environ Int. 2019;133(Pt B):105268.
90. Klompmaker JO, Hoek G, Bloemsma LD, Marra M, Wijga AH, van den Brink C, et al. Surrounding green, air pollution, traffic noise exposure and non-accidental and cause-specific mortality. Environ Int. 2020;134:105341.
91. Puisieux A, Pommier RM, Morel AP, Lavial F. Cellular pliancy and the multistep process of tumorigenesis. Cancer Cell. 2018;33:164–72.
92. Yu XJ, Yang MJ, Zhou B, Wang GZ, Huang YC, Wu LC, et al. Characterization of somatic mutations in air pollution-related lung cancer. EBioMedicine. 2015;2(6):583–90.
93. Brauer M, Avila-Casado C, Fortoul TI, Vedal S, Stevens B, Churg A. Air pollution and retained particles in the lung. Environ Health Perspect. 2001;109:1039–43.
94. Castano-Vinyals G, D'Errico A, Malats N, Kogevinas M. Biomarkers of exposure to polycyclic aromatic hydrocarbons from environmental air pollution. Occup Environ Med. 2004;61:e12.

95. Moorthy B, Chu C, Carlin DJ. Polycyclic aromatic hydrocarbons: from metabolism to lung cancer. Toxicol Sci. 2015;145:5–15.
96. Dunn BP. Carcinogen adducts as an indicator for the public health risks of consuming carcinogen-exposed fish and shellfish. Environ Health Perspect. 1991;90:111–6.
97. Peluso M, Airoldi L, Armelle M, Martone T, Coda R, Malaveille C, et al. White blood cell DNA adducts, smoking, and NAT2 and GSTM1 genotypes in bladder cancer: a case-control study. Cancer Epidemiol Biomark Prev. 1998;7(4):341–6.
98. Tang D, Santella RM, Blackwood AM, Young TL, Mayer J, Jaretzki A, et al. A molecular epidemiological case-control study of lung cancer. Cancer Epidemiol Biomark Prev. 1995;4(4):341–6.
99. Birnbaum LS. The mechanism of dioxin toxicity: relationship to risk assessment. Environ Health Perspect. 1994;102 Suppl 9(Suppl 9):157–67.
100. Meng Z, Qin G, Zhang B. DNA damage in mice treated with sulfur dioxide by inhalation. Environ Mol Mutagen. 2005;46(3):150–5.
101. Bai J, Meng Z. Effect of sulfur dioxide on expression of proto-oncogenes and tumor suppressor genes from rats. Environ Toxicol. 2010;25:272–83.
102. Abbas I, Verdin A, Escande F, Saint-Georges F, Cazier F, Mulliez P, et al. In vitro short-term exposure to air pollution PM2.5-0.3 induced cell cycle alterations and genetic instability in a human lung cell coculture model. Environ Res. 2016;147:146–58.
103. Reyes-Caballero H, Rao X, Sun Q, Warmoes MO, Lin P, Sussan TE, et al. Air pollution-derived particulate matter dysregulates hepatic Krebs cycle, glucose and lipid metabolism in mice. Sci Rep. 2019;9(1):17423. Erratum in: Sci Rep. 2020;10(1):5082
104. Santibanez-Andrade M, Quezada-Maldonado EM, Osornio-Vargas A, Sanchez-Perez Y, Garcia-Cuellar CM. Air pollution and genomic instability: the role of particulate matter in lung carcinogenesis. Environ Pollut. 2017;229:412–22.
105. Taghizadeh S, Najmabadi H, Kamali K, Behjati F. Evaluation of chromosomal aberrations caused by air pollutants in some taxi drivers from two polluted districts of urban Tehran and its comparison with drivers from rural areas of Lahijan: a pilot study. J Environ Health Sci Eng. 2014;12:144.
106. Dagher Z, Garçon G, Billet S, Gosset P, Ledoux F, Courcot D, et al. Activation of different pathways of apoptosis by air pollution particulate matter (PM2.5) in human epithelial lung cells (L132) in culture. Toxicology. 2006;225(1):12–24.
107. Novack L, Yitshak-Sade M, Landau D, Kloog I, Sarov B, Karakis I. Association between ambient air pollution and proliferation of umbilical cord blood cells. Environ Res. 2016;151:783–8.
108. Liu C, Guo H, Cheng X, Shao M, Wu C, Wang S, et al. Exposure to airborne PM2.5 suppresses microRNA expression and deregulates target oncogenes that cause neoplastic transformation in NIH3T3 cells. Oncotarget. 2015;6(30):29428–39.
109. Ding X, Wang M, Chu H, Chu M, Na T, Wen Y, Wu D, et al. Global gene expression profiling of human bronchial epithelial cells exposed to airborne fine particulate matter collected from Wuhan, China. Toxicol Lett. 2014;228(1):25–33.
110. Zhou W, Tian D, He J, Wang Y, Zhang L, Cui L, et al. Repeated PM2.5 exposure inhibits BEAS-2B cell P53 expression through ROS-Akt-DNMT3B pathway-mediated promoter hypermethylation. Oncotarget. 2016;7(15):20691–703.
111. Gondalia R, Baldassari A, Holliday KM, Justice AE, Méndez-Giráldez R, Stewart JD, et al. Methylome-wide association study provides evidence of particulate matter air pollution-associated DNA methylation. Environ Int. 2019;132:104723.
112. Sanchez-Guerra M, Zheng Y, Osorio-Yanez C, Zhong J, Chervona Y, Wang S, et al. Effects of particulate matter exposure on blood 5-hydroxymethylation: results from the Beijing truck driver air pollution study. Epigenetics. 2015;10(7):633–42.
113. Clark SJ, Melki J. DNA methylation and gene silencing in cancer: which is the guilty party? Oncogene. 2002;21:5380–7.

114. Zhang W, Klinkebiel D, Barger CJ, Pandey S, Guda C, Miller A, et al. Global DNA Hypomethylation in epithelial ovarian cancer: passive demethylation and association with genomic instability. Cancers (Basel). 2020;12(3):764.
115. Heßelbach K, Kim GJ, Flemming S, Häupl T, Bonin M, Dornhof R, et al. Disease relevant modifications of the methylome and transcriptome by particulate matter (PM2.5) from biomass combustion. Epigenetics. 2017;12(9):779–92.
116. Swanton C, Hill W, Lim E, Lee C, Weeden CE, Augustine M, et al. Mechanism of action and an actionable inflammatory axis for air pollution induced non-small cell lung cancer: towards molecular cancer prevention. Ann Oncol. 2022;33(suppl_7):S808–69.
117. Cohen AJ, Brauer M, Burnett R, Anderson HR, Frostad J, Estep K, et al. Estimates and 25-year trends of the global burden of disease attributable to ambient air pollution: an analysis of data from the Global Burden of Diseases Study 2015. Lancet. 2017;389(10082):1907–18. Erratum in: Lancet. 2017;389(10087):e15. Erratum in: Lancet. 2018;391(10130):1576
118. Burnett R, Chen H, Szyszkowicz M, Fann N, Hubbell B, Pope CA 3rd, et al. Global estimates of mortality associated with long-term exposure to outdoor fine particulate matter. Proc Natl Acad Sci U S A. 2018;115(38):9592–7.
119. Eckel SP, Cockburn M, Shu YH, Deng H, Lurmann FW, Liu L, et al. Air pollution affects lung cancer survival. Thorax. 2016;71(10):891–8.
120. Cheng I, Yang J, Tseng C, Wu J, Conroy SM, Shariff-Marco S, et al. Outdoor ambient air pollution and breast cancer survival among California participants of the multiethnic cohort study. Environ Int. 2022;161:107088.
121. Wong CM, Tsang H, Lai HK, Thomas GN, Lam KB, Chan KP, et al. Cancer mortality risks from long-term exposure to ambient fine particle. Cancer Epidemiol Biomark Prev. 2016;25(5):839–45.
122. Coleman NC, Ezzati M, Marshall JD, Robinson AL, Burnett RT, Pope CA 3rd. Fine particulate matter air pollution and mortality risk among US cancer patients and survivors. JNCI Cancer Spectr. 2021;5(1):pkab001.
123. Kaufman JD, Adar SD, Barr RG, Budoff M, Burke GL, Curl CL, et al. Association between air pollution and coronary artery calcification within six metropolitan areas in the USA (the Multi-Ethnic Study of Atherosclerosis and Air Pollution): a longitudinal cohort study. Lancet. 2016;388(10045):696–704. Erratum in: Lancet. 2016;388(10045):660
124. Pope CA 3rd, Burnett RT, Thurston GD, Thun MJ, Calle EE, Krewski D, et al. Cardiovascular mortality and long-term exposure to particulate air pollution: epidemiological evidence of general pathophysiological pathways of disease. Circulation. 2004;109(1):71–7.
125. Yang BY, Qian Z, Howard SW, Vaughn MG, Fan SJ, Liu KK, et al. Global association between ambient air pollution and blood pressure: a systematic review and meta-analysis. Environ Pollut. 2018;235:576–88.
126. DeVries R, Kriebel D, Sama S. Outdoor air pollution and COPD-related emergency department visits, hospital admissions, and mortality: a meta-analysis. COPD. 2017;14:113–21.
127. Zhu RX, Nie XH, Chen YH, Chen J, Wu SW, Zhao LH. Relationship between particulate matter (PM2.5) and hospitalizations and mortality of chronic obstructive pulmonary disease patients: a meta-analysis. Am J Med Sci. 2020;359:354–64.
128. Li X, Wang M, Song Y, Ma H, Zhou T, Liang Z, et al. Obesity and the relation between joint exposure to ambient air pollutants and incident type 2 diabetes: a cohort study in UK biobank. PLoS Med. 2021;18(8):e1003767.
129. Wolf K, Popp A, Schneider A, Breitner S, Hampel R, Rathmann W, et al. KORA-Study Group. Association Between Long-term Exposure to Air Pollution and Biomarkers Related to Insulin Resistance, Subclinical Inflammation, and Adipokines. Diabetes. 2016;65(11):3314–26. Erratum in: Diabetes. 2017;66(10):2725
130. Bowe B, Xie Y, Li T, Yan Y, Xian H, Al-Aly Z. Associations of ambient coarse particulate matter, nitrogen dioxide, and carbon monoxide with the risk of kidney disease: a cohort study. Lancet Planet Health. 2017;1:e267–76. https://doi.org/10.1016/S2542-5196(17)30117-1.

131. Bernatsky S, Smargiassi A, Barnabe C, Svenson LW, Brand A, Martin RV, et al. Fine particulate air pollution and systemic autoimmune rheumatic disease in two Canadian provinces. Environ Res. 2016;146:85–91.
132. Di D, Zhang L, Wu X, Leng R. Long-term exposure to outdoor air pollution and the risk of development of rheumatoid arthritis: a systematic review and meta-analysis. Semin Arthritis Rheum. 2020;50:266–75.
133. Merk R, Heßelbach K, Osipova A, Popadić D, Schmidt-Heck W, Kim G-J, et al. Particulate matter (PM2.5) from biomass combustion induces an anti-oxidative response and cancer drug resistance in human bronchial epithelial BEAS-2B cells. Int J Environ Res Public Health. 2020;17:8193.
134. Lagunas-Rangel FA, Liu W, Schiöth HB. Can exposure to environmental pollutants be associated with less effective chemotherapy in cancer patients? Int J Environ Res Public Health. 2022;19(4):2064.
135. Altman MC, Kattan M, O'Connor GT, Murphy RC, Whalen E, LeBeau P, et al. Associations between outdoor air pollutants and non-viral asthma exacerbations and airway inflammatory responses in children and adolescents living in urban areas in the USA: a retrospective secondary analysis. Lancet Planet Health. 2023;7(1):e33–44.
136. Calderón-Garcidueñas L, Torres-Jardón R, Kulesza RJ, et al. Alzheimer disease starts in childhood in polluted metropolitan Mexico City. A major health crisis in progress. Environ Res. 2020 Apr;183:109137. https://doi.org/10.1016/j.envres.2020.109137.
137. Chandra M, Rai CB, Kumari N, Sandhu VK, Chandra K, Krishna M, et al. Air pollution and cognitive impairment across the life course in humans: a systematic review with specific focus on income level of study area. Int J Environ Res Public Health. 2022;19(3)
138. Livingston G, Huntley J, Sommerlad A, Ames D, Ballard C, Banerjee S, et al. Dementia prevention, intervention, and care: 2020 report of the lancet commission. Lancet. 2020;396(10248):413–46.
139. Vohra K, Vodonos A, Schwartz J, Marais EA, Sulprizio MP, Mickley LJ. Global mortality from outdoor fine particle pollution generated by fossil fuel combustion: results from GEOS-Chem. Environ Res. 2021;195:110754.
140. Vohra K, Marais EA, Bloss WJ, Schwartz J, Mickley LJ, Van Damme M, et al. Rapid rise in premature mortality due to anthropogenic air pollution in fast-growing tropical cities from 2005 to 2018. Sci Adv. 2022;8(14):eabm4435.
141. Tran KV, Casey JA, Cushing LJ, Morello-Frosch R. Residential proximity to oil and gas development and birth outcomes in California: a retrospective cohort study of 2006-2015 births. Environ Health Perspect. 2020;128(6):67001.
142. Guxens M, Lubczyńska MJ, Muetzel RL, Dalmau-Bueno A, Jaddoe VWV, Hoek G, et al. Air pollution exposure during fetal life, brain morphology, and cognitive function in school-age children. Biol Psychiatry. 2018;84(4):295–3.
143. Choi YJ, Cho J, Hong YC, Lee DW, Moon S, Park SJ, et al. DNA methylation is associated with prenatal exposure to sulfur dioxide and childhood attention-deficit hyperactivity disorder symptoms. Sci Rep. 2023;13(1):3501.

Leaving No Stone Unturned: Unraveling the Path to Maximizing the Potential for Discovery of Novel Antineoplastics

3

Solmaz Karimi, Godsfavour Umoru, and Cynthia El Rahi

Introduction

The discovery of anti-microbials such as penicillin led to a paradigm shift in mortality patterns in the 1950s as evidenced by the significant decline in infectious disease related mortality while heart disease and cancer, conversely, became leading causes of human mortality. Consequently, interest in cancer research advancement intensified. In 1955, the National Cancer Institute (NCI) created the Cancer Chemotherapy National Service Center (CCNSC) to mitigate the rise in cancer-related mortality through research focused on the evaluation of known and synthetic compounds [1]. In 1960, Jonathan L. Hartwell of the NCI expanded the center by implementing a screening program for agents with potential anti-cancer properties [2]. Plant samples were collected at random and supplied to the NCI by the U.S Department of Agriculture (USDA) through a partnership [9–10]. Between 1960 and 1981, a total of 30,000 plant samples were collected and tested under this program [1]. The first part of the chapter will highlight the chemotherapy agents that were discovered from plants to advance cancer research while the latter part will aim to unravel the potential impact that deforestation and destruction of rainforest portends for drug discovery.

Vinca Alkaloids

Discovery and Origin

Vinca alkaloids were originally discovered in the 1950s by Canadian scientists, Robert Noble, MD, PhD and Charles Beer, PhD [3]. Charles Beer, born in England,

S. Karimi · G. Umoru (✉) · C. El Rahi
Department of Pharmacy, Houston Methodist Hospital, Houston, TX, USA

Houston Methodist Cancer Center, Houston, TX, USA
e-mail: goumoru@houstonmethodist.org

E. H. Bernicker (ed.), *Environmental Oncology*,
https://doi.org/10.1007/978-3-031-33750-5_3

first came to Canada in 1954 as a British Empire Cancer Campaign Fellow after spending three years at New York's Sloan Kettering Institute for cancer research [3]. Robert Noble, son of a Toronto physician, grew up in Toronto surrounded by scientists and physicians driven by discovery, fame, fortune, and serendipity [3]. After completing his Doctor of Medicine degree at the University of Toronto in 1934 and his Doctor of Philosophy degree at the University of London in 1937, Noble took a position working at the J. B. Collip's endocrinology lab [3]. The chain of events that led to the discovery of vinca alkaloids began in 1949, a time where researchers were keen on finding an oral antidiabetic agent to eliminate insulin injections. Historically, the extracts of the leaves of the Madagascar periwinkle were known for their antidiabetic properties, although Australian researchers had previously investigated and refuted the antidiabetic potential of periwinkle in the late 1920s [3]. However, Noble's team found that oral administration of periwinkle extracts in rats had no effect on blood sugar or glucagon levels and decided to inject the periwinkle water extracts to further investigate its antidiabetic potential [3]. Interestingly, the injections proved to be lethal with autopsy results revealing multiple abscesses associated with a pseudomonas infection akin to rats that received high doses of cortisone [3]. Additional studies revealed that the injected periwinkle water extracts resulted in a rapid decrease in white blood counts, granulocytopenia, and profoundly depressed bone marrow in rats [3]. These findings prompted further investigation into the purification and isolation of the active compound responsible for bone marrow suppression in hopes of discovering an anti-tumor agent. Beer is responsible for isolation, purification, and crystallization of the active agent from the periwinkle extract, which he and Noble named "vincaleukoblastine"—later shortened to vinblastine owing to its origin from periwinkle (vinca) plant and its effects on the bone marrow [3].

Noble and Beer were respectively recognized for the discovery and isolation of vinblastine in 1958 [3]. The Robert L. Noble Prize is awarded each year in honor of Robert Noble by the Canadian Cancer Society to researchers whose contributions have led to a significant advancement in basic biomedical cancer research [4]. In 1997, the late researchers Charles Beer (1915–2010) and Robert Noble (1910–1990) were inducted into the Canadian Medical Hall of Fame for their "chance" discovery of vinblastine from the subtropical plant *Catharanthus roseus* G. Don (also known as Madagascar periwinkle or *vinca rosea*) [3].

Pharmacology

Vinca alkaloids are naturally occurring or semi-synthetic nitrogenous bases that have played a pivotal role in cancer therapy. There are four FDA-approved vinca alkaloids available in the United States: Vinblastine (VBL), vincristine (VCR), vinorelbine (VRL), and liposomal vincristine [5]. Vinorelbine is a semi-synthetic agent derived from catharanthine and vindoline, which are natural products of the Madagascar periwinkle plant [6].

Vinca alkaloids are cell-cycle specific agents that specifically bind to β-tubulin and block its interaction with α-tubulin to form microtubules [6]. Inhibiting the formation of microtubules blocks the formation of the mitotic spindle and ultimately leads to cell cycle arrest in metaphase [6]. Cells blocked in mitosis undergo changes characteristic of apoptosis [6]. Of note, vinca alkaloids do not distinguish between malignant and non-malignant cells; therefore, some of their adverse effects are related to direct effects on microtubules found in different parts of the body including high concentrations in the brain [4–7]. Vinca alkaloids are lethal if given intrathecally. Furthermore, these agents can cause gastrointestinal toxicity including constipation, bloating, abdominal pain, ileus, and perforations. They are also associated with neurotoxicity primarily peripheral neuropathy (dose-limiting toxicity), but can also cause paresthesia, loss of reflexes, and ataxia [4–7]. Neurotoxicity is more commonly seen with VCR [7]. Other class toxicities reported include myelosuppression (dose-limiting toxicity), mucositis, pharyngitis, stomatitis, hyperuricemia, pulmonary toxicity, and extravasation [4–7]. Vinca alkaloids are primarily metabolized by hepatic cytochrome P450 isoenzymes in the CYP 3A family and can accumulate in patients with renal insufficiency, leading to increased risk of toxicity [5–7]. Although the three vinca alkaloid agents have structural similarities, they have unique pharmacology and clinical efficacy.

Vinblastine

Pharmacokinetic studies in cancer patients have shown that VBL displays a triphasic plasma concentration decline following intravenous injection [8]. The initial, middle, and terminal half-lives are 3.7 mins, 1.6 h, and 24.8 h, respectively [8]. VBL exhibits extensive reversible tissue binding [8]. Its major route of elimination is through the biliary system; therefore, increased toxicity is seen in patients with hepatic insufficiency [8]. VBL has significant bone marrow suppression among the vinca alkaloids [8]. Figure 3.1 depicts the chemical structure of vinblastine.

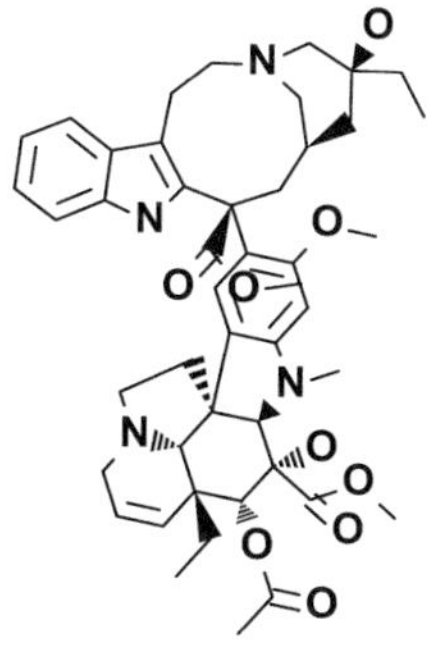

Fig. 3.1 Chemical structure of vinblastine

Vincristine

Pharmacokinetic studies in cancer patients have shown a triphasic plasma concentration decline following rapid intravenous injection of VCR. The initial, middle, and terminal half–lives are 5 minutes, 2.3 h, and 85 h, respectively [7]. The range of terminal half-life can be anywhere from 19 h to 155 h [7]. VCR is widely distributed following intravenous administration and exhibits extensive reversible tissue binding. Like VBL, VCR is also eliminated through the biliary system with 80% of the injected dose recovered in feces and 10–20% in urine [7]. VCR is often part of a polychemotherapy due to its lack of significant bone marrow suppression as well as its risk of neuropathy [7]. Figure 3.2 depicts the chemical structure of vincristine.

Vinorelbine

Like the other two agents, pharmacokinetic studies have revealed triphasic plasma decay with a mean terminal phase half-life of 27.7–43.6 h with VRL [9]. Following intravenous administration of of VRL, 18% and 46% of the administered drug is recovered in urine and feces, respectively [9]. VRL carries a boxed warning for myelosuppression [9]. Figure 3.3 depicts the chemical structure of vinorelbine.

Clinical Use

Vinca alkaloids are widely used in combination chemotherapy regimens for a variety of malignancies. VBL is an integral part of the treatment regimen for testicular carcinoma, Hodgkin lymphoma, and non-Hodgkin lymphoma [6]. Other FDA-approved indications for use include Kaposi sarcoma, Letterer-Siwe disease, mycosis fungoides, and breast cancer that is unresponsive to appropriate surgery and hormonal therapy [8]. VCR is commonly used as a chemotherapy agent in leukemia, lymphoma, myeloma, breast, head, and neck cancer [10]. It is FDA-approved

Fig. 3.2 Chemical structure of vincristine

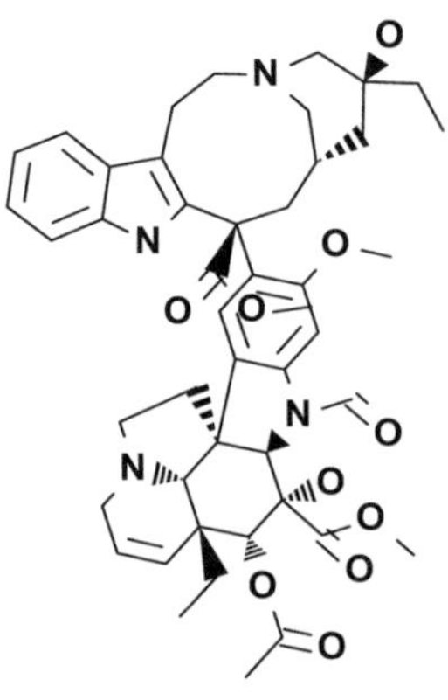

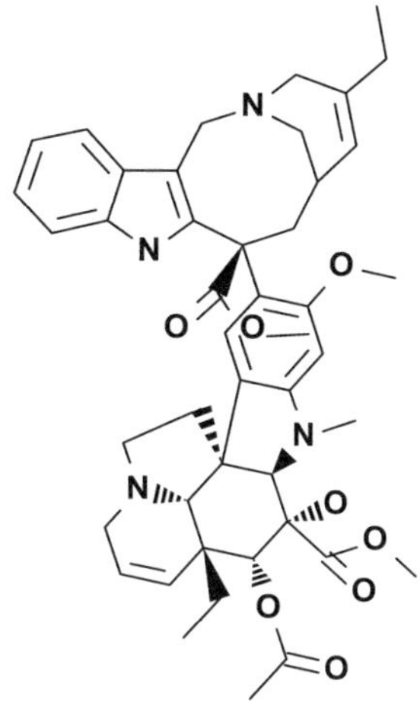

Fig. 3.3 Chemical structure of vinorelbine

for use in acute leukemia, Hodgkin lymphoma, non-Hodgkin lymphoma, rhabdomyosarcoma, neuroblastoma, and Wilm's tumor [7]. Moreover, VCR is part of many chemotherapy protocols including non-Hodgkin lymphoma treatment with CHOP (cyclophosphamide, doxorubicin, vincristine, and prednisone), Hodgkin lymphoma treatment with MOPP (mechlorethamine, vincristine, procarbazine, and prednisone), COPP (cyclophosphamide, vincristine, procarbazine, and prednisone), BEACOPP (bleomycin, etoposide, doxorubicin, cyclophosphamide, vincristine, procarbazine, and prednisone), acute lymphoblastic leukemia (ALL) treatment with the Stanford V (mechlorethamine, doxorubicin, vinblastine, vincristine, bleomycin, etoposide and prednisone), and nephroblastoma treatment [10]. Vincristine is also utilized as an immunosuppressant in the treatment of thrombotic thrombocytopenic purpura and chronic idiopathic thrombocytopenic purpura [6, 10]. Lastly, VRL is FDA-approved for use in combination with cisplatin for first-line treatment of patients with locally advanced or metastatic non-small cell lung cancer or as a single agent for first-line treatment of patients with metastatic non-small cell lung cancer [9].

Taxanes

Discovery and Origin

In 1962, USDA botanist Arthur S. Barclay, PhD, and three college student field assistants collected 650 plant samples across Oregon, Washington, and California, including samples from Taxus *brevifolia* (Pacific or Western Yew) in Washington State, which had never been investigated for medicinal use [2]. Barclay's samples of *T. brevifolia* were sent to Wisconsin Alumni Research Foundation where the initial crude extracts showed cytotoxicity against 9 KB cell cultures that had been derived from a human cancer of the nasopharynx [2]. On the basis of this evidence, the plant was assigned by Dr. Hartwell of the NCI to the Research Triangle Institute (RTI) in North Carolina, where Mansukh C. Wani, PhD and Monroe E. Wall, PhD had noted a correlation between L1210 (lymphoid leukemia in mice) in vivo activity and 9 KB cytotoxicity assay when studying camptothecin [2]. Walls and Wani extracted paclitaxel, the most cytotoxic compound from the bark of the tree, through laborious isolation steps [9–10]. In 1977, NCI confirmed paclitaxel's antitumor activity through efficacy shown in mouse models and the drug became a candidate for clinical development [1].

In 1979, Susan Band Horwitz, PhD, discovered the mechanism of cytotoxicity of paclitaxel through a grant from NCI. She demonstrated that paclitaxel prevented cell division through a mechanism that was different than previous antimitotic drugs, which was remarkable in the era of drug resistance and non-response [1, 2]. Early clinical trials showed promising results; however, due to the slow-growing nature of the Pacific yew trees and the large amount of bark needed to produce the therapeutic doses of paclitaxel, further research was delayed [1, 2]. Furthermore, the molecule proved to be extremely hydrophobic which posed a bottle neck with formulation attempts for clinical use [2]. To circumvent this issue, paclitaxel was prepared in Cremophor EL and ethanol solvents. Notably, the first patients to receive the drug experienced severe hypersensitivity reactions to paclitaxel due to the Cremophor EL solvent [2]. These hypersensitivity reactions halted the use of the drug in clinical trials for 5 years (1983–1988) [2]. During this time, researchers began to find a solution to the severe allergic reactions experienced by patients when given paclitaxel and found that premedication with antihistamines, steroids, and administration as a 24-h infusion in lieu of a bolus were strategies for preventing and mitigating the allergic reactions [2].

Moreover, there were delays with the production of paclitaxel for clinical use due to high ecological and manufacturing costs. The delay encouraged the search for other similar anti-tumor drugs. In 1981, a French group discovered a new semi-synthetic taxane called docetaxel [11]. Docetaxel is produced by esterification of 10-deacetylbaccatin III, which does not have cytotoxic activity on its own and is isolated from the needles of the European yew tree *Taxus baccata* [11]. Development of docetaxel proved to be more renewable than paclitaxel [11]. To overcome the availability problems with paclitaxel, Bristol-Myers Squibb (BMS) developed a semisynthetic form of paclitaxel in 1991, Taxol, in partnership with NCI [1]. The following year, Taxol was FDA approved for treatment of ovarian cancer and breast cancer in 1994 [1].

Pharmacology

Taxanes are chemically similar and share a taxane ring with a four-membered oxetane side ring at positions C4 and C5 and an ester side chain at C13, which plays a pivotal role in binding to microtubules [2]. Taxanes exert their cytotoxic effects by binding to the β-tubulin subunit on the inner surface of microtubules, leading to the formation of stable tubulin bundles, and disruption of their physiological disassembly. This disruption ultimately halts cell cycle at metaphase and inhibits cell proliferation resulting in apoptosis [2]. There are four FDA-approved drugs under the taxane category, and these include paclitaxel, docetaxel, nab-paclitaxel and cabazitaxel. Docetaxel and paclitaxel are different in their molecular pharmacology which could explain their different activity and toxicity profiles [11]. Docetaxel has a greater affinity to β-tubulin subunit and targets centrosome organization and acts on cells in three phases of cell cycle including S, G_2, and M phase, while paclitaxel acts in the G_2 and M phases of cell cycle [11]. Docetaxel also induces the phosphorylation of BCL-2 gene that encodes the oncoprotein that prevents apoptosis thereby promoting apoptotic cell death, at concentrations 100-fold less than those required by paclitaxel [11]. Additionally, docetaxel has a higher uptake into tumor cells and a slower efflux from tumor cells leading to a longer retention of the drug in cells, which provides an explanation for the incomplete cross-resistance between the two drugs [11]. Taxanes require. Taxanes share some class toxicities that differ in intensity from one agent to the other, these toxicities include myelosuppression (dose-limiting toxicity), myalgia and arthralgia, peripheral neuropathy, hand-foot syndrome, oncolysis, total body alopecia, increased liver function tests and extravasation risk [12–14]. Taxanes are metabolized by the liver and thus require dose adjustment in patients with hepatic dysfunction.

Paclitaxel

Pharmacokinetic studies have revealed that paclitaxel exhibits a biphasic plasma concentration decline [12]. The initial rapid decline occurs due to distribution to the peripheral compartment and the elimination of the drug, while the second phase occurs as a result of the relatively slow efflux from the peripheral compartment [12]. Following administration of 225 or 250 mg/m^2 dose of radiolabeled paclitaxel as a 3-h infusion in 5 patients, a mean of 71% radioactivity was excreted in the feces in 120 h and only 14% was recovered in the urine. In vitro studies with human liver microsomes and tissue slices revealed that the drug is primary metabolized to its major metabolite, 6α-hydroxypaclitaxel, by CYP2C8 and two minor metabolites, 3′-*p*-hydroxypaclitaxel and 6α, 3′-*p*-dihydroxypaclitaxel, by CYP3A4. Paclitaxel carries a boxed warning for anaphylaxis and severe hypersensitivity occurring in 2–4% of patients receiving the drug, and is characterized by dyspnea, hypotension requiring treatment, angioedema, and generalized urticaria [12]. This hypersensitivity reaction is the result of its formulation with a vehicle known as polyoxyethylated castor oil (Cremophor EL) [12]. There is a nanoparticle albumin-bound formulation of paclitaxel, more commonly known as nab-paclitaxel, that does not contain Cremophor EL [15]. Figure 3.4 depicts the chemical structure of paclitaxel.

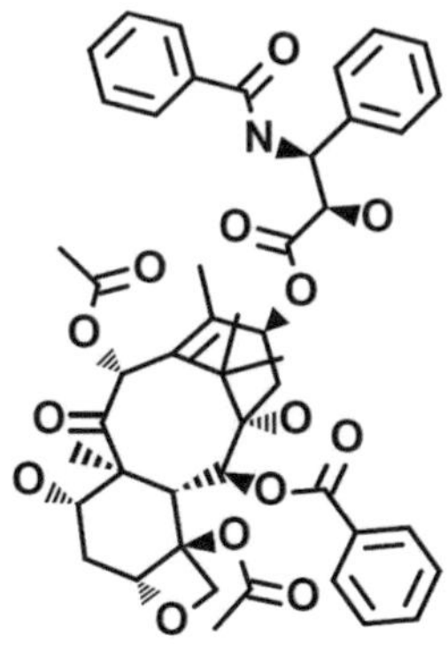

Fig. 3.4 Chemical structure of paclitaxel

Docetaxel

The pharmacokinetic profile of docetaxel is one of three compartment model, with half-lives for α, β, and γ phases being 4 min, 36 min, and 11.1 h, respectively [13]. The initial rapid decline is due to the representation of distribution to the peripheral compartments while the terminal phase is due to a relatively slow efflux from the peripheral compartment [13]. In vitro studies have revealed that docetaxel is metabolized by CYP3A4 enzyme, and its metabolism is affected by concomitant administration of drugs that act on the CYP3A4 enzyme [13]. Like paclitaxel, docetaxel is primarily excreted in the feces. After intravenous administration of a radioactive docetaxel, about 80% of radioactivity was recovered in feces during the first 48 h as one major and 3 minor metabolites [13]. Docetaxel carries several boxed warnings. Treatment-related mortality is reported to increase with abnormal liver function, at higher doses, and in patients with non-small cell lung cancer and prior platinum-based therapy receiving docetaxel at 100 mg/m^2 [13]. Another boxed warning is that docetaxel should not be given to patients with bilirubin higher than upper limit of normal, or if AST and/or ALT is one and a half times higher than the upper limit of normal concomitant with alkaline phosphatase that is two and a half time more than upper limit of normal, as these can increase risk of severe or life-threatening complications [13]. Docetaxel is given in conjunction with dexamethasone to combat the risk of fluid retention. Even in the setting of the recommended 3-day course of dexamethasone therapy, docetaxel can still cause severe fluid retention. Lastly, docetaxel can cause severe hypersensitivity reaction including very rare fatal anaphylaxis as the drug is formulated with polysorbate 80 and use in patients with hypersensitivity to polysorbate 80 is contraindicated. Figure 3.5 depicts the chemical structure of docetaxel.

Cabazitaxel

Following intravenous injection, cabazitaxel plasma concentrations exhibit a three-compartment pharmacokinetic model with half-lives for α, β, and γ phases being 4 min, 2 h, and 95 h, respectively [14]. Binding of cabazitaxel to serum proteins is

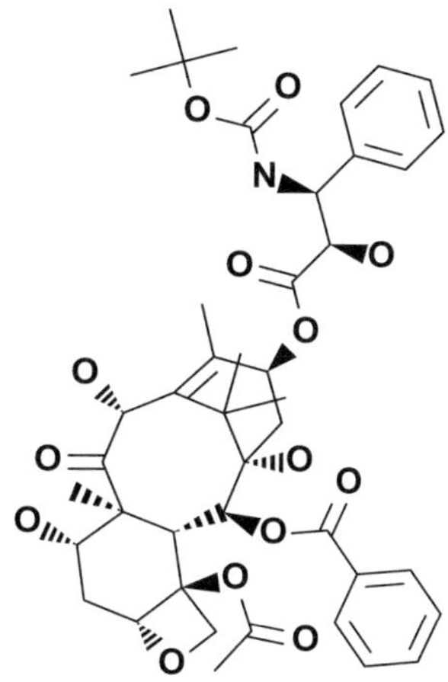

Fig. 3.5 Chemical structure of docetaxel

89% to 92% in vitro [14]. The drug is primarily bound to serum albumin (82%) and lipoproteins (88% for HDL, 70% for LDL, and 56% for VLDL) [14]. Cabazitaxel is extensively metabolized by CYP3A4/5 isoenzyme (80–90%) and to a lesser extent by CYP2C8 [14]. There are seven metabolites that have been detected in plasma (including the 3 active metabolites) and around 20 metabolites are excreted into urine and feces [14]. Like the other taxanes, cabazitaxel metabolites are primarily excreted in the feces (76% of the dose), while renal excretion of cabazitaxel and its metabolites accounts for only 3.7% of the dose [14]. After intravenous administration of a radioactive cabazitaxel, about 80% of administered dose was eliminated within 2 weeks [14]. Cabazitaxel has boxed warning for neutropenic deaths; therefore, use is contraindicated in patients with neutrophil counts of ≤1500 cells/mm^3 [14]. Primary prophylaxis with granulocyte colony stimulating factor (G-CSF) is typically recommended in patients who are at high-risk, and it can be considered in those receiving a dose of 25 mg/m^2 [14]. Cabazitaxel is formulated with polysorbate 80 and as such severe hypersensitivity reactions can occur, including generalized rash, hypotension, and bronchospasm [14]. Figure 3.6 depicts the chemical structure of cabazitaxel.

Clinical Use and Impact

Taxanes are used in a myriad of solid tumors and play an integral role in combination regimens as well as treatment-resistant solid tumors. Paclitaxel is FDA-approved for use in ovarian cancer, breast cancer, non-small cell lung cancer, and AIDS-related Kaposi sarcoma [12]. Docetaxel is FDA-approved for use in breast cancer, non-small cell lung cancer, hormone refractory prostate cancer, gastric adenocarcinoma, and squamous cell carcinoma of the head and neck cancer [13]. Cabazitaxel is indicated for use in combination with prednisone for treatment of

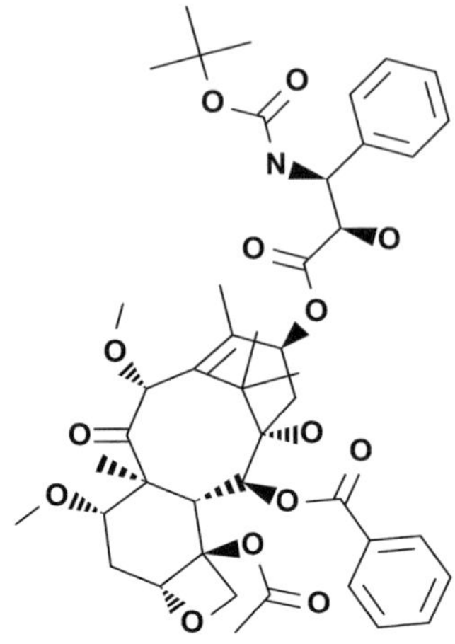

Fig. 3.6 Chemical structure of cabazitaxel

patients with metastatic castration-resistant prostate cancer previously treated with a docetaxel-containing treatment regimen [14].

Camptothecin Derivatives

Discovery and Origin

Camptothecin was discovered and isolated from the bark and stem of a Chinese tree, *Camptotheca acuminata*, by Mansukh C. Wani, PhD and Monroe E. Wall, PhD in 1966 [16]. At the time, the bark was commonly used to treat psoriasis, stomach ailments, and common cold according to Chinese traditional medicine [16]. This was another discovery that was the fruit of the NCI plant program led by Jonathan Hartwell, PhD, an organic chemist [17]. At the time, USDA was providing plants to NCI and ultimately sent samples to different laboratories to evaluate their anticancer potential [17]. Samples were initially screened for cytotoxicity against KB cell culture (cell line from human cancer) and subsequently tested in three tumor xenograft mouse models for S-180 (a sarcoma), CA-755 (an adenocarcinoma), and L1210 (a lymphoid leukemia) [17]. Crude extracts that showed anticancer potential were then fractionated to three different laboratories, one of them being the RTI center where Wani and Wall worked.

One of the thousand ethanolic plant extracts screened was the *Camptotheca acuminata*, which showed very cytotoxic potency in L1210 mouse leukemia cell assay. Wani and Wall were eager to isolate the active compound responsible for the anticancer activity and began activity-based fractionation, where fractions of the crude extract were tested for bioactivity and fractions showing potent activity were carried on to the next phase of purification [16–17]. They were able to isolate the pure compound that caused cell death in L1210 leukemic cells and termed the compound as camptothecin [17]. Camptothecin was not only active against the L1210 cells but was also found to be cytotoxic against p338 leukemic cells [17]. RTI scientists Keith Palmer and Harold Taylor were initially responsible for the isolation of camptothecin and Ed Cook for determining the structure of the compound [17]. In the 1970s, it was clinically approved to treat stomach cancer, bladder cancer, and few types of leukemia [16]. Its efficacy is largely limited by its poor water solubility, high toxicity including off-target toxicity, in vivo rapid hydrolysis of the lactone rings, and acquired resistance [16]. Due to its poor water solubility and its off-target toxicities, further investigation with camptothecin was paused [18]. In the 1980s, its molecular target was identified as topoisomerase 1, which is responsible for relieving the stress of DNA supercoiling [16]. It was not until the discovery of its mechanism of action that interest in this compound began to surface again. Chemical modifications of the rings of the parent camptothecin compound improved its toxicity and pharmacokinetic profiles which led to the commercial manufacturing of derivatives of camptothecin: topotecan and irinotecan [17].

Pharmacology

Camptothecin consists of a planar pentacyclic ring system. It has three fused rings: pyrrolo-(3, 4-β)-quinoline part (rings A, B and C), fused to a pyridine (ring D) [16]. The active form of camptothecin has a chiral center that is within the α-hydroxy lactone ring (ring E) with an (S)-configuration [16]. Hydrolysis of the lactone ring occurs at physiologic pH which leads to an equilibrium between the inactive carboxylate form and the active lactone form [16]. Camptothecin and its derivates act on topoisomerase I, which is one of two enzymes that relieves the torsional strain or supercoils associated with the unwinding of DNA during replication and transcription [17]. Topoisomerase I enzymes induce reversible single strand breaks. Camptothecin and its derivatives bind to topoisomerase I-DNA complex through hydrogen bonds and form a ternary complex, which prevents the re-ligation of nicked DNA [16–19]. The ternary complex causes accumulation of DNA strand breaks leading to programmed cell death during the S phase [16–19]. Camptothecin derivates, topotecan and irinotecan, have improved solubility and lactone stability through addition of solubilizing groups. Topotecan has a tertiary amine at the 9-position while irinotecan owes its solubility to the 10-hydroxyl moiety [18]. The common toxicities among camptothecin derivates are the risk of myelosuppression and gastrointestinal toxicities including diarrhea, constipation, nausea, and vomiting [18–19].

Topotecan

The pharmacokinetics of the drug has been evaluated in cancer patients following 0.5–1.5 mg/m^2 dose administered as a 30-minute infusion [19]. It has been unraveled that topotecan exhibits multi-exponential pharmacokinetics with a terminal half-life of 2–3 h. Topotecan is about 35% plasma protein bound. As previously

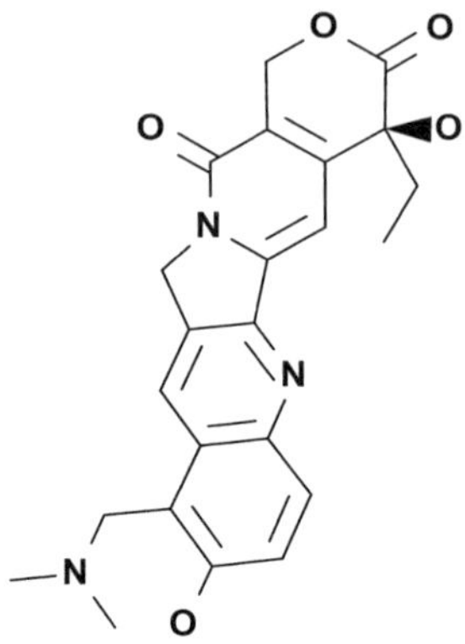

Fig. 3.7 Chemical structure of topotecan

mentioned, topotecan undergoes a reversible pH dependent hydrolysis of the lactone moiety. At pH ≤ 4, the lactone is exclusively present and at physiologic pH the ring-opened hydroxy-acid form is present [19]. In vitro studies have revealed that topotecan is metabolized to an N-demethylated metabolite. Renal clearance is the major route of elimination; however, topotecan is also eliminated through feces [19]. Topotecan has a boxed warning against use in patients with bone marrow suppression, neutrophil counts <1500 cells/mm^3 [19]. The most common adverse effects associated with topotecan use include, neutropenia, anemia, thrombocytopenia, nausea, alopecia, vomiting, sepsis or infection, diarrhea, constipation, fatigue, and pyrexia [19]. Figure 3.7 depicts the chemical structure of topotecan.

Irinotecan

Irinotecan is a water-soluble prodrug of the lipophilic metabolite, SN-38 [20–22]. SN-39 is formed by carboxylesterase-mediated cleavage of the carbamate bond between the dipiperidino side chain and the camptothecin moiety [20]. SN-39 is 1000-fold more potent as a topoisomerase I inhibitor than irinotecan in human and rodent cell lines and in vitro cytotoxicity studies show that SN-38 is 2- to 2000-fold more potent than irinotecan [20]. Both irinotecan and SN-38 have a pH-dependent equilibrium between their active lactone form that exists in acidic pH and their inactive acid anion form that exists in a basic pH [20]. Following intravenous administration of irinotecan, its plasma concentrations decline in a multi-exponential pharmacokinetic model with a mean terminal elimination half-life of about 6–12 h for irinotecan and 10–20 h for SN-38 [20]. Irinotecan is moderately bound to plasma proteins (30–68%) whereas SN-38 is highly protein-bound (95%), primarily bound to albumin [20]. Irinotecan is extensively metabolized by various enzymes including esterases to the active metabolite SN-38, UGT1A1 to form an inactive metabolite, and CYP3A4 oxidation to several inactive metabolizes. Figure 3.8 depicts the chemical structure of irinotecan.

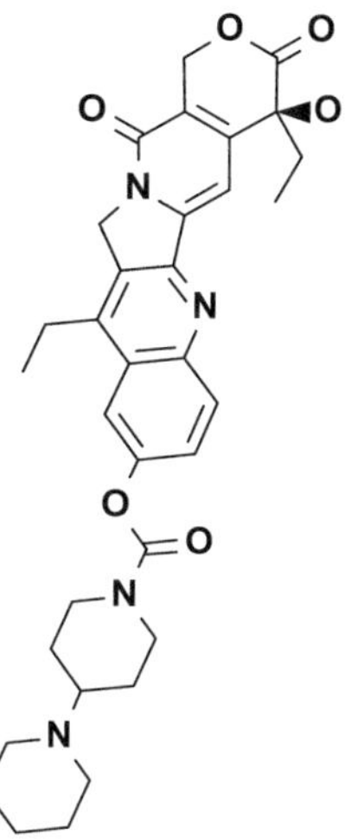

Fig. 3.8 Chemical structure of irinotecan

Irinotecan and its metabolites are cleared through both the urinary and the biliary system and carries two boxed warnings for myelosuppression and early and late forms of diarrhea. Acute diarrhea occurs immediately after drug administration while delayed diarrhea occurs >24 h up to 14 days after administration of irinotecan [21]. The immediate-onset diarrhea is caused by cholinergic excess and is characterized by salivation, lacrimation, abdominal cramping, and rhinitis [20, 21]. Mean duration of these symptoms is 30 minutes and is relieved rapidly by atropine [21]. Delayed diarrhea is characterized as secretory diarrhea and can occur at all dose levels [21]. Delayed or late-onset diarrhea can be life threatening and requires prompt treatment with loperamide [20]. The mechanism of irinotecan-induced diarrhea can be explained by enzymatic conversion of the inactive metabolite to the active metabolite in the intestinal lumen. SN38 is inactivated by UGT1A1 to SN38-glucuronide (SN38G), which is deconjugated by bacterial B-glucuronidase to SN38 in the intestinal lumen [21, 22]. This active form is responsible for the irinotecan-induced diarrhea [21, 22]. SN38 induces direct mucosal damage with water and electrolyte malabsorption, and also induces the production of prostaglandin E_2 and thromboxane A_2, inflammatory cytokines and TNF-alpha causing additional mucosal damage [21, 22]. Irinotecan causes colonic damage along with excessive mucous secretion which can lead to changes in absorption rates and diarrhea. The UGT1A1 activity is reduced in individuals who harbor the UGT1A1*28 polymorphism, which can lead to a higher exposure to SN-38 compared to individuals with the wild-type UGT1A1 allele [20].

Clinical Use and Impact

Clinical uses of camptothecin derivatives have been well-established. Research investigating further derivatives has always been around and there is potential for more camptothecin derivatives in the future. These agents are crucial in treatment-resistant or recurrent cancers. Topotecan is FDA-approved for the treatment of small cell lung cancers in patients with chemotherapy-sensitive disease after failure of first-line chemotherapy and in combination therapy with cisplatin for the treatment of stage IVB, recurrent, or persistent carcinoma of the cervix which is not amenable to curative treatment with surgery or radiation therapy [19]. Irinotecan is FDA-approved for first-line therapy in combination with 5-fluorouracil and leucovorin for patients with metastatic carcinoma of the colon or rectum and patients with metastatic carcinoma of the colon or rectum whose disease has recurred or progressed following initial fluorouracil-based therapy [20].

Epipodophyllotoxins

Discovery and Origin

Podophyllum emodi Wall., which grows in the Himalayan region and its American counterpart, *Podophyllum peltatum* L. are old medicinal plants that were used by the natives as cathartic and anthelminthics [23]. In the 1940s, Kaplan demonstrated an alcoholic extract, podophyllin, from the *Podophyllum* rhizomes exhibits curative effects in condylomata cuminata [23]. In the early 1950s, chemists from Sandoz, Ltd. postulated that *podophyllum* lignans might be present as glycosides in the plant, and that they might exhibit pharmacological properties. Utilizing experiences learned from extraction of *Digitalis* glycosides that exhibited cardiac pharmacological properties, podophyllum roots were extracted using procedures that would preserve glycosides [23]. These chemists were able to isolate podophyllotoxin glycoside and it's 4′-demethyl derivatives [23]. In their attempt to find useful drugs, Hartmann F. Stahelin and Albert von Wartburg along with their team, prepared a large series of derivatives of glucosides through chemical modifications. Two preparations were then selected as potential anticancer agents for more extensive testing in vitro for both animals and humans. These two preparations were SP-G, which was the condensation product of crude *podophyllum* glucoside fraction with benzaldehyde, and SP-I, which was podophyllinic acid ethyl hydrazide [23]. SP-G molecule was shown to increase the life span of mice inoculated with L1210 leukemia cells [23]. After 2 years of chemical modifications, a compound was found in SP-G which was responsible for prolonging survival time of leukemic cells. The compound was coined benzylidene lignan P. In the 1960s, further chemical modifications led to discovery of teniposide (VM-26) and etoposide (VP-16) [23]. In 1983, the FDA approved etoposide and a decade later, its analog, teniposide was approved for use in the United States [24].

Pharmacology

Teniposide and etoposide are semisynthetic derivatives of podophyllotoxin, which are increasingly utilized in a variety of cancers. Both are phase-specific cytotoxic drugs acting in the late S or early G_2 phase of the cell cycle, preventing developing cells from entering mitosis [23–26]. These medications exert their cytotoxic effects through inhibition of topoisomerase II. Topoisomerase II makes double-strand breaks to alleviated DNA supercoiling [24]. Etoposide and teniposide prevent topoisomerase II from re-ligating the cleaved DNA leading to cell cycle arrest and apoptosis [24]. In vitro, teniposide is about ten-fold more potent than etoposide in terms of cytotoxicity and this difference is thought to arise from better cellular uptake of teniposide [24]. The dose limiting toxicity associated with these agents are risk of myelosuppression. The most common toxicities associated with etoposide are gastrointestinal toxicities, hypersensitivity reactions, secondary malignancies, and hypotension associated with rapid infusion. The most common toxicities associated with teniposide are acute central nervous system depression, metabolic acidosis, and hypotension associated with rapid infusion.

Etoposide

Following intravenous administration, etoposide follows a biphasic pharmacokinetic model with a distribution half-life of about 1.5 h and terminal half-life ranging between 4 and 11 h [25]. Following intravenous injection, the area under the concentration curve and the maximum plasma concentrations increases linearly [25]. Etoposide is highly protein bound (97%), primarily to albumin. The drug is widely distributed but has a poor cerebrospinal fluid penetration [25]. Etoposide is metabolized by opening of the lactone ring, glucuronidation, sulfation, and O-demethylation [25]. O-demethylation is done through CYP3A4 isoenzyme that produces the active catechol metabolite [25]. Figure 3.9 depicts the chemical structure of etoposide.

Fig. 3.9 Chemical structure of etoposide

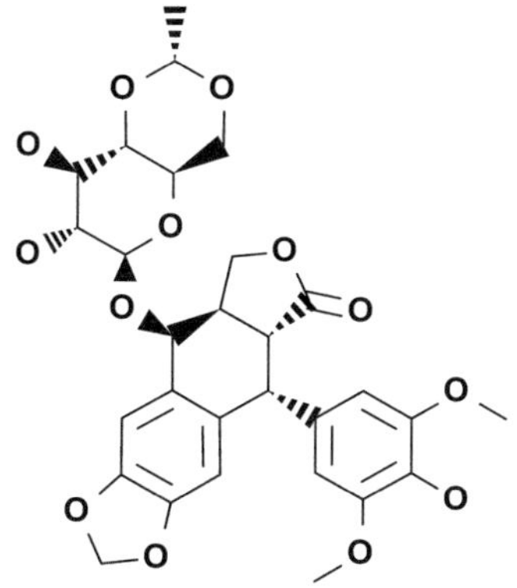

Following administration of a radioactive etoposide, the mean recovering of radioactivity in the urine was 56% (with less than 8% as metabolites) after 120 h [25]. Due to its poor water solubility, etoposide is dissolved in a solubilizer composed of polyethylene glycol, alcohol, and polysorbate 80 [24–25]. These additives are believed to be responsible for the hypersensitivity reactions occasionally seen during etoposide infusion [24]. Etoposide has a boxed warning for severe myelosuppression resulting in infection or bleeding [25].

Teniposide

Following intravenous injection and at doses 100–333 mg/m^2/day in adults, plasma drug levels of teniposide increases linearly with dose [26]. In pediatric patients after infusions of 137–203 mg/m^2 over a period of 1–2 h, maximum plasma concentrations exceeded 40 mcg/mL and by 20–24 h post infusion plasma levels were < 2 mcg/mL [26]. In adults, after intravenous administrations of 10 mg/kg or 67 mg/m^2 of radiolabeled teniposide, 44% of the radiolabeled drug was recovered in urine as parent (4–12%) and metabolites within 120 h [26]. Fecal excretion of the radioactivity was much lower (0–10%) at 72 h after dosing [26]. Teniposide is poorly absorbed across the blood-brain barrier. Compared to etoposide, teniposide is more extensively bound to plasma proteins (>99%), has a higher cellular uptake, lower systemic clearance, a longer elimination half-life, and is excreted in the urine as parent drugs to a lesser extent [26]. Teniposide carries a boxed warning for severe myelosuppression and hypersensitivity reactions, including anaphylaxis-like symptoms, which may occur with initial or repeated exposure to teniposide. Figure 3.10 depicts the chemical structure of teniposide.

Clinical Use and Impact

Etoposide is FDA-approved for use in combination with other approved chemotherapeutic agents in patients with refractory testicular tumors who have already

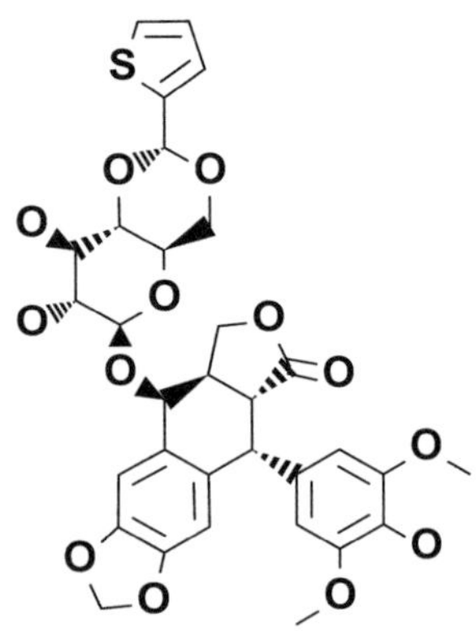

Fig. 3.10 Chemical structure of teniposide

received appropriate surgical, radiotherapeutic, and chemotherapeutic therapy [25]. It is also approved for use in combination with other agents as first line treatment in patients with small cell lung cancer [25]. Etoposide has been widely used off-label for other malignancies including leukemias, non-Hodgkin lymphomas, Kaposi sarcoma, neuroblastoma, and soft tissue sarcomas [24]. Teniposide is FDA-approved for induction therapy in patients with refractory childhood acute lymphoblastic leukemia [26]. Although, teniposide is not a major chemotherapeutic agent used in adult cancers, it has demonstrated antitumor activity in small cell lung cancer, bladder cancer, leukemias, lymphomas, and Kaposi sarcomas [24].

Potential Impact of Deforestation on Cancer Drug Discovery

Humans have relied on natural resources since the beginning of time for their basic needs including food, shelter, and medicine [27]. Remarkably, harnessing natural resources has historically proven to be one of the most successful strategies for the development of new drugs [27]. Consequently, nearly half of therapeutic medications were derived primarily from plants or are synthetic products derived from

Table 3.1 Summary of cancer drugs discovered from plant-based resources

Drug class	Origin	Place in therapy
Vinca alkaloids	*Catharanthus roseus* (Madagascar periwinkle)	Vinblastine: Testicular carcinoma, Hodgkin lymphoma, non-Hodgkin lymphoma, Kaposi sarcoma, letterer-Siwe disease, mycosis fungoides, and breast cancer
		Vincristine: Acute leukemia, Hodgkin lymphoma, non-Hodgkin lymphoma, rhabdomyosarcoma, neuroblastoma, and Wilm's tumor
		Vinorelbine: Locally advanced or metastatic non-small cell lung cancer
Taxanes	*Taxus brevifolia* (Pacific yew tree)	Paclitaxel: Ovarian cancer, breast cancer, non-small cell lung cancer, and AIDS-related Kaposi sarcoma
		Docetaxel: Breast cancer, non-small cell lung cancer, hormone refractory prostate cancer, gastric adenocarcinoma, and squamous cell carcinoma of the head and neck cancer
		Cabazitaxel: Metastatic castration-resistant prostate cancer
Camptothecin derivatives	*Camptotheca acuminate* (Chinese happy tree)	Topotecan: Small cell lung cancer and stage IVB, recurrent, or persistent carcinoma of the cervix
		Irinotecan: Metastatic carcinoma of the colon or rectum
Epipodophyllotoxins	*Podophyllum*	Etoposide: Refractory testicular tumors, small cell lung cancer, and off-label uses in leukemias, non-Hodgkin lymphomas, Kaposi sarcoma, neuroblastoma, and soft tissue sarcomas
		Teniposide: Refractory childhood acute lymphoblastic leukemia

nature. This is noteworthy given the thought that perhaps only 1% of tropical trees and plants have been tested for scientific purposes. With regards to cancer drug discovery, Table 3.1 is a summary of antineoplastic agents which were either directly or indirectly derived from the plant ecosystem and their respective utilization for various malignant indications. However, for natural resources to remain as a potential source for new drug discovery, sustainable access to plant, marine, and microbial ecosystems is required and as such efforts in the preservation of these ecosystems is paramount [27]. At present, the continued need for preservation of tropical rainforests and other plant-based ecosystems through minimization of deforestation and encouragement of afforestation is a pivotal factor that is often dismissed or viewed as a task for only nature conservationists. Particularly, these strategies aimed at preserving forests and plant-based habitats should take priority on a global scale as these ecosystems not only play a critical role in mitigating climate change by serving as the carbon dioxide sink but also provide us with untapped opportunities for expansion of our treatment armamentaria and potential discovery of antineoplastic agents that ultimately improve survival outcomes for oncology patients.

Deforestation is defined as the conversion of forest land to other land uses irrespective of whether it is human-induced or otherwise caused by natural causes [28]. Deforestation can take many forms including fires, cutting down trees for agricultural and/or living purposes, and degrading due to climate change. To better understand the potential impact of forest loss on cancer drug discovery, it is vital to review net forest changes that have occurred over the past several decades. The total forest area in the world is estimated to be about 4.06 billion hectares (ha), which accounts for approximately 31% of the total land area [28]. The tropical domain has the largest proportion of the world's forests area (45%), followed by boreal (27%), temperate (16%), and subtropical (11%) domains respectively [28]. Fifty-four percent of the world's forests are in only five countries: Russian Federation (815 million ha), Brazil (497 million ha), Canada (347 million ha), United States of America (310 million ha), and China (220 million ha) [28]. Many of our forests across the world are currently under environmental pressure and deforestation poses a serious

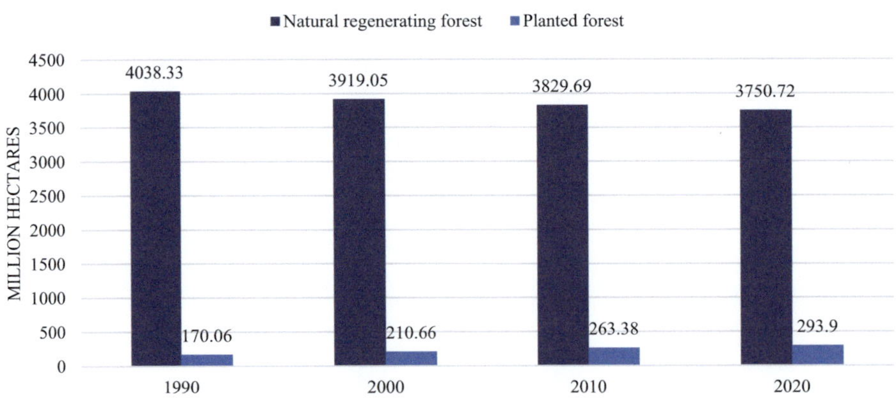

Fig. 3.11 Naturally regenerating and planted forests (1990–2020) based on UN FAO 2020

problem for preservation of our plant-based ecosystems. However, it is of absolute importance to view the global impact of deforestation through the prism of overall net forest loss [29]. Unlike deforestation, the net forest loss accounts for the impact of forest gains over a given period; therefore, net change, can be positive or negative depending on whether gains exceed losses or vice versa.

The Food and Agriculture Organization of the United Nations (UN FAO) provided data in 2020 which estimated that the world has lost 178 million hectares of forest (approximately the size of Libya) since 1990 [28]. Figure 3.11 depicts the naturally regenerating and planted forests over the past four decades, emphasizing the depth of forest loss since 1990 and the importance of forest preservation through afforestation. Afforestation can be defined as a process of establishing and rebuilding a forest in an area that has either lost its natural plants or an area where no forests and plant-based ecosystems existed previously [27–29].

Notably, the rate of net forest loss has substantially decreased over the past four decades owing to both a reduction in deforestation in some countries and significant increases in forest area in other countries through afforestation and the natural expansion of forests [28]. Naturally regenerating forests account for 93% (3.75 billion ha) of the forest area worldwide while planted forests account for 7% (290 million ha) [28]. It is estimated that 420 million ha of forest has been lost since 1990 worldwide because of deforestation [28]. Figure 3.12 depicts the rate of net forest loss over the period from 1990 through 2020 and highlights the slow rate of decline of net forest loss in the most recent decade which could be attributed to the reduction in the rate of forest expansion [28]. Given the reduction in the rates of forest expansion, it is crucial that we maximize efforts aimed at the preservation of rainforest and other forest biomes to not only protect the biodiversity within the forest habitat but also ensure continuous access to unlimited natural resources for drug discovery.

It is important to emphasize that net loss of forest over the years is drastically different depending on the region and the decade. For instance, Africa has had the largest annual rate of net forest loss in 2010–2020, followed by South America while Asia has had the highest net gain of forest area in 2010–2020, followed by Oceania and Europe [28]. Figure 3.13 depicts the net change in annual forest area in various regions across the world between 1990 and 2020 [28].

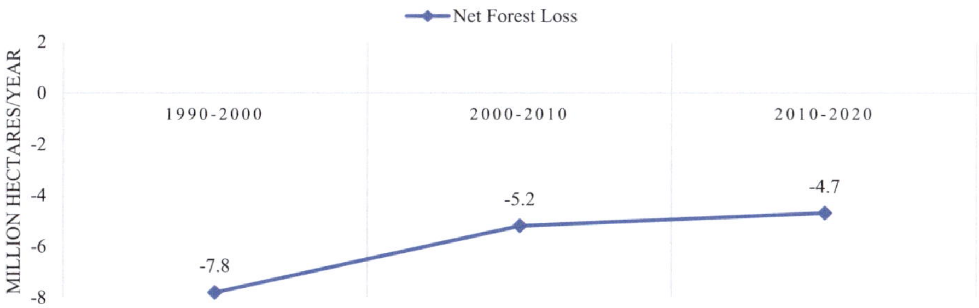

Fig. 3.12 Rate of net forest loss between 1990 through 2020 based on UN FAO 2020

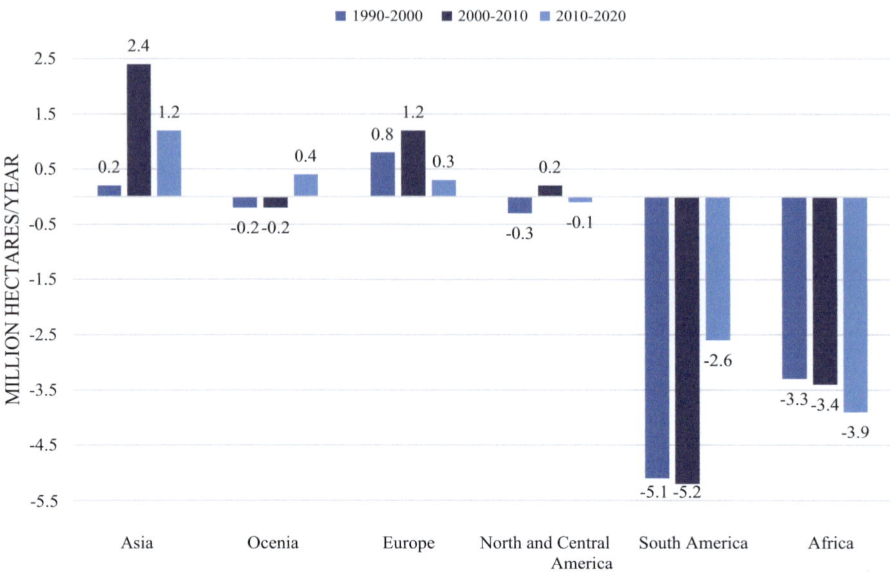

Fig. 3.13 Annual net forest change based on region obtained from UN FAO 2020

Although the rate of deforestation has declined in the most recent years (2015–2020; roughly ten million ha/year) as compared to prior years (2010–2015; roughly 12 million ha/year), we are still losing ten million ha of forest land annually [28]. Agriculture is by far the largest driver of deforestation and while most environmental concerns are modern issues, deforestation is more of a chronic long-term issue. Humans have been cutting trees for millennia and we are just now realizing the impact that years of deforestation and agricultural pressures have had on a global scale [29]. If one takes a closer look at historical data, it will be evident that the earth's surface has changed drastically throughout the past 10,000 years with the loss of forest lands and expansion of agricultural and urban lands [29]. Half of total forest loss occurred well before any of our times during 8000 BC to 1900 and the other half occurred in last century alone. This drives home two important factors; the fact that deforestation is not by any means a new concept and the fact that deforestation has accelerated over the last century [29]. In just over 100 years, the world has lost as much forest as it had in the previous 9000 years, with the primary driver being expansion of agricultural lands [29]. However, with the advancement of technological innovations like lab-grown meat and use substitute products, our per capita agricultural land has decreased significantly [29]. Data from UN FAO 2020 is promising as it shows a steady decline in rates of deforestation and total net loss of forests.

Based on the most recent statistics, it is evident that deforestation has led to a net forest loss over the years. Conversely, the rate of afforestation and natural regeneration has increased over the years but has not been significant enough to allow for complete recovery and preservation of our forests on a global scale. While to date, there is no strong causal relationship between deforestation and loss of drug discovery, one can undoubtedly predict that there is lost opportunity with the loss of rainforests and other plant-based ecosystems. An often-unintended impact of deforestation is this potential threat of loss of discovery of medical cures and remedies for some of the life-threatening illnesses such as cancer. Over the past decades, efforts in preserving rainforests have made a huge impact on the net change in forest loss. Continued preservation efforts are needed to limit our environmental footprint on the rainforests and to sustain our medical discovery endeavors.

Acknowledgements All chemical structures were drawn using PubChem Sketcher V2.4. No special permission required.

References

1. National Cancer Institute. A Story of Discovery: Natural Compound Helps Treat Breast and Ovarian Cancers. 2015. Retrieved from https://www.cancer.gov/research/progress/discovery/taxol
2. Wani MC, Horwitz SB. Nature as a remarkable chemist: a personal story of the discovery and development of Taxol. Anti-Cancer Drugs. 2014;25(5):482–7. https://doi.org/10.1097/CAD.0000000000000063. PMID: 24413390; PMCID: PMC3980006
3. Duffin J. Poisoning the spindle: serendipity and discovery of the anti-tumor properties of the Vinca alkaloids. Can Bull Med Hist. 2000 Nov;17(1–2):155–92. https://doi.org/10.3138/cbmh.17.1.155.
4. Awards for Excellence in Cancer Research. Canadian Cancer Society. Retrieved from https://cancer.ca/en/research/for-researchers/awards-for-excellence
5. Moudi M, Go R, Yien CY, Nazre M. Vinca alkaloids. Int J Prev Med. 2013;4(11):1231–5. PMID: 24404355; PMCID: PMC3883245
6. Keglevich P, Hazai L, Kalaus G, Szántay C. Modifications on the basic skeletons of vinblastine and vincristine. Molecules 2012 17(5):5893–5914. doi: https://doi.org/10.3390/molecules17055893. PMID: 22609781; PMCID: PMC6268133.
7. Vincristine [package insert]. Lake Forest, IL: Hospira, Inc., 2014.
8. Vinblastine [package insert]. Bedford, OH: Bedford Laboratories™, 2012.
9. Vinorelbine [package insert]. Parsippany, NJ: Pierre Fabre, 2020.
10. Dhyani P, Quispe C, Sharma E, Bahukhandi A, Sati P, Attri DC, Szopa A, Sharifi-Rad J, Docea AO, Mardare I, Calina D, Cho WC. Anticancer potential of alkaloids: a key emphasis to colchicine, vinblastine, vincristine, vindesine, vinorelbine and vincamine. Cancer Cell Int. 2022;22(1):206. https://doi.org/10.1186/s12935-022-02624-9. PMID: 35655306; PMCID: PMC9161525
11. de Weger VA, Beijnen JH, Schellens JH. Cellular and clinical pharmacology of the taxanes docetaxel and paclitaxel--a review. Anticancer Drugs. 2014;25(5):488–94. https://doi.org/10.1097/CAD.0000000000000093. Erratum in: Anticancer Drugs. 2015 Feb;26(2):240. PMID: 24637579
12. Paclitaxel [package insert]. Princeton, NJ: Bristol-Myers Squibb Co., 2011.
13. Docetaxel [package insert]. Princeton, NJ: Ebewe Pharma, 2012.
14. Cabazitaxel [package insert]. Bridgewater, NJ: Sanofi-aventis U.S. LLC, 2021.
15. Abraxane [package insert]. Summit, NJ: Abraxis BioScience, LLC, 2020.
16. Khaiwa N, Maarouf NR, Darwish MH, Alhamad DWM, Sebastian A, Hamad M, Omar HA, Orive G, Al-Tel TH. Camptothecin's journey from discovery to WHO essential medicine: fifty years of promise. Eur J Med Chem. 2021;5(223):113639. https://doi.org/10.1016/j.ejmech.2021.113639. Epub 2021 Jun 17

17. Nat R, Chem P, Raveendran VV. Camptothecin-discovery, clinical perspectives and biotechnology. Nat Prod Chem Res. 2015;3(3)
18. Venditto VJ, Simanek EE. Cancer therapies utilizing the camptothecins: a review of the in vivo literature. Mol Pharm. 2010;7(2):307–49. https://doi.org/10.1021/mp900243b. PMID: 20108971; PMCID: PMC3733266
19. Topotecan [package insert]. Sellersville, PA: Teva Pharmaceuticals USA. 2014.
20. Irinotecan [package insert]. New York, NY: Pfizer. 2014.
21. Stein A, Voigt W, Jordan K. Chemotherapy-induced diarrhea: pathophysiology, frequency and guideline-based management. Ther Adv Med Oncol. 2010;2(1):51–63. https://doi.org/10.1177/1758834009355164. PMID: 21789126; PMCID: PMC3126005
22. Koselke E, Kraft S. Chemotherapy-induced diarrhea: options for treatment and prevention. J Hematol Oncol Pharm. 2012;2(4):143–51.
23. Stähelin HF, von Wartburg A. The chemical and biological route from podophyllotoxin glucoside to etoposide: ninth Cain memorial award lecture. Cancer Res. 1991;51(1):5–15.
24. Hande KR. Topoisomerase II inhibitors. Update on Cancer Ther. 2008;3(1):13–26.
25. Etoposide [package insert]. Princeton, NJ: Bristol-Myers Squibb Company. 2004.
26. Teniposide [package insert]. Paramus, NJ: WG Critical Care, LLC. 2015.
27. Cao S, Kingston DG. Biodiversity conservation and drug discovery: can they be combined? The Suriname and Madagascar experiences. Pharm Biol. 2009;47(8):809–23. https://doi.org/10.1080/13880200902988629. PMID: 20161050; PMCID: PMC2746688
28. FAO. Global Forest resources assessment 2020 – key findings. Rome; 2020.
29. Ritchie H, Roser M. Forests and deforestation. Published online at OurWorldInData.org. 2021. Retrieved from https://ourworldindata.org/forests-and-deforestation [Online Resource].

Part II

Food, Lifestyle and Cancer

Meat and Alcohol Consumption: Diet and Lifestyle Choice and Cancer

4

Renee Stubbins

Introduction

Colorectal cancer (CRC) is the third most common cancer and the second most deadly and its incidence is increasing drastically, especially in developed countries. It is estimate that by 2040, there will be 3.2 million cases [1]. It has been well established that our lifestyle choices heavily influence the risk to cancer [2, 3]. The American Cancer Society estimates that at least 18% of all cancers and about 16% of cancer deaths in the United States of America (USA) are related to excess body weight, physical inactivity, alcohol consumption, and/or poor nutrition [4]. The USA follows a predominately western dietary pattern which is characterized by increased red and processed meat, refined grains and added sugar intake. A western diet has been strongly associated with an increased risk for CRC [5–7] compared to a diet that is rich in fiber from fruits, vegetables, whole grains and lean proteins (i.e. poultry and fish) [8, 9]. A recent study suggests that a western diet coupled with binge alcohol drinking (even short-term) can result in an increased risk of liver dysfunction [10]. Despite continuous educational efforts to encourage Americans to eat more plant-based foods and minimize alcohol intake, our meat consumption has shifted, and our alcohol consumption as increased dramatically especially during the pandemic [11]. For the purposes of this chapter, we are going to focus on meat and alcohol consumption and cancer risk.

R. Stubbins (✉)
Neal Cancer Center, Houston Methodist Hospital, Houston, TX, USA
e-mail: restubbins@houstonmethodist.org

E. H. Bernicker (ed.), *Environmental Oncology*,
https://doi.org/10.1007/978-3-031-33750-5_4

Meat Consumption and Cost

The average American male consumes approximately 4.8 oz of meat daily and in comparison, the average American female consumes approximately 3.3 oz of meat daily. Americans on average consume more meat than other countries and despite the health and environmental costs, meat consumption is on the rise. Meat consumption in the U.S. increased by 40% between 1961 and 2017, and globally, meat consumption increased by 58% between 1998 and 2018. It is expected that the U.S. meat consumption will increase by 1% each year through 2023 [12]. It is estimated that 70% of Americans have at least tried a plant-based meat alternative, and by 2040 approximately 60% of the meat eaten globally will be from plant-based or lab-grown alternatives due to changes in the preference of consumers, but limited long-term health information is known of the plant-based meat alternatives [12].

The health costs of meat vary on the type of meat, amount of meat and how the meat is prepared. The American Institute for Cancer Research (AICR) says there's "strong evidence" that eating an excessive amount of red or processed meat may increase the risk of colorectal cancer (CRC) and may also be linked to prostate, nasopharyngeal, lung, gastric and pancreatic cancer. In 2015, the IARC classified processed meat as a human carcinogen (Group 1), which means there is enough evidence to conclude that it can cause cancer in humans. The evidence for red meat was less definitive, so IARC classified it as a probable carcinogen (Group 2A) [13]. Additionally, the World Health Organization says red meat is "probably carcinogenic to humans." Interestingly, there is no established link between white meat (poultry) and fish and increased cancer risk [14–16]. Recently, it has been observed that there is an increased incidence of CRC in adults under the age of 50, which is unfortunately projected to increase over 140% by 2030 [17–19]. It is prudent to gain a better understanding of the mechanisms behind the relationship between red/process meat and increased cancer risk. Some of the mechanisms that have been explored to explain the relationship between red meat and CRC are as follows: Heme-iron in red meat, heterocyclic amines (HCAs), formation of N-glycolylneuraminic acid (Neu5Gc), polycyclic aromatic hydrocarbons (PAHs) formed when cooking meat and N-nitroso compounds (NOCs) and more recently the effects on the intestinal microbiota.

The Role of Heme Iron in Colon Cancer

Heme-iron is found in red and processed meat in the form of hemoglobin and myoglobin. Red meat (beef, veal, lamb, pork) is unprocessed mammalian muscle meat, while processed meat is meat that has been transformed through salting, curing, fermentation, smoking, or other processes to add flavor and improve preservation. Several epidemiological and experimental studies have published a strong association between red or processed meat consumption and colorectal carcinoma (CRC) [20–24]. Studies have demonstrated that consuming 25 g (approximately 1 oz) of processed meat or 50–100 g (approximately 2–4 oz) of red meat daily increases the risk to CRC [23, 24]. Processed meat contains N-nitroso compounds (NOC) from the nitrites or nitrates used to preserved (cure/salt) the meat [25]. Additionally, studies have shown that red meat consumption promotes the formation of N-nitroso compounds (NOC) in a dose dependent manner; NOCs can cause various forms of DNA damage as well increasing the risk to carcinogenesis [26]. In addition to the amount of red meat consumed, the cooking of the red meat also poses as a risk factor for CRC. Specifically, when red meat is barbequed, smoked or cooked a high temperature, two carcinogens can form heterocyclic amines (HCAs) and polycyclic aromatic hydrocarbons (PAHs) [25, 27].

In a recent review, Seiwert et al. discusses multiple pathways in which heme iron can act as a carcinogen: formation of DNA-damaging compounds, changes in intestinal epithelium and microbiota [28]. Specifically, it has been established that heme iron forms reactive oxygen species (ROS) through the Fenton's reaction producing hydroxyl radicals [29]; however, studies on a direct correlation between the ROS produced by dietary iron causing damage to the epithelium are few and only indirect evidence suggests that dietary heme causes oxidative DNA damage. Hydroxyl radicals can also oxidize lipids causing lipid peroxidation and a series of further reactions ultimately creating reactive aldehydes: malondialdehyde (MDA) and 4-hydroxynonenal (4-HNE); these reactive aldehydes have been found to generate mutagenic DNA adducts [30]. Seiwert et al. also highlights recent findings that indicate that heme iron causes hyperproliferation of the colon epithelium through the WNT pathway and the growing knowledge of the gut microbiota and its potential role in CRC risk [28].

Gut Microbiome and Colon Cancer

The gut microbiota continues to be a growing field of interest especially its role in colon carcinogenesis. The gut microbiota is a term that includes the microorganisms (bacteria, fungi, protozoans, archaea and viruses) that reside in our gastrointestinal tract (GIT), predominately in the colon [31]. The host-microbe relationship is actively being investigated and is heavily influenced by our diet, lifestyle choices and medications, which explains why each individual microbiome is unique and diverse. A primary function of the gut microbiome is to aid in the digestion of our food; specifically breaking down nutrients that the host is unable to metabolize such as fiber. A high fiber diet has been shown to increase the diversity of the gut microbiome [32, 33]. Both soluble and insoluble fiber have an important role in the gut microbiome, so both are important to have in the diet. Soluble fiber is soluble in water and fermentable and slows the GI-transit time; bananas, oats, barley, beans, peeled apples or pears, white rice and potatoes are examples of soluble fiber. Insoluble fiber is not soluble in water and adds bulk to our stool and prevents constipation; whole grains, bran, raw vegetables, peels of fruit, and berries are examples of insoluble. Soluble fiber is essential to the health of the gut microbiota since it serves has a prebiotic and can be fermented to short chain fatty acids (SCFAs): acetate, propionate, and butyrate by the intestinal microbiota [34]. Short chain fatty acids have anti-inflammatory properties and are formed from dietary fiber, plant-based proteins, and microbiota accessible carbohydrates (MAC) [35–37]. Short chain fatty acids, specifically butyrate and propionate have been shown to modulate gene expression by acting as histone deacetylases (HDACs) inhibitors [38]. HDAC inhibitors have been explored for their anti-inflammatory properties and potential use in cancer therapy. Specifically, studies have shown that butyrate can act as a HDAC inhibitor protecting against colon cancer [39]. Thus, if the gut microbiome has a healthy amount of SCFAs, then one could argue that the gut microbiome environment is better protected from cancer. It should be noted that butyrate can act as a bifunctional metabolite with different outcomes in both normal colonocytes and cancerous colonocytes. Specifically, in normal colonocytes, butyrate can stimulate cell proliferation but in cancerous colonocytes it inhibits cell proliferation due to the Warburg effect [40].

Research supports that diet has a significant impact on our gut microbiome and that our diet can contribute up to 20% in microbial variation [41–43]. However, there is a surprisingly lack of research investigating how red meat affects the gut microbiome. In Albracht-Schulte et al., they reviewed 15 articles studying a possible link between beef consumption and its impact on the microbiota and found that more research is needed to determine the mechanisms involved. Specifically, the need to discover potential interactions with other dietary compounds and how the beef is cooked or processed and its affect in the microbiome [44]. As previously mentioned, it has been established that dietary fiber intake can have a significant impact on the gut microbiome and The Academy of Nutrition and Dietetics recommends 25–35 g of fiber daily; however, most Americans only consume 10–15 g of fiber. Thus, consistent with the Western diet, high intake of red meat and lack of fiber intake create multiple pathways to colon carcinogenesis. The food industry has considered meat alternatives to curve the American appetite for red meat, but we do not yet know if these alternatives are necessarily a healthier option. However, it has been well documented that substituting plant protein for animal protein (red meat) is associated a reduced risk of CRC [45].

Immunotherapy and Colorectal Cancer

Studies have shown that a high fiber intake is associated with improved response to immunotherapy [46, 47]. Immunotherapy includes a class of drugs that act as immune checkpoint inhibitors (ICIs). ICIs are antibodies (anti-PDL1, anti-PD1, and anti-CTLA4) that enhance the anti-tumor response from our immune system. It has been shown that Immunotherapy works best in a subset of CRC tumors that have high levels of DNA microsatellite instability, which are formally classified as microsatellite instability-high (MSI-H) CRCs and represent approximately 3% of all CRC cases [48, 49]. MSI-H tumors have an overexpression immune checkpoint proteins (like PD-1 and PD-L1), which are targets for certain classes of ICIs. Again, as of now, research only supports the use of immunotherapy in a small percentage of CRC cases. Research efforts to increase the efficacy of ICIs in CRC is much needed. Future exploration of modulating various dietary factors (i.e., dietary fiber) during cancer treatment and its effect on cancer prognosis and survival should be explored.

Societal Costs and Colorectal Cancer

As previously mentioned, a healthy balanced diet with high fiber foods such as fruits, vegetables and whole grains is modifiable lifestyle factor that can reduce the risk to CRC possibly through modifying the gut microbiome. However, there is not equitable access to these healthy foods and studies have shown that not only does our food affect the gut microbiome, but our environment (i.e., exposure to pollutants) can have a substantial impact as well. Thus, health inequality can impact our gut microbiome and by extension the risk to chronic diseases. Studies have shown that changes in the microbiome in both early and late stages of life can alter metabolism and increase the risk to obesity, cancer, and diabetes [50–52]. Furthermore, it has been suggested that health inequity in the form of a more compromised gut microbiome can be passed through generations from mother to baby [53, 54]. Despite being a hot topic of recent discussions, healthy inequalities persist and only changes to healthy policy will initiate this needed change. In a recent paper, by Amato et al., they discuss host–gut microbe interactions and their impact on health inequities and argue that health policy is a crucial step to understanding the link to our environment and population health [55]. There are multiple barriers to address when targeting health inequalities and it is important to take note that it is not only access to healthy food that creates disparity but also the environment (access to health care and/or exposures to carcinogens) that unfortunately can have a greater impact on lower socioeconomic classes and minorities [56, 57]. It has been well documented that there is an inverse relationship between socioeconomic status and overall health as well as exposure to environmental injustice [58, 59]. Environmental injustice is described as "avoidance of hazards and acquisition of benefits through relationships that negatively impact the environment of others" [60]. Environmental injustice is augmented by factory farming, which can indirectly be linked to increased risk of colorectal cancer. The formal definition for factory farming is "a system of rearing livestock using intensive methods, by which poultry, pigs, or cattle are confined indoors under strictly controlled conditions." Animals kept in these tight spaces are fed in excess to make them grow faster than their bodies can support and the waste output from these conditions is harmful to animals, workers, and our environment [61, 62]. The term "factory farming" is socially driven, but the term concentrated animal feeding operations (CAFOs) is a term from the USDA, which is defined as an intensive animal feeding operation (AFO) in which over 1000 animal units are confined for over 45 days a year. Pollutants from CAFOs are just as impactful on our health as pollutants from factory farms. CAFOs regulated by the EPA but they still produce a substantial amount of air and water containments that can affect our health, including cancer. Studies have suggested that livestock and poultry farmers are at increased risks to certain cancer (colon cancer and multiple myeloma) [63, 64]. Current industrial farming is not sustainable and come with heavy societal and healthy costs to not only our generation but future generations as well.

Alcohol Intake and Cancer Risk

Both IACR and AICR, have shown that alcohol consumption increases a person's risk of developing at least six different cancer types: breast, nasopharyngeal, mouth, laryngeal, esophageal, and colon [65, 66]. Not only does alcohol increase the risk to cancer, but it is also associated with increased cancer mortality [67]. Alcohol consumption has a dose-dependent effect on cancer risk and evidence suggests that three or more drinks per day increases the risk to stomach and liver cancer [65, 68]. Alcohol cessation is a modifiable lifestyle behavior that we could choose to adopt to reduce our cancer risk, but many Americans are unaware that alcohol consumption is associated with increased cancer risk [69]. Recent national surveys have found that only 38–45% of Americans are aware that alcohol intake increases the risk to cancer; however, approximately 89% of Americans acknowledge that tobacco use was associated with cancer risk [70, 71]. Suggesting that public health education could potentially be an avenue to improve awareness and decrease alcohol consumption thus potentially decreasing cancer incidence and mortality. An important part of public education is understanding the mechanisms by which alcohol increases carcinogenesis and mortality and then disseminating the results to the public. Understanding how alcohol increases our risk to cancer is complex and the mechanisms behind the causal relationship between alcohol and cancer are not well elucidated. However, explored explanations suggests that alcohol increases damage to our cells, alters hormone signals and changes cellular physiology to easier absorb other carcinogens (i.e., tobacco) [72–74].

Alcohol is rapidly absorbed through the stomach and small intestines and metabolized by the liver. During alcohol metabolization, a hydrogen is transferred from the ethanol molecule to nicotinamide dinucleotide (NAD+) by alcohol dehydrogenase (ADH) forming acetaldehyde (AA). The formation of AA increases the risk to carcinogenesis by directly and indirectly effecting DNA damage. It has been shown in animal studies, that acetaldehyde is carcinogenic [75, 76]. AA can directly bind to DNA causing errors in oncogenes or tumor suppressor genes [77]. AA can indirectly effect DNA damage by increasing the production of reactive oxygen species (ROS) and reactive nitrogen species (RNS) through multiple pathways in the liver [78]. ROS and RNS molecules in an altered state cause cellular damage that increases the risk to carcinogenesis [79, 80]. Redox homeostasis is part of our normal biology, but if this balanced is disturbed it can be pathogenic and lead to malignancies. Alcohol and AA have multiple pathophysiological mechanisms by which they increase carcinogenesis. As mentioned previously, alcohol and AA can alter DNA and impact its ability to repair damage; additionally it has been suggested that alcohol and AA can also impact our DNA expression by affecting DNA methylation and ultimately potentially impacting the expression of oncogenes and tumor suppressor genes [81, 82]. Alcohol and AA have multiple pathways that can be linked to increased cancer risk, including disrupting hormone signals and endocrine function.

Alcohol and Breast Cancer

Epidemiological studies have shown a clear link between alcohol and breast cancer [83–85]. Alcohol-induced hormonal dysregulations can affect the entire body along with its metabolite, (Acetaldehyde, AA). Studies have shown that even one drink per day can significantly increase the risk of breast cancer in women [86]. The mechanisms by which alcohol and its metabolite (AA) increase breast cancer risk are constantly under review. One proposed hypothesis is that alcohol can increase endogenous estrogen, higher levels of estrogen interact with estrogen receptor (ER) and increase breast cancer incidence [87–89]. Additionally, alcohol might affect endogenous estrogen levels by interacting with aromatase activity and potentially menstrual cycles, thus increasing lifetime exposure to estrogen [90, 91]. Recently, a proposed mechanism suggests that alcohol might alter Brf1 (TFIIIB-related factor 1) expression and initiate changes that support tumorigenesis [92]. Another pattern that has emerged is alcohol consumption increases mammary density, which is also correlated with higher levels of endogenous estrogen and ER+ breast cancers [93]. Acetaldehyde has also been shown to accumulate in breast mammary after even a single dose of ethanol and since AA is a known mutagen it could be another possible mechanism by which alcohol increases the risk to breast cancer [94, 95]. Overall, researchers tend to agree that it is the lifelong exposure of alcohol and estrogen that influence the risk of breast cancer.

Alcohol and Head and Neck Cancer

Head and neck cancer is the seventh most common cancer in the world, with 1.1 million new diagnoses each year [96]. Furthermore, most head and neck cancer patients are diagnosed at an advanced stage so the patients have an unfortunate poor prognosis [97]. It is well established that alcohol, tobacco, and human papilloma virus 16 (HPV 16) are the main risk factors for head and neck cancer. However, studies have found that when alcohol and tobacco are used together, they synergistically increase the risk to head and neck cancer [98–100]. Alcohol also impacts the oral microbiota; the enzyme catalase starts metabolizing alcohol to AA in the oral cavity but further metabolism to acetate is minimal thus there is increased exposure to AA and its harmful effects [101]. It has been shown that heavy drinkers are more prone to oral microbiota dysbiosis and that there is a decrease in beneficial microbiota (Lactobacilli) [102]. Lastly, it has been argued that alcohol increases the exposure of oral mucosa epithelial cells to other carcinogens further augmenting the incidence of carcinogenesis [103].

Alcohol and Upper and Lower GI Cancer

In a recent report, the authors demonstrate that alcohol increases the risk of esophageal cancer [104]. The causation of the increased risk is comparable to the aforementioned mechanisms (oxidative stress, inflammation, DNA damage, epigenetics etc.). However, there is additional support that when alcohol and tobacco are consumed together, the risk to esophageal cancer increases drastically [105]. Alcohol consumed in higher amounts (>45 g per day, >4 drinks per day) has been shown to increase the link to both stomach and liver cancer [65, 106]. On the contrary, the same report found an increased risk to colorectal cancer (CRC) with only 10 g (1 drink per day) of ethanol consumption [65]. In our previous section, we discussed the significance of CRC and the gut microbiome, it should not surprise that alcohol consumption significantly impacts our intestines microbiome and thus providing another avenue for carcinogenesis [107]. Animal studies have shown alcohol consumption (up to 40% of calories) induced bacterial overgrowth and dysbiosis [108]. Epidemiological studies have shown that increased alcohol consumption increases the amount of AA, and the colonic microbes are unable to metabolize further leading to an accumulation of AA; augmenting the chance of carcinogenesis [109, 110].

In conclusion, the IARC classifies alcohol has a class I carcinogen, which is the same class as tobacco. However, alcohol and tobacco do not have the same warning labels. Tobacco products clearly warn consumers of the hazards of the tobacco smoke and their health; but no such label currently exists on alcohol products. It is a disservice to public health to not provide transparent and factual evidence on the risks associated with certain lifestyle behaviors. If our goal is to improve public health by decreasing cancer incidence and mortality, it should start with public health education and changes in labeling requirements for alcohol.

Conclusion

The incidence of colon cancer is on the rise along with alcohol consumption; specifically in the younger population (<50 years old). Our diet and lifestyle offer a unique opportunity to help reduce our risk by changing our behaviors and actions. Following a high fiber diet that is plentiful in fruits, vegetables, whole grains, and plant-proteins and limiting our red and processed meat intake is a proven sustainable intervention that not only reduces the costs to our health but also our environment. Creating equitable access to healthy food and safe environments for all socioeconomic classes is an essential step in changing health policy and impacting population health. Change starts with education; educating the public on the costs of a meat-based diet vs plant-based diet and how it is impacts not only their health and risk to colon cancer but also the long-term impact the health of our planet. Lastly, encouraging policy makers to take a more transparent educational effort on the impact of alcohol on cancer risk, could greatly impact overall cancer incidence and mortality in future generations.

References

1. Xi Y, Xu P. Global colorectal cancer burden in 2020 and projections to 2040. Transl Oncol. 2021;14(10):101174.
2. Murphy N, et al. Lifestyle and dietary environmental factors in colorectal cancer susceptibility. Mol Asp Med. 2019;69:2–9.
3. Keum N, Giovannucci E. Global burden of colorectal cancer: emerging trends, risk factors and prevention strategies. Nat Rev Gastroenterol Hepatol. 2019;16(12):713–32.
4. Rock CL, et al. American Cancer Society guideline for diet and physical activity for cancer prevention. CA Cancer J Clin. 2020;70(4):245–71.
5. Terry P, et al. Prospective study of major dietary patterns and colorectal cancer risk in women. Am J Epidemiol. 2001;154(12):1143–9.
6. Wu K, et al. Dietary patterns and risk of colon cancer and adenoma in a cohort of men (United States). Cancer Causes Control. 2004;15(9):853–62.
7. Magalhães B, Peleteiro B, Lunet N. Dietary patterns and colorectal cancer: systematic review and meta-analysis. Eur J Cancer Prev. 2012;21(1):15–23.
8. Kim MK, et al. Dietary patterns and subsequent colorectal cancer risk by subsite: a prospective cohort study. Int J Cancer. 2005;115(5):790–8.
9. Koushik A, et al. Fruits, vegetables, and colon cancer risk in a pooled analysis of 14 cohort studies. J Natl Cancer Inst. 2007;99(19):1471–83.
10. Skinner RC, Hagaman JA. The interplay of Western diet and binge drinking on the onset, progression, and outlook of liver disease. Nutr Rev. 2022;80(3):503–12.
11. Pollard MS, Tucker JS, Green HD. Changes in adult alcohol use and consequences during the COVID-19 pandemic in the US. JAMA Netw Open. 2020;3(9):e2022942.
12. Christen C. Meat consumption in the U.S. is growing at an alarming rate. 2021 [cited 2022 10/20/2022].
13. Bouvard V, et al. Carcinogenicity of consumption of red and processed meat. Lancet Oncol. 2015;16(16):1599–600.
14. Kobayashi M, et al. Fish, long-chain n-3 polyunsaturated fatty acids, and risk of colorectal cancer in middle-aged Japanese: the JPHC study. Nutr Cancer. 2004;49(1):32–40.
15. Tiemersma EW, et al. Meat consumption, cigarette smoking, and genetic susceptibility in the etiology of colorectal cancer: results from a Dutch prospective study. Cancer Causes Control. 2002;13(4):383–93.
16. Willett WC, et al. Relation of meat, fat, and fiber intake to the risk of colon cancer in a prospective study among women. N Engl J Med. 1990;323(24):1664–72.
17. Patel SG, Ahnen DJ. Colorectal cancer in the young. Curr Gastroenterol Rep. 2018;20(4):15.
18. Mauri G, et al. Early-onset colorectal cancer in young individuals. Mol Oncol. 2019;13(2):109–31.
19. Bailey CE, et al. Increasing disparities in the age-related incidences of colon and rectal cancers in the United States, 1975-2010. JAMA Surg. 2015;150(1):17–22.
20. Bernstein AM, et al. Processed and unprocessed red meat and risk of colorectal cancer: analysis by tumor location and modification by time. PLoS One. 2015;10(8):e0135959.
21. Norat T, et al. Meat, fish, and colorectal cancer risk: the European prospective investigation into cancer and nutrition. J Natl Cancer Inst. 2005;97(12):906–16.
22. English DR, et al. Red meat, chicken, and fish consumption and risk of colorectal cancer. Cancer Epidemiol Biomark Prev. 2004;13(9):1509–14.
23. Sandhu MS, White IR, McPherson K. Systematic review of the prospective cohort studies on meat consumption and colorectal cancer risk: a meta-analytical approach. Cancer Epidemiol Biomark Prev. 2001;10(5):439–46.
24. Smolińska K, Paluszkiewicz P. Risk of colorectal cancer in relation to frequency and total amount of red meat consumption. Systematic review and meta-analysis. Arch Med Sci. 2010;6(4):605–10.
25. Jägerstad M, Skog K. Genotoxicity of heat-processed foods. Mutat Res. 2005;574(1–2):156–72.

26. Bingham SA, Hughes R, Cross AJ. Effect of white versus red meat on endogenous N-nitrosation in the human colon and further evidence of a dose response. J Nutr. 2002;132(11 Suppl):3522S–5S.
27. Cross AJ, Sinha R. Meat-related mutagens/carcinogens in the etiology of colorectal cancer. Environ Mol Mutagen. 2004;44(1):44–55.
28. Seiwert N, et al. Mechanism of colorectal carcinogenesis triggered by heme iron from red meat. Biochim Biophys Acta Rev Cancer. 2020;1873(1):188334.
29. Stohs SJ, Bagchi D. Oxidative mechanisms in the toxicity of metal ions. Free Radic Biol Med. 1995;18(2):321–36.
30. Guéraud F, et al. Chemistry and biochemistry of lipid peroxidation products. Free Radic Res. 2010;44(10):1098–124.
31. Ley RE, Peterson DA, Gordon JI. Ecological and evolutionary forces shaping microbial diversity in the human intestine. Cell. 2006;124(4):837–48.
32. Daniel CR, McQuade JL. Nutrition and cancer in the microbiome era. Trends Cancer. 2019;5(9):521–4.
33. Menni C, et al. Gut microbiome diversity and high-fibre intake are related to lower long-term weight gain. Int J Obes. 2017;41(7):1099–105.
34. Koh A, et al. From dietary fiber to host physiology: short-chain fatty acids as key bacterial metabolites. Cell. 2016;165(6):1332–45.
35. Topping DL, Clifton PM. Short-chain fatty acids and human colonic function: roles of resistant starch and nonstarch polysaccharides. Physiol Rev. 2001;81(3):1031–64.
36. Hryckowian AJ, et al. Microbiota-accessible carbohydrates suppress Clostridium difficile infection in a murine model. Nat Microbiol. 2018;3(6):662–9.
37. Cummings JH, et al. Short chain fatty acids in human large intestine, portal, hepatic and venous blood. Gut. 1987;28(10):1221–7.
38. Johnstone RW. Histone-deacetylase inhibitors: novel drugs for the treatment of cancer. Nat Rev Drug Discov. 2002;1(4):287–99.
39. Flint HJ, et al. The role of the gut microbiota in nutrition and health. Nat Rev Gastroenterol Hepatol. 2012;9(10):577–89.
40. Donohoe DR, et al. The Warburg effect dictates the mechanism of butyrate-mediated histone acetylation and cell proliferation. Mol Cell. 2012;48(4):612–26.
41. Wilson AS, et al. Diet and the human gut microbiome: an international review. Dig Dis Sci. 2020;65(3):723–40.
42. Singh RK, et al. Influence of diet on the gut microbiome and implications for human health. J Transl Med. 2017;15(1):73.
43. Johnson AJ, et al. A guide to diet-microbiome study design. Front Nutr. 2020;7:79.
44. Albracht-Schulte K, et al. Systematic review of beef protein effects on gut microbiota: implications for health. Adv Nutr. 2021;12(1):102–14.
45. Liao LM, et al. Substitution of dietary protein sources in relation to colorectal cancer risk in the NIH-AARP cohort study. Cancer Causes Control. 2019;30(10):1127–35.
46. Malczewski AB, et al. Enhancing checkpoint inhibitor therapy in solid tissue cancers: the role of diet, the microbiome & microbiome-derived metabolites. Front Immunol. 2021;12:624434.
47. Spencer CN, et al. Dietary fiber and probiotics influence the gut microbiome and melanoma immunotherapy response. Science. 2021;374(6575):1632–40.
48. Nosho K, et al. Tumour-infiltrating T-cell subsets, molecular changes in colorectal cancer, and prognosis: cohort study and literature review. J Pathol. 2010;222(4):350–66.
49. Golshani G, Zhang Y. Advances in immunotherapy for colorectal cancer: a review. Ther Adv Gastroenterol. 2020;13:1756284820917527.
50. Larsen N, et al. Gut microbiota in human adults with type 2 diabetes differs from non-diabetic adults. PLoS One. 2010;5(2):e9085.
51. Kitten AK, et al. Gut microbiome differences among Mexican Americans with and without type 2 diabetes mellitus. PLoS One. 2021;16(5):e0251245.
52. Turnbaugh PJ, et al. An obesity-associated gut microbiome with increased capacity for energy harvest. Nature. 2006;444(7122):1027–31.

53. Kapourchali FR, Cresci GAM. Early-life gut microbiome-the importance of maternal and infant factors in its establishment. Nutr Clin Pract. 2020;35(3):386–405.
54. Ferretti P, et al. Mother-to-infant microbial transmission from different body sites shapes the developing infant gut microbiome. Cell Host Microbe. 2018;24(1):133–145.e5.
55. Amato KR et al. The human gut microbiome and health inequities. Proc Natl Acad Sci USA. 2021;118(25).
56. Henderson S, Wells R. Environmental racism and the contamination of black lives: a literature review. J Afr Am Stud. 2021;25(1):134–51.
57. Bailey ZD, et al. Structural racism and health inequities in the USA: evidence and interventions. Lancet. 2017;389(10077):1453–63.
58. Evans GW, Kantrowitz E. Socioeconomic status and health: the potential role of environmental risk exposure. Annu Rev Public Health. 2002;23(1):303–31.
59. Kelly-Reif K, Wing S. Urban-rural exploitation: an underappreciated dimension of environmental injustice. J Rural Stud. 2016;47:350–8.
60. Wing S. Environmental injustice connects local food environments with global food production. Local Food Environments: Food Access in America. CRC Press; 2014.
61. Ferguson DD, et al. Detection of airborne methicillin-resistant Staphylococcus aureus inside and downwind of a swine building, and in animal feed: potential occupational, animal health, and environmental implications. J Agromedicine. 2016;21(2):149–53.
62. Parris K. Impact of agriculture on water pollution in OECD countries: recent trends and future prospects. Int J Water Resour Dev. 2011;27(1):33–52.
63. Beane Freeman LE, et al. Poultry and livestock exposure and cancer risk among farmers in the agricultural health study. Cancer Causes Control. 2012;23(5):663–70.
64. Eriksson M, Karlsson M. Occupational and other environmental factors and multiple myeloma: a population based case-control study. Br J Ind Med. 1992;49(2):95–103.
65. Research, W.C.R.F.A.I.f.C., Continuous update project expert report 2018. Alcoholic drinks and the risk of cancer. London: World Cancer Research Fund/American Institute for Cancer Research; 2018.
66. Stewart BW, W.C., World cancer report 2014. 2014, International Agency for Research on Cancer: Lyon.
67. Goding Sauer A, et al. Proportion of cancer cases and deaths attributable to alcohol consumption by US state, 2013-2016. Cancer Epidemiol. 2021;71(Pt A):101893.
68. Bagnardi V, et al. Alcohol consumption and site-specific cancer risk: a comprehensive dose-response meta-analysis. Br J Cancer. 2015;112(3):580–93.
69. Scheideler JK, Klein WMP. Awareness of the link between alcohol consumption and cancer across the world: a review. Cancer Epidemiol Biomark Prev. 2018;27(4):429–37.
70. Wiseman KP, Klein WMP. Evaluating correlates of awareness of the association between drinking too much alcohol and cancer risk in the United States. Cancer Epidemiol Biomark Prev. 2019;28(7):1195–201.
71. Research, A.I.f.C., 2017 AICR cancer risk awareness survey report. 2017, American Institute for Cancer Research.
72. Hirano T. Alcohol consumption and oxidative DNA damage. Int J Environ Res Public Health. 2011;8(7):2895–906.
73. Rachdaoui N, Sarkar DK. Effects of alcohol on the endocrine system. Endocrinol Metab Clin N Am. 2013;42(3):593–615.
74. Pöschl G, Seitz HK. Alcohol and cancer. Alcohol Alcohol. 2004;39(3):155–65.
75. Woutersen RA, et al. Inhalation toxicity of acetaldehyde in rats. III Carcinogenicity study. Toxicology. 1986;41(2):213–31.
76. Feron VJ, Kruysse A, Woutersen RA. Respiratory tract tumours in hamsters exposed to acetaldehyde vapour alone or simultaneously to benzo(a)pyrene or diethylnitrosamine. Eur J Cancer Clin Oncol. 1982;18(1):13–31.
77. Fang J-L, Vaca CE. Development of a 32P-postlabelling method for the analysis of adducts arising through the reaction of acetaldehyde with 2′-deoxyguanosine-3′-monophosphate and DNA. Carcinogenesis. 1995;16(9):2177–85.

78. Galicia-Moreno M, Gutiérrez-Reyes G. The role of oxidative stress in the development of alcoholic liver disease. Rev Gastroenterol Mex. 2014;79(2):135–44.
79. Ames BN. Endogenous oxidative DNA damage, aging, and cancer. Free Radic Res Commun. 1989;7(3–6):121–8.
80. Cerutti PA. Oxy-radicals and cancer. Lancet. 1994;344(8926):862–3.
81. Seitz HK, Stickel F. Molecular mechanisms of alcohol-mediated carcinogenesis. Nat Rev Cancer. 2007;7(8):599–612.
82. Varela-Rey M, et al. Alcohol, DNA methylation, and cancer. Alcohol Res. 2013;35(1):25–35.
83. Seitz HK, et al. Epidemiology and pathophysiology of alcohol and breast cancer: update 2012. Alcohol Alcohol. 2012;47(3):204–12.
84. Singletary KW, Gapstur SM. Alcohol and breast cancer: review of epidemiologic and experimental evidence and potential mechanisms. JAMA. 2001;286(17):2143–51.
85. McDonald, J.A., A. Goyal, and M.B. Terry, Alcohol intake and breast cancer risk: weighing the overall evidence. Curr Breast Cancer Rep, 2013. 5(3), 208, 221.
86. Allen NE, et al. Moderate alcohol intake and cancer incidence in women. J Natl Cancer Instit. 2009;101(5):296–305.
87. Ginsburg ES. Estrogen, alcohol and breast cancer risk. J Steroid Biochem Mol Biol. 1999;69(1–6):299–306.
88. Suzuki R, et al. Alcohol intake and risk of breast cancer defined by estrogen and progesterone receptor status—a meta-analysis of epidemiological studies. Int J Cancer. 2008;122(8):1832–41.
89. Deandrea S, et al. Alcohol and breast cancer risk defined by estrogen and progesterone receptor status: a case-control study. Cancer Epidemiol Biomark Prev. 2008;17(8):2025–8.
90. Castro GD, Castro JA. Alcohol drinking and mammary cancer: pathogenesis and potential dietary preventive alternatives. World J Clin Oncol. 2014;5(4):713–29.
91. Cooper GS, et al. Association of physical and behavioral characteristics with menstrual cycle patterns in women age 29-31 years. Epidemiology. 1996;7:624–8.
92. Huang C, Zhang Y, Zhong S. Alcohol intake and abnormal expression of Brf1 in breast cancer. Oxidative Med Cell Longev. 2019;2019:4818106.
93. Frydenberg H, et al. Alcohol consumption, endogenous estrogen and mammographic density among premenopausal women. Breast Cancer Res. 2015;17(1):103.
94. Fanelli SL, et al. Further studies on the potential contribution of acetaldehyde accumulation and oxidative stress in rat mammary tissue in the alcohol drinking promotion of breast cancer. J Appl Toxicol. 2011;31(1):11–9.
95. Castro GD, et al. Acetaldehyde accumulation in rat mammary tissue after an acute treatment with alcohol. J Appl Toxicol. 2008;28(3):315–21.
96. Mody MD, et al. Head and neck cancer. Lancet. 2021;398(10318):2289–99.
97. de França GM, et al. Five-year survival and prognostic factors for oropharyngeal squamous cell carcinoma: retrospective cohort of a cancer center. Oral Maxillofac Surg. 2022;26(2):261–9.
98. Prabhu A, Obi KO, Rubenstein JH. The synergistic effects of alcohol and tobacco consumption on the risk of esophageal squamous cell carcinoma: a meta-analysis. Off J Am Coll Gastroenterol. 2014;109(6):822–7.
99. Giraldi L, et al. Alcohol and cigarette consumption predict mortality in patients with head and neck cancer: a pooled analysis within the international head and neck cancer epidemiology (INHANCE) consortium. Ann Oncol. 2017;28(11):2843–51.
100. Talamini R, et al. Combined effect of tobacco and alcohol on laryngeal cancer risk: a case–control study. Cancer Causes Control. 2002;13(10):957–64.
101. Cao Y, et al. Light to moderate intake of alcohol, drinking patterns, and risk of cancer: results from two prospective US cohort studies. BMJ. 2015:351.
102. Fan X, et al. Drinking alcohol is associated with variation in the human oral microbiome in a large study of American adults. Microbiome. 2018;6(1):1–15.
103. Howie NM, et al. Short-term exposure to alcohol increases the permeability of human oral mucosa. Oral Dis. 2001;7(6):349–54.

104. Rumgay H, et al. Global burden of cancer in 2020 attributable to alcohol consumption: a population-based study. Lancet Oncol. 2021;22(8):1071–80.
105. Steevens J, et al. Alcohol consumption, cigarette smoking and risk of subtypes of oesophageal and gastric cancer: a prospective cohort study. Gut. 2010;59(01):39–48.
106. World Cancer Research Fund, I., Diet, nutrition, physical activity and cancer: a global perspective: a summary of the Third Expert Report. 2018: World Cancer Research Fund International.
107. Rossi M, et al. Colorectal cancer and alcohol consumption—populations to molecules. Cancers. 2018;10(2):38.
108. Engen PA, et al. The gastrointestinal microbiome: alcohol effects on the composition of intestinal microbiota. Alcohol Res. 2015;37(2):223–36.
109. Na H-K, Lee JY. Molecular basis of alcohol-related gastric and colon cancer. Int J Mol Sci. 2017;18(6):1116.
110. Siddiqui R, et al. The pivotal role of the gut microbiome in colorectal cancer. Biology. 2022;11. https://doi.org/10.3390/biology11111642

Tobacco, Second-Hand Smoke and Cancer

5

Decha Pinkaew, Tarek Dammad, Mohamad Bitar, Sandeep Sahay, and Rodney J. Folz

Introduction

Tobacco smoking kills more than eight million people worldwide each year. Over seven million of these deaths were directly caused by smoking, and around 1.2 million were nonsmokers exposed to second-hand smoke [1]. Although smoking is the leading cause of chronic obstructive pulmonary disease (COPD) in the United States, one out of four Americans with COPD is a never-smoker [2]. Exposure to second-hand smoke, smoke exhaled by a smoker and smoke from the lighted end of a burning cigarette, is associated with the development of COPD [3]. Physiologically speaking, COPD is characterized by chronic airflow restriction resulting from an inflammatory and injurious response to a variety of inhaled pollutants present in tobacco smoke [4]. Patients with COPD could have either chronic bronchitis with inflammation and excess mucus or emphysema with destruction of the alveolus, or more commonly a combination of both [5]. Numerous chemical molecules in tobacco smoke can lead to the production of reactive oxygen species (ROS) that directly or indirectly damage lung cells [6]. An imbalance of ROS and antioxidants in the body, called oxidative stress, usually results in damage to lipids, proteins, and DNA [7]. DNA and protein modification by ROS can cause DNA mutations and impaired protein function, respectively. Moreover, ROS can cause mitochondrial DNA (mtDNA) mutations and mitochondrial dysfunction and activate pro-inflammatory nuclear factor-κB (NF-κB) signaling [8, 9]. Inflammation in the lungs of patients with COPD involves innate and adaptive immunity, as well as structural airway cells, which contribute to the pathogenesis and disease severity of COPD [10].

D. Pinkaew · T. Dammad · M. Bitar · S. Sahay · R. J. Folz (✉)
Division of Pulmonary, Critical Care and Sleep Medicine, Houston Methodist Hospital, Houston, TX, USA
e-mail: dpinkaew@houstonmethodist.org; tdammad@houstonmethodist.org; mzbitar@houstonmethodist.org; ssahay@houstonmethodist.org; rfolz@houstonmethodist.org

E. H. Bernicker (ed.), *Environmental Oncology*,
https://doi.org/10.1007/978-3-031-33750-5_5

Numerous studies have demonstrated a link between the pathogenesis of COPD and lung cancer. Accumulating evidence shows that oxidative stress, mitochondrial dysfunction, inflammation, and epigenetic regulation have a role in the development of COPD and lung cancer. Oxidative stress–induced DNA damage and impaired DNA repair can lead to deleterious mutations. Furthermore, mtDNA is particularly prone to mutations when exposed to ROS. In addition, ROS-mediated mitochondrial dysfunction drives cancer progression toward chemoresistance or metastasis [11]. Inflammatory cells and cytokines are more likely to promote lung tumor growth, proliferation, and metastasis [12]. ROS also activates NF-κB, which induces inflammation and carcinogenesis by promoting angiogenesis and preventing cancer cell apoptosis [13]. Moreover, DNA methylation and dysregulated miRNA expression of various genes, including immune genes, oncogenes, and tumor suppressor genes, increase the risk of developing lung cancer in patients with COPD. In this chapter, we will review the risk for the development of COPD from tobacco smoke exposure, the role of oxidative stress in COPD, and how oxidative stress and potentially related pathways link COPD and lung cancer.

Tobacco Smoking and Second-Hand Smoke

Tobacco smoking is still the main cause of disease and death in the United States that is completely preventable [14]. Annually, nearly half a million people die from smoking in the United States [15]. In 2020, 47.1 million American adults aged 18 years and older used tobacco products, with 30.8 million people still smoking cigarettes [16]. Smoke from tobacco contains many harmful substances that damage the airways and lungs. Smoking is the primary risk factor driving the pathogenesis and progression of hypertension [17], cardiovascular diseases [18, 19], cancer [15, 20], and pulmonary diseases [21, 22]. Smoking is usually the leading cause of COPD [23, 24]. As many as eight out of ten COPD-related fatalities are caused by smoking [15]. However, one out of four Americans with COPD is a never-smoker [25].

Second-hand smoke (SHS) is defined as smoke exhaled by a smoker and smoke from the lit end of a burning cigarette or tobacco [26]. Breathing in other people's tobacco smoke is called passive smoking. SHS contains the same harmful chemicals that smokers inhale, and at least 250 chemicals in SHS are toxic or carcinogenic [27]. The concentrations of these dangerous compounds in the body increase as more SHS is breathed in. A recent study showed that SHS-exposed student nurses were 4.45 times more likely than non-exposed student nurses to test positive for SHS, as indicated by urine cotinine levels [28]. Another study found that the serum cotinine levels of smokers residing in homes where others smoke indoors were significantly higher than those of smokers who reported no SHS exposure, regardless of individual smoking intensity [29].

SHS is a risk factor for cardiovascular diseases [27], lung cancer [30], hypertension [31], childhood asthma [32], and COPD [33]. Whether a person is exposed to second-hand smoking as a child, an adult, or both, the exposure raises their chance of developing respiratory conditions like asthma and COPD [34]. Never-smokers exposed to SHS at home and at a previous workplace had a greater prevalence of COPD than non-exposed participants (8.0% and 8.3% vs. 4.2%, respectively). When never-smokers were exposed to SHS at home, previous work, and current work, the prevalence of COPD was highest (18.8% vs 4.2%) [35].

Smoke-Induced Oxidative Stress in the Pathogenesis of COPD

COPD is characterized by airflow restriction resulting from an inflammatory response to inhaled pollutants, most commonly tobacco smoke [4]. According to recent data, COPD was the sixth leading cause of death in the United States in 2020, with over 150,000 deaths per year [36, 37]. During the COVID-19 pandemic, patients with COPD and COVID-19 had a higher mortality rate than those without COPD (15% vs. 4%) [38]. In 2015, an estimated 15.5 million Americans over the age of 18 were diagnosed with COPD [39]. It was estimated that COPD would cost $800.9 billion in direct medical costs over 20 years, from 2019 to 2038, in the United States [40].

Smoking is a well-known risk factor for COPD and is responsible for approximately 75% cases of COPD [41] (Fig. 5.1). Exposure to air pollution, second-hand smoke, and dust, fumes and chemicals can cause COPD [42, 43]. A small number of people have alpha-1 antitrypsin deficiency (Alpha-1)-related COPD, an uncommon form of COPD that affects 2–3% of all COPD patients in the US [44]. This

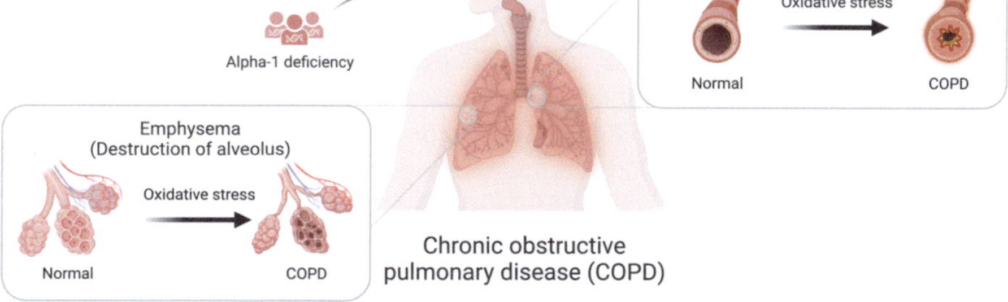

Fig. 5.1 Chronic obstructive pulmonary disease (COPD). COPD is characterized by airflow restriction resulting from an inflammatory response to inhaled pollutants, most commonly tobacco smoke. COPD can be also caused by long-term exposure to air pollution, second-hand smoke and dust, gases, and chemicals, as well as alpha-1 antitrypsin deficiency. A person with COPD could have either chronic bronchitis (inflammation and excess mucus) or emphysema (destruction of the alveolus). The graphic was created with BioRender.com

type of COPD is caused by a genetic disease that impairs the body's ability to produce circulating alpha-1 antitrypsin protein (ATT), a serine protease inhibitor that protects lung tissues against damage caused by neutrophil elastase (NE) [45]. Excessive NE can destroy alveolar structures and the extracellular matrix of the lung if AAT is considerably diminished, inactivated or deficient [46].

A person with COPD can have chronic bronchitis, emphysema, or both [5]. Chronic obstructive bronchitis is a disorder in which long-term swelling and inflammation cause the airways to become abnormally narrow and to include goblet cells that produce and secrete excessive amounts of mucus, which reduces airflow [5, 47] (Fig. 5.1). Emphysema is defined by the loss of alveolar epithelial cells and impairment in re-epithelialization. This destroys the alveolar wall and reduces the lung parenchyma's surface area for gas exchange, resulting in hyperinflation, air trapping, and diaphragm flattening [48, 49] (Fig. 5.1).

COPD's pathogenesis involves many inflammatory cells, inflammatory mediators, and cell signaling pathways. Also, COPD regulates the expansion of goblet cells, the formation of mucoproteins, and the discharge of mucus [50]. Tobacco smoke contains a variety of oxidants and prooxidants, which can produce ROS that are harmful to lung cells [6, 51] (Fig. 5.2). In addition, oxidants that are present in cigarette smoke have the potential to activate alveolar macrophages, causing them to generate ROS and to release a variety of mediators that both attract and activate neutrophils and other inflammatory cells into the lungs [52, 53]. Neutrophils and macrophages migrate into the lungs of cigarette smokers in greater numbers than into the lungs of nonsmokers and can generate ROS by the reduced nicotinamide adenine dinucleotide phosphate oxidase system [52, 53]. The excessive levels of ROS can overwhelm cellular and extracellular antioxidant capacity which in turn can cause harmful modifications of DNA, proteins, and lipids in cells, leading to cell damage [54]. An imbalance of ROS and antioxidants in the body is called oxidative stress (Fig. 5.2). Numerous studies have demonstrated that oxidative stress plays an important role in the development of COPD [55].

The level of oxidative stress in patients with COPD is elevated, particularly during acute exacerbations of the disease. Several biomarkers of oxidative stress have been studied, including malondialdehyde (MDA), 4-Hydroxy-2-nonenal (4-HNE) and 8-isoprostane. The levels of these oxidative stress biomarkers increase concomitant with a decrease in endogenous antioxidants in patients with COPD. MDA is generated as a byproduct of ROS-mediated polyunsaturated fatty acid degradation [56]. The serum MDA concentrations are significantly higher in patients with COPD than in healthy controls [57, 58]. In addition, plasma MDA levels increase in a severity of COPD–dependent manner, from mild to very severe stages, according to Global Initiative for Chronic Obstructive Lung Disease (GOLD) categories [59]. Another oxidative biomarker—4-HNE, an end-product of lipid peroxidation—is very reactive and can attack protein targets, generating 4-HNE protein adducts [60]. Previous studies found that 4-HNE protein adduct was elevated in COPD patients' lungs, and serum from those patients with acute exacerbation had 4-HNE protein adduct levels higher than those of healthy subjects [61, 62]. One of the biomarkers that is most often used to measure oxidative stress is 8-isoprostane [63]. The levels

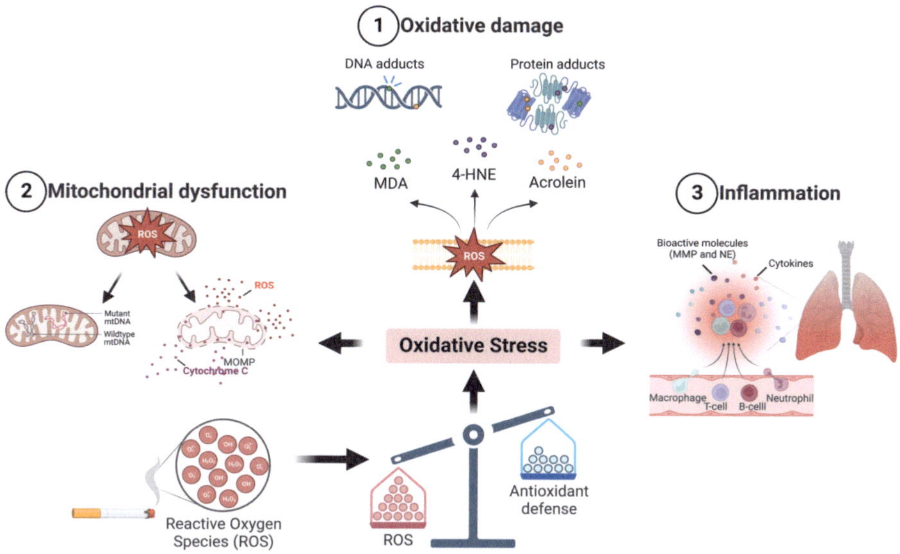

Fig. 5.2 Oxidative stress–induced damage, mitochondrial dysfunction and inflammation in COPD. Tobacco smoke contains a variety of oxidants and prooxidants, which produce reactive oxygen species (ROS) that are harmful to lung cells. An imbalance of ROS and antioxidants in the body is called oxidative stress. ROS oxidizes polyunsaturated fatty acids (PUFAs), components of cell membranes, and subsequently produces lipid peroxyl radicals and hydroperoxides, which can be broken down further into MDA, 4-HNE, and acrolein. These compounds cause DNA modification and interact with proteins to form protein adducts, resulting in DNA mutations and impaired protein function, respectively. Moreover, ROS can cause mitochondrial DNA mutations and mitochondrial dysfunction, which leads to inflammation. Inflammation in the lungs of COPD patients involves innate immunity (e.g. neutrophils and macrophages) and adaptive immunity (T and B cells) as well as structural airway cells, contributing to the pathogenesis and disease severity of COPD. The graphic was created with BioRender.com

of 8-isoprostane in sputum were higher in healthy smokers than in nonsmokers and correlated with the severity of COPD [64]. The amount of 8-isoprostane in the exhaled breath condensate of patients with COPD was almost four times higher than in healthy controls. This amount had a positive relationship with COPD GOLD stage and a negative relationship with lung function [65].

Superoxide dismutase (SOD), catalase, glutathione peroxidases (GPX), and glutathione S-transferases (GST) are the primary enzymatic antioxidants found in the lungs [66, 67]. Superoxide anion, the primary ROS, is converted to hydrogen peroxide by SOD. The hydrogen peroxide produced is then reduced to water by catalase and GPX [66, 67]. GST contributes to the detoxification of both endogenous and exogenous toxic agents, including ROS [68]. Glutathione, a non-enzymatic antioxidant, is an important antioxidant defense mechanism that protects the lungs against oxidative damage [69]. It neutralizes ROS either directly or indirectly through processes that are mediated by GPX and GST [70, 71]. Several studies have found that SOD, catalase, and GPX levels are significantly lower in patients with COPD than in healthy controls [59, 72, 73]. Moreover, patients diagnosed with COPD have significantly lower reduced glutathione levels than healthy controls [59, 74]. These findings indicate that inadequate antioxidant defense plays a crucial role in the development of COPD.

Oxidative Stress–Induced Damage in COPD

Among the effects of excessive oxidative stress is oxidative damage to cells, tissues, and organs. There are three main mechanisms by which oxidative stress damages cells: lipid peroxidation of membranes, oxidative alteration of proteins, and DNA damage. High levels of free radicals or ROS have been known to cause direct damage to lipids [56]. Lipids are important components of cell membranes, which preserve cell structure and regulate cellular function [75]. Polyunsaturated fatty acids (PUFAs) of cellular membranes, like arachidonic and linoleic acids, are the primary targets of lipid peroxidation [76, 77]. ROS oxidizes PUFAs and subsequently produces lipid peroxyl radicals and hydroperoxides. As a result, the hydroperoxides can be broken down further into aldehydes, yielding MDA, 4-HNE, and acrolein [77, 78] (Fig. 5.2).

MDA is mutagenic and causes covalent modification of DNA, resulting in increased mutations in mammalian cells, mostly insertions and deletions. MDA also promotes UV-induced mutagenesis and prevents nucleotide excision repair by directly interacting with cellular repair proteins [79]. Additionally, MDA can form highly stable and immunogenic malondialdehyde-acetaldehyde (MAA) adducts by covalently interacting with the epsilon-amino group of lysine within proteins [80, 81]. A Previous study found that MAA-surfactant protein D adducts induced lung inflammation through scavenger receptor A [82]. In animal models of cardiovascular disease, MAA-low density lipoprotein adducts, and MAA-human serum albumin adducts appeared to affect cellular metabolism, causing an increase in triglyceride buildup and further development of the inflammatory response [81]. Recently, MAA-extracellular matrix protein adducts produced by exposure to alcohol and cigarette smoke increased endoplasmic reticulum stress and suppressed X-box-binding protein-1 expression, resulting in acinar cell death [83].

During oxidative stress, 4-HNE preferentially interacts with cysteine, histidine, and lysine residues of proteins to form stable 4-HNE-protein adducts [84] (Fig. 5.2). These HNE-bound proteins have the potential to disrupt protein structure and function [85]. When compared to patients without COPD, the levels of 4-HNE-modified proteins were significantly higher in the lungs, bronchial epithelial and endothelial cells, and alveolar epithelial cells of patients with COPD, as well as in serum from COPD rats [62, 86, 87]. A recent study of right ventricle myocardium collected from patients indicated that 4-HNE bound to metabolic and mitochondrial proteins, sarcomeric proteins, and cytoskeletal proteins. Increased levels of these 4-HNE-protein adducts were linked to decreased oxidative phosphorylation and mitochondrial structural damage [88]. In addition, 4-HNE prevents nucleotide excision repair by directly altering DNA repair proteins in human cells [89]. Another study discovered an increase in 4-HNE-thioredoxin reductase (4-HNE-Trx1) adducts in serum from patients with COPD with acute exacerbation compared to healthy controls. Adducts of 4HNE-Trx1 may alter the expression of proteins involved primarily in RNA metabolism in human bronchial epithelial cells [61].

Acrolein is one of the hazardous substances found in cigarette smoke. Acrolein can also be produced through lipid peroxidation. According to a recent study, the levels of acrolein in lung tissue of COPD patients and non-COPD smokers were significantly higher than in non-COPD never-smokers. Acrolein levels in lung tissue were also positively correlated with oxidant status in patients with COPD [90]. Acrolein can form both DNA and protein adducts (Fig. 5.2). The sulfhydryl group of cysteine residues in proteins is the main target for acrolein adduct formation [91]. As cysteine residues are frequently involved in enzyme catalysis and are found at the active sites of many proteins, the formation of acrolein-cysteine adducts has broad functional implications [92]. Conjugation of the catalytic cysteine residue in protein tyrosine phosphatase 1B to acrolein results in enzyme inactivation [93]. Acrolein interacts rapidly with DNA to generate well-defined adducts. A recent study demonstrated that e-cigarette users had higher levels of acrolein-DNA adducts in oral buccal cells than non-users [94]. Moreover, acrolein mutagenicity depended on the formation of DNA adducts [95]. In human lung cells, acrolein preferentially bound to mutational hotspots, CpG-containing codons, in the *p53* gene. More than half of the base substitution mutations caused by acrolein-DNA adducts are G to T transversions [96]. These findings suggest that oxidative products from lipid, protein, and DNA modifications caused by oxidative stress are harmful to cells.

Mitochondrial Dysfunction in COPD

Mitochondria are essential cellular organelles that are responsible for energy production, cellular metabolism, intracellular signaling, calcium transport, and the regulation of apoptotic cell death [97–99]. Oxidative stress can cause mtDNA mutations and disrupt the respiratory chain, membrane permeability, and Ca^{2+} homeostasis [8]. Disruptions in mitochondrial morphology, function, membrane potential, reactive oxygen species generation, and intracellular calcium flow are all hallmarks of mitochondrial dysfunction [100]. Injured mitochondria produce ROS and release soluble mediators, which trigger subsequent processes like inflammation and cell death [100]. Mitochondrial dysfunction generates excessive ROS through a vicious cycle that amplifies the release of ROS and mtDNA into the cytosol, triggering inflammation [101]. An increase in mitochondrial ROS and oxidative stress causes inflammation by activating NF-κB [102]. Multiple pro-inflammatory genes, including those producing cytokines and chemokines, are induced by NF-κB. Furthermore, NF-κB regulates the survival, activation, and differentiation of innate immune cells and inflammatory T cells [103]. When mitochondrial damage is beyond repair, it can cause apoptotic cell death. During apoptosis, the mitochondrial outer membrane is permeabilized, leading to the release of cytochrome c and other apoptogenic substances, which in turn activate caspases [101, 104] (Fig. 5.2). In addition, permeabilization of the mitochondrial inner membrane during apoptosis allows mtDNA release into the cytoplasm, which activates the cyclic GMP-AMP synthase–stimulator of interferon genes (cGAS-STING) signaling pathway, resulting in inflammation [105].

In bronchial epithelial cells, cigarette smoke extract lowered cellular ATP, triggered a fast depolarization of mitochondrial membrane potential, and enhanced mitochondrial ROS generation [106, 107]. Animals exposed to cigarette smoke exhibited aberrant mitochondrial morphology, including cristae loss and mitochondrial enlargement, in lung epithelial cells. In addition, mitochondria isolated from lungs of animals exposed to cigarette smoke displayed a reduced membrane potential [108]. Another study found that long-term exposure to cigarette smoke extract caused significant alterations in mitochondrial structure, including branching, fragmentation, and the number of cristae. Increased concentrations of the pro-inflammatory IL-6, IL-8, and IL-1β were present along with these alterations [109]. In patients with COPD, mitochondria in bronchiolar epithelial cells were swollen and fragmented more than in non-COPD smokers [110]. Moreover, endobronchial biopsies of COPD GOLD stage II patients showed a significant decrease in mitochondrial membrane potential and an increase in cellular and mitochondrial ROS when compared to ex-smokers or never-smokers [111]. A recent study also showed that cell-free mtDNA was higher in the plasma of patients with COPD and in the serum of a mouse model of emphysema induced by cigarette smoke exposure than in their controls [112]. In bronchial epithelial cells, cigarette smoke extract reduced mitochondrial membrane potential, increased superoxide anion generation, and promoted apoptosis and mtDNA release into extracellular vesicles, as determined by a mechanistic investigation. Furthermore, cigarette smoke extract increased the expression of DNA-sensing receptors (cGAS and NLRP3), as well as pro-inflammatory cytokines (IL-1β, IL-6, IL-8, IL-18, and CXCL2) involved in inflammation [112]. These findings suggest that mitochondrial dysfunction may play an important role in COPD pathogenesis and may provide a novel therapeutic target for COPD treatment.

Inflammation in COPD

COPD is a chronic inflammation that mostly affects the lung parenchyma and peripheral airways, mainly linked to cigarette smoke exposure. This causes airflow obstruction that is typically progressive and irreversible [113]. Inflammation in the lungs of patients with COPD involves not only innate immunity (e.g. neutrophils, macrophages, and dendritic cells) and adaptive immunity (T and B lymphocytes), but also structural airway and alveolar epithelial cells, endothelial cells, and fibroblasts [10].

Patients with COPD have higher levels of active neutrophils in their sputum and blood, which is correlated with the severity of their condition [114, 115]. Neutrophils are recruited to the airways of patients with COPD and release proteases, such as matrix metalloproteinases (MMPs) and NE, which may contribute to the deterioration of alveoli (Fig. 5.2). Several studies have demonstrated elevated MMPs in the plasma and bronchoalveolar lavage fluid of patients with COPD, which contribute to airway constriction by degrading the structural elements of the extracellular matrix [116–118]. Mucus hypersecretion is associated with airway neutrophilia because NE is a potent stimulant of mucus secretion from submucosal glands and goblet cells by upregulating *MUC5AC* expression [119–121]. Furthermore, NE activates pro-enzyme MMPs to fully active MMPs and inhibits tissue inhibitor of metalloproteinase-1 (TIMP-1), promoting extracellular matrix degradation [122–124].

Macrophages are important cells that cause inflammation in COPD. They play a role in airway remodeling by secreting pro-inflammatory mediators and proteinase enzymes (Fig. 5.2). In patients with emphysema, macrophages are found at sites of alveolar wall destruction, and there is a link between macrophage numbers in the parenchyma and the severity of the disease [10]. Patients with COPD have a significant increase (5- to 10-fold) in the number of macrophages in their sputum, bronchoalveolar lavage fluid, and lung parenchyma [125]. In response to cigarette smoke, these lung macrophages release pro-inflammatory mediators such as TNF-α, IL-6, IL-1β, and MCP-1, as well as MMPs and ROS [126, 127]. Monocyte chemoattractant protein-1 (MCP-1) attracts macrophages, neutrophils, and lymphocyte inflammatory cells to the lungs, which have been linked to greater COPD disease severity [128]. Also, MCP-1 is involved in mucus hypersecretion by upregulating the expression of *MUC5AC* in human airway epithelium [129]. TNF-α promotes inflammation by recruiting inflammatory cells, stimulating the production of inflammatory mediators, increasing oxidative stress, and causing airway hyperresponsiveness [130]. The emphysema-like changes that are observed in TNF-transgenic mice are completely reversed when the TNF-α signal is blocked by deletion of TNF receptor 1 [131]. Moreover, TNF-α knockdown attenuates pulmonary function and reduces inflammatory damage in rats with COPD [132].

Long-term cigarette smoke exposure causes inflammation in the airways, involving both innate and adaptive immune cells, such as T cells and B cells (Fig. 5.2). CD8+ T cells play a critical role in the pathogenesis of COPD. The level of circulating CD8+ T lymphocytes is considerably higher in COPD patients than in healthy controls and is inversely associated with COPD severity. Additionally, cigarette smoke extract enhances the proliferation of CD8+ T lymphocytes and inhibits the apoptosis of CD8+ T cells [133]. Reduced CD8+ T-lymphocyte apoptosis may be a major contributor to the buildup of these cells in the airway submucosa of smokers with mild to moderate COPD [134]. According to a study in peripheral T cells, CD8+ T cells from patients with COPD produced more IFN-γ and IL-4 than those from healthy controls. IFN-γ production by these circulating CD8+ T cells from patients with COPD was also linked to disease severity [135]. Another study found that COPD patients had significantly higher proportions of IFN-γ + and TNF-α + CD8+ T cells than healthy controls [136]. Based on these results, it seems likely that lung CD8+ T cells contribute to the progression of COPD through IFN-γ production. Extensive evidence has suggested critical roles for T helper 17 (Th17) cells and regulatory T cells (Tregs) in COPD. Evidence from several studies showed that patients with COPD had higher proportions of Th17 cells and lower proportions of Tregs than healthy controls [137, 138]. The increased ratio of Th17 to Treg was inversely correlated with lung function in patients with COPD, with greater serum IL-17A levels and lower expression of Foxp3 in peripheral cells than in healthy controls [137]. Additionally, COPD patients with acute exacerbation had considerably higher amounts of released IL-17 and TGF-β1 in their sputum than control groups. In patients with COPD, Th17 cell frequencies were negatively related to Treg frequencies and positively related to IL-17 concentrations [138]. Another study discovered peripheral Th17 and Treg imbalance in COPD patients, and an altered TGF-β/BAMBI pathway may enhance Th17 and Treg imbalance in smokers with COPD [139]. Moreover, restoration of Th17 and Treg balance through STAT3 inhibition and STAT5 activation improved COPD [140]. Recent research indicated that Th17/Th1 cells were resistant to glucocorticoid-mediated T cell suppression, which may have therapeutic implications in glucocorticoid-resistant inflammation [141]. These results suggest that an imbalance of Th17 cells and Tregs may contribute to pathogenesis and disease severity in patients with COPD.

Progression of COPD results in an elevated number and volume of B cells in lymphoid follicles [142]. CXCL13, a B cell–attracting chemokine, is elevated in the lymphoid follicles of cigarette smoke–exposed mice and COPD patients' lungs. CXCL13 is required for cigarette smoke–induced lymphoid follicle formation in an animal model, and neutralizing CXCL13 reduces bronchoalveolar lavage inflammation and alveolar wall destruction [143]. Autoantibodies from B cells that attack lung cells, cell components, and extracellular matrix proteins are also found in the lungs of patients with COPD. In these patients, these autoantibodies may cause inflammation and damage to the lungs by activating complement components [144]. The survival, maturation, differentiation, and maintenance of B cell lymphoid follicles require B cell–activating factor of tumor necrosis factor family (BAFF) [145]. BAFF-positive B cells were higher, but apoptotic B cells were less numerous, in the lymphoid follicles of COPD GOLD stage IV patients than in smoker controls. Also, patients with COPD had higher BAFF levels in their blood and bronchoalveolar lavage B cells than non-smokers and smoker controls. These data suggest that B cell BAFF may contribute to the development of COPD by promoting B cell survival [146].

Bronchial epithelial cells are composed of various cell types including basal, ciliated and mucous (goblet) cells. It is believed that basal cells are the primary stem cells that can differentiate into mucous and ciliated epithelial cells [147, 148]. Additionally, they can secrete bioactive molecules such as cytokines. Ciliated epithelial cells contain numerous mitochondria that direct mucus transport from the lung to the throat. Mucous cells produce mucus to trap foreign particles in the airway lumen [147, 148]. Cigarette smoke is one of the leading causes of both COPD and lung cancer. The epithelial barrier function in the lungs can be severely impaired by cigarette smoke, resulting in an increase in epithelial permeability and a decrease in mucus clearance [148]. Mucus hyperplasia is a characteristic of many patients with COPD in response to chronic airway irritation by cigarette smoke. The transcription of *MUC* genes, which encode mucin, a major component of mucus, is regulated by a variety of inflammatory mediators and signaling pathways [148].

IL-1β and IL-17A promote mucus formation via activating NF-κB in airway epithelial cells [149]. IL-1β and TNF-α also induce mucin production in human airway epithelial cells through cyclooxygenase-2, prostaglandin E2 receptors, and cyclic AMP-protein kinase A signaling [150]. Additionally, EGF receptors (EGFR) regulate mucus hyperplasia and secretion and are activated by NE secretion [10]. Studies have demonstrated that cigarette smoke increases the availability of soluble EGFR ligands, which then bind to and activate EGFR, causing mucin overproduction in airway epithelial cells [151, 152]. When bronchial epithelial cells are exposed to cigarette smoke, several redox-associated intracellular signaling pathways are activated, including mitogen-activated protein kinases (MAPKs) and NF-κB [153, 154]. The activation of MAPKs and NF-κB pathways by oxidative stress leads to the release of pro-inflammatory cytokines such as TNF-α, IL-6 and IL-8, which triggers an inflammatory response [155, 156].

The Relationship Between COPD and Lung Cancer

During the twentieth century, the total cancer death rate climbed, primarily due to substantial increases in lung cancer fatalities among males as a result of the tobacco epidemic [157]. Lung cancer is the most prevalent cancer in both men and women, accounting for 14% and 12%, respectively, of all anticipated new cases. Also, lung and bronchus cancers are the leading causes of cancer death in both men and women, accounting for 27% and 25%, respectively, of all predicted cancer deaths in 2017 [158]. Cigarette smoking causes COPD and lung cancer, and there is growing evidence linking the two diseases [157].

COPD is a significant risk factor for lung cancer development. In patients with lung cancer, the prevalence of COPD is 50%, compared to 8% in matched smokers [159]. A cohort study found that 32.6% of patients with lung cancer also had COPD, which was strongly linked to a shorter overall survival time for lung cancer patients. This study also discovered that COPD with an emphysema-predominant phenotype was a distinct predictive risk factor for squamous carcinoma [160]. Furthermore, two large cohort studies found that COPD not only contributed to lung cancer development regardless of smoking status but was also a risk factor for lung cancer in people who had never smoked [161, 162]. In addition, patients with early-stage non-small cell lung cancer (NSCLC) who had anatomic resection with COPD had a greater risk of tumor recurrence [163]. The Bergen COPD cohort study also found that the risk of developing lung cancer was considerably greater in COPD patients with emphysema, and smoking did not increase the risk of lung cancer among patients with COPD [164]. The most recent systematic review and meta-analysis revealed that both visual and quantitative chest computed tomography (CT) emphysema assessments have been linked to an elevated chance of developing lung cancer, which increased with emphysema severity [165]. These results indicate that patients with COPD, regardless of smoking status, are at high risk of lung cancer. Among these patients, emphysema was associated with a higher cancer rate.

Oxidative Stress–Induced DNA Damage Links COPD to Lung Cancer

DNA damage and carcinogenesis result from increased oxidative stress in COPD. DNA damage triggers a number of cellular responses, allowing the cell to rectify the damage or to initiate programmed cell death, presumably to eliminate cells with mutations that could be dangerous [166]. Oxidative stress–mediated DNA damage induces cancer initiation via point mutations, single-stand breaks, and double-strand breaks. If this damage is incorrectly repaired, it results in mutations [167] (Fig. 5.3). Studies have shown that oxidative stress marker MDA and DNA strand breaks are significantly higher in patients with COPD who smoke than in healthy non-smokers [168]. Furthermore, 38% of patients with COPD showed microsatellite DNA instability (MSI), a marker of acquired somatic mutations, compared to 0% in healthy controls. The 8-OHdG levels, a marker of oxidative stress, in the sputum of MSI-positive COPD patients were also significantly higher than those of non-COPD smokers and healthy controls [169]. Another study also demostrated that patients with COPD had higher levels of phosphorylated histone 2AX, a marker

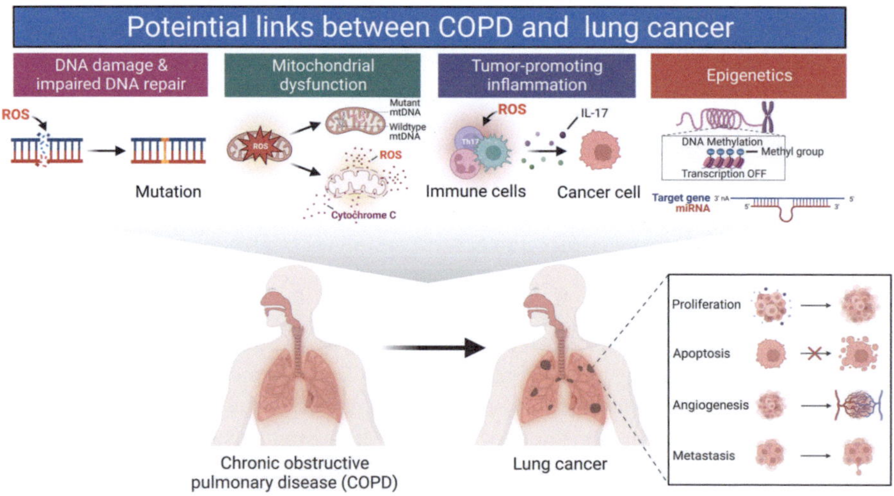

Fig. 5.3 Potential links between COPD and lung cancer. Numerous studies have demonstrated a link between the pathogenesis of COPD and lung cancer. There is accumulating evidence that oxidative stress, mitochondrial dysfunction, inflammation, and epigenetic regulation have a role in the development of COPD and lung cancer. Oxidative stress–induced DNA damage and impaired DNA repair can cause mutations in lung cells. Furthermore, mitochondrial DNA is particularly prone to mutations when exposed to reactive oxygen species (ROS). In addition, ROS-mediated mitochondrial dysfunction drives cancer progression toward chemoresistance or metastasis. Inflammatory cells and cytokines are more likely to promote cancer formation, proliferation, and metastasis. ROS also activates NF-κB signaling, which induces inflammation and carcinogenesis by promoting angiogenesis and preventing cancer cell apoptosis. Moreover, epigenetic regulation, such as DNA methylation and dysregulated miRNA expression of various genes, plays a role in the development of lung cancer in COPD patients. The graphic was created with BioRender.com

of double-strand breaks, than non-COPD smokers and non-smokers [170]. A recent study demonstrated that peripheral blood mononuclear cells from patients with lung cancer had higher levels of both single- and double-strand breaks but lower levels of oxidative stress markers GSH and GSSG than those from healthy controls [171]. These findings indicate that oxidative stress causes DNA damage, a common pathogenic mechanism in COPD, which may contribute to the development of cancer.

Another possible mechanism connecting COPD and lung cancer is a deficiency in DNA repair [172]. Cigarette smoke contains ROS and carcinogens that lead to DNA adduct formation, as described earlier. A failure to fix DNA damage by DNA repair mechanisms results in mutations [173]. Transcriptomic analysis of lung tissue from patients with severe COPD identified 15 genes involved in DNA damage and repair and showed that the nucleotide excision repair mechanism was down-regulated as COPD severity increased [174]. Because nucleotide excision repair is mainly responsible for eliminating DNA adducts [175], these findings have implications for the higher risk of lung cancer in COPD patients.

Mitochondrial Dysfunction Links COPD to Lung Cancer

Both somatic and germline mitochondrial mutations can cause cancer. Histones and introns protect most DNA, but mtDNA lacks these crucial defenses [176]. Additionally, mitochondria lack the DNA repair mechanisms found in the nucleus [177]. As a result, mtDNA is highly susceptible to mutations when exposed to ROS. The two most frequent mtDNA alterations in cancers are mutations and copy number changes [178]. Lung cancer progression in patients with lung cancer who have never smoked is influenced by mtDNA mutations in the respiratory Complex I [179]. It was reported that 92% of NSCLC samples (24 out of 26) showed somatic mtDNA mutations, with a maximum of nine alterations per sample. These mutations were widely dispersed over the whole mitochondrial genome, and most mutations were discovered in the non-coding displacement *(D) Loop*, *12S rRNA*, *16S rRNA*, *ND4*, and *ND5* [180]. Exhaled breath condensate from patients with lung cancer also had significantly higher mutation rates in the *D-loop* region of mtDNA than in healthy controls [181]. Moreover, recent research indicated that somatic mutations in the D-loop region were associated with longer relapse-free survival in patients with stage 1 lung adenocarcinoma, but somatic mutations in the mitochondrial Complex IV and Complex V genes were associated with shorter relapse-free survival [182]. According to a different study, patients with lung cancer who have never smoked have greater rates of mtDNA mutations than those who currently smoke. Since most coding mtDNA mutations in this study targeted the respiratory Complex I, the overexpression of one of these variants causes lung cancer cells to proliferate, invade, and produce more superoxide [179]. It has been postulated that mtDNA mutations may be a marker of lung carcinogenesis.

Several studies have demonstrated the significance of mtDNA copy number in lung cancer. The development of both COPD and NSCLC was substantially correlated with an increase in cell-free mtDNA copy numbers, along with an increase in IL-6. Also, cell-free mtDNA copy numbers in patients with NSCLC were inversely correlated with overall survival but positively correlated with metastasis [183]. In addition, patients with advanced T4 NSCLC had lower DNA copy numbers and expression of mitochondrial transcription factor A, which regulates both replication and transcription of mtDNA, than those with T1 and T2/T3 NSCLC [184]. Another study found that patients with lung cancer had lower mtDNA copy numbers than healthy controls and that the lower plasma mtDNA copy number in lung cancer patients was also associated with tumor metastatic potential and a poor prognosis [185].

Furthermore, mitochondrial dysfunction drives cancer progression toward a chemoresistant or invasive phenotype [11]. In drug-resistant triple-negative breast cancer, mitochondrial oxidative phosphorylation and ROS production were enhanced through MYC and MCL1. This elevated ROS increased hypoxia inducible factor-1α (HIF-1α), which promoted cancer stem cells and tumorigenesis [186]. Also, mitochondrial dysfunction caused by mitochondrial respiration inhibitors led to cell migration and invasion by inducing epithelial-to-mesenchymal transition through the activation of AKT and AMPK pathways in lung cancer cells [187]. Another study identified top gene sets related to metastasis-specific lethality as mitochondrial ribosome and mitochondria-associated genes in cells. When compared to non-metastatic primary tumors, metastatic tumors from the lung cancer mouse model showed lower mitochondrial membrane potential and diminished mitochondrial functionality [188]. These findings indicate that mitochondrial dysfunction may be a promising determinant in the development of lung cancer (Fig. 5.3).

Inflammation Links COPD to Lung Cancer

As described above, COPD is a chronic inflammatory disease that involves neutrophils, macrophages, T cells, and B cells. Inflammatory cell infiltration is common to both COPD and lung cancer [189]. The infiltration of leukocytes is observed in malignant tissues, suggesting that cancers may originate from chronic inflammation sites. Tumor development, proliferation, and immunosuppression are more likely to be facilitated by these inflammatory cells and cytokines [190] (Fig. 5.3). The severity of COPD is associated with increased IL-17 expression produced by Th17 cells, as previously stated. In a mouse model of lung cancer, COPD-induced inflammation promoted lung adenocarcinoma and increased Th17 cells in the tumor tissue. Also, IL-17 deficiency reduced tumor cell growth and angiogenesis, as well as the production of pro-inflammatory molecules and myeloid cell recruitment in this mouse model [191]. Another recent study found that deleting IL-17C, which was expressed by non-hematopoietic cells and modulated innate immune responses, in a lung cancer animal model with COPD-like inflammation reduced tumor development and neutrophil infiltration in tumor lesions [192]. Moreover, COPD-like airway inflammation induced by *Hemophilus influenzae* promoted lung tumor development and upregulated the expression of HIF-1α in a model of lung cancer. Deleting HIF-1α in airway epithelial cells reduced tumor growth, angiogenesis and COPD-like airway inflammation in these animals [193, 194].

Oxidative stress activates NF-κB, which in turn induces inflammation, as described previously. TNFα, IL-6, IL-1, and IL-8 are all regulated by NF-κB, and several of them are powerful activators of NF-κB [195]. NF-κB is also a potential mediator of inflammation-induced carcinogenesis. According to a meta-analysis, patients with NSCLC had poorer overall survival rates when NF- κB expression was higher, and its expression was closely linked to tumor stage [196]. It is known that NF-κB promotes tumor survival, whereas it inhibits tumor apoptosis. NF-κB activation decreased the stability of the tumor suppressor protein p53 by increasing the expression of the p53 inhibitor E3 ligase MDM2 [197, 198]. NF-κB also inhibited the transcription of tumor suppressor phosphatase and tensin homolog (PTEN) by binding directly to the *PTEN* promoter, while increasing the expression of Snail 1, a transcriptional suppressor of PTEN [199–201]. PTEN inhibits the PI3K/AKT signaling cascade, which is one of the most important cancer-promoting pathways [202]. In addition, tobacco-specific carcinogen 4-(methylnitrosamino)-1-(3-pyridyl)-1-butanone promoted bronchial cell proliferation by activating NF-κB, which increased cyclin D1 expression [203]. Deletion of I kappa B kinase (IKK) 2 in tumor cells significantly reduced tumor proliferation and extended survival in mice [204]. On the other hand, mice expressing IKK in airway epithelial cells developed chronic airway inflammation and significantly enhanced lung tumor formation after urethane-induced lung tumorigenesis [205]. Tobacco smoke–induced lung tumor growth and inflammation were also inhibited by IKK deletion in myeloid cells [206]. Additionally, NF-κB signaling cascade contributed to angiogenesis in cancer cells by upregulating vascular endothelial growth factor, IL-8, and microRNAs. Deletion of IKKβ in KRAS-positive lung cancer cells reduced the expression and secretion of IL-8 and VEGF, as well as angiogenesis [207]. In a recent study, IL-1β secreted from NSCLC activated NF-κB, which in turn repressed the expression of miR-22, a potent angiogenesis inhibitor [208]. IL-6/STAT3 transactivated miR-135b in NSCLC, which in turn activated NF-κB signaling by targeting the deubiquitinase CYLD, a negative regulator of NF-κB activation and angiogenesis [209]. These findings suggest the significance of ROS-mediated NF-κB in promoting tumor growth and angiogenesis.

Epigenetics Link COPD to Lung Cancer

The development of COPD and lung cancer has also been linked to epigenetic alterations, such as DNA methylation and microRNA expression [210, 211] (Fig. 5.3). Oxidative stress and the inflammatory response change the redox state of the cells, which causes the DNA to become unstable and ultimately results in epigenetic changes [212]. DNA methylation is defined as the covalent addition of a methyl group by DNA methyltransferase enzymes to the carbon-5 position of a cytosine base. DNA methylation happens preferentially in DNA abundant with CpG dinucleotides, known as CpG islands, which are typically found near or within gene promoters [210]. A previous study from two cohorts found that the sputum samples of ex-smokers with chronic mucous hypersecretion (CMH) had significantly higher total methylation in the promoters of eleven lung cancer tumor suppressor genes than in those without CMH, indicating that former smokers with persistent CMH may have a higher chance of developing lung cancer [213]. Non-cancerous lung tissue from patients with lung adenocarcinomas appeared to have DNA methylation profiles reflecting smoking and COPD, which may influence the aggressiveness of tumors in patients [214]. An epigenome-wide association analysis of NSCLC found that DNA methylation in lung tumors from NSCLC patients with COPD was considerably higher than in NSCLC patients without COPD. In this study, *CCDC37* and *MAP1B* methylation were substantially more common, and their expressions were significantly lower in lung cancers from patients with COPD than in lung cancers from non-COPD patients. Additionally, analysis of the two cohorts revealed that *CCDC37* was methylated more frequently in the sputum of COPD patients than in controls [215]. These findings suggest that *CCDC37* methylation may be a potential biomarker for lung cancer risk in patients with COPD. Another study observed *IL-12Rβ2* and *Wif-1* methylation more frequently in NSCLC patients with COPD than in those without COPD, suggesting that COPD may impact *IL-12Rβ2* and *Wif-1* methylation during COPD-related NSCLC pathogenesis [216]. Another study of DNA methylation profiles in NSCLC discovered that COPD had a greater impact on the DNA methylation profile in lung tumors than age and smoking, both of which are key risk factors for NSCLC. In addition, methylation of immune genes was substantially enriched in NSCLC patients with COPD but not in those without COPD, according to an ontology analysis of genes containing differentially methylated CpGs [217]. Alpha-1 antitrypsin, or AAT, which is encoded by the human serine protease inhibitor A1 (*SERPINA1*) gene, has a role in the pathophysiology of emphysema and raises the risk of COPD in both active and passive smokers [218, 219]. Recently, it was found that lung cancer patients with COPD had significantly greater *SERPINA1* methylation levels than patients with COPD or lung cancer, indicating that *SERPINA1* methylation is associated with lung cancer development in patients with COPD [220]. MicroRNA-7 (miR-7) is a tumor suppressor non-coding RNA that suppresses the development of lung cancer by inhibiting lung cancer cell growth, proliferation, and migration while promoting apoptosis [221]. Compared to smokers with unrestricted airways, patients with COPD had greater levels of miR-7 methylation. Among patients with different clinical COPD phenotypes, hypermethylation of miR-7 was increased in COPD patients with emphysema [222]. Another study found that the overall survival of patients with lung adenocarcinoma was lower in those with low miR-7 expression than in those with high miR-7 expression, indicating that low miR-7 expression may be biomarker that prognosticates poor outcomes for lung cancer. Furthermore, miR-7 overexpression improves the sensitivity of lung adenocarcinoma cells to cisplatin by inducing apoptosis through targeting Bcl-2 [223]. These findings may help to explain why these patients have a higher risk of lung cancer.

MicroRNAs (miRNAs), endogenous 19–24 nucleotide RNAs, play key regulatory roles in animals by modulating protein-coding gene transcription and/or translation [224]. According to genome-wide microRNA expression profiles in patients with COPD, nine miRNAs had significantly lower expression in blood samples from patients who were developing cancer than in those who were not [225]. This study also found that miRNAs with the most upregulated and downregulated expression were associated with regulation of MAP kinase, integrin signaling, and focal adhesion pathways that are involved in cancer development [225]. These findings suggest that several miRNAs in COPD patients' blood may be used as potential biomarkers to identify patients with COPD who are at increased risk of cancer. A study comparing the tumors of lung cancer patients with and without COPD discovered that expression of miR-21, miR-200b, miR-210, and miR-let7c were dramatically upregulated, whereas their downstream gene targets, such as *PTEN*, *MARCKs*, and *k-RAS*, were downregulated in the COPD group [226]. *PTEN*, *MARCKs*, and *k-RAS* have been described as tumor suppressor genes [227–229]. In addition, a previous study used pathway enrichment analysis to identify various common microRNAs—such as miR-15b, miR-106a, miR-17, miR-103, and miR-107—between COPD and NSCLC [230]. Another study found 14 significantly dysregulated miRNAs in plasma extracellular vesicles that could discriminate between patients with lung cancer and stable COPD, with miR-27a-3p (AUC 0.803) being the single most important miRNA. The combination of miR-106b-3p and miR-361-5p also enhanced discriminatory power (AUC 0.870) [231]. Furthermore, a comprehensive biological and computational approach discovered that the genes *RGS6* (Regulator of G-protein signaling 6) and *PARK2* (parkin RBR E3 ubiquitin protein ligase) were substantially linked to the risk of developing lung cancer and COPD. This study also indicated that miR-137, miR-221-3p, and miR-222-3p, as well as miR-320a, miR-320b, miR-320c, and miR-320d, were computationally predicted to regulate *RGS6* and *PARK2*, respectively [232]. These results provide insight into the miRNA biology behind the observed increased risk of developing lung cancer in patients with COPD.

Smoking and the Management of Cancer

The difficulty and discomfort arising during the process of smoking cessation, coupled with the stresses of a cancer diagnosis, mean that many cancer patients choose to continue to smoke. This was described well in a cross-sectional study that assessed the smoking behaviors among US adult cancer survivors. It was found that survivors had a higher prevalence of current smoking, especially those with smoking-related cancers compared to non-smoking-related cancers [233]. In another study evaluating the awareness of 172 patients seeking care at a urology clinic, only 36% were aware that current smoking increased their risk of renal cancer [234]. In a study evaluating the post-operative complications of simple mastectomy, modified radical mastectomy, and breast-conserving surgery in breast cancer, continued smoking was strongly associated with wound infection, skin flap necrosis, and epidermolysis, regardless of the number of cigarettes smoked per day [235]. In another retrospective analysis of tobacco smoking during the management of head and neck cancer, it was found that active smokers had poorer control of cancer, lower-disease-free survival, and an inferior 5-year overall survival. The incidence of

Grade 3 or greater late complications from radiotherapy was increased among active smokers [236]. Smoking cessation before the diagnosis of cancer has been associated with a lower likelihood of developing all major types of lung cancer, with the most reduction seen in the incidence of small and squamous cell lung cancer as well as heavy smokers, especially women [237]. In addition, smoking cessation at diagnosis decreases cancer-related morbidity and mortality and the incidence of other co-morbid conditions such as coronary artery disease, stroke, and COPD [238]. A meta-analysis that looked at the influence of smoking cessation after diagnosis of early-stage lung cancer on prognosis found that people who continued to smoke had a higher risk of recurrence, a second primary tumor, and a doubling in all-cause mortality than those who stopped smoking at that time. They also found that the benefits of smoking cessation were considerable, and the effect on all-cause mortality was driven mainly by the reduced progression of cancer rather than cardiorespiratory disease [239]. Therefore, research should be geared toward strategies that improve patient awareness about the consequences of smoking during and after cancer treatment, as well as strategies that can aid smoking cessation.

Conclusions

Tobacco smoke is the most prevalent risk factor for chronic obstructive pulmonary disease (COPD), one of the most common chronic respiratory disorders, which has a high morbidity and mortality rate. Long-term second-hand smoke exposure also increases the risk of developing COPD. Tobacco smoke contains reactive oxygen species, which can cause oxidative stress–induced damage to lipids, DNA, and proteins, and can trigger mitochondrial dysfunction and inflammation, all of which contribute to the development and progression of COPD. Moreover, COPD is a substantial risk factor for the development of lung cancer. Although the precise mechanism driving the elevated incidence of lung cancer in patients with COPD remains unknown, numerous studies have found compelling evidence that oxidative stress–induced DNA damage, mitochondrial dysfunction, inflammation, and epigenetic regulation all play a role in the development of COPD and lung cancer.

References

1. World Health Organization [Internet]. Geneva, Switzerland: World Health Organization. Tobacco Fact sheets; [updated 2022 May 24; cited 2022 Sep 12]. Available from https://www.who.int/news-room/fact-sheets/detail/tobacco.
2. Wheaton AG, Liu Y, Croft JB, VanFrank B, Croxton TL, Punturieri A, et al. Chronic obstructive pulmonary disease and smoking status–United States, 2017. MMWR Morb Mortal Wkly Rep. 2019;68(24):533–8.
3. Flexeder C, Zock JP, Jarvis D, Verlato G, Olivieri M, Benke G, et al. Second-hand smoke exposure in adulthood and lower respiratory health during 20 year follow up in the European Community respiratory health survey. Resp Res. 2019;20(1):33.
4. Barnes PJ, Burney PG, Silverman EK, Celli BR, Vestbo J, Wedzicha JA, et al. Chronic obstructive pulmonary disease. Nat Rev Dis Primers. 2015;1:15076.

5. Lareau SC, Fahy B, Meek P, Wang A. Chronic obstructive pulmonary disease (COPD). Am J Respir Crit Care Med. 2019;199(1):P1–2.
6. Goel R, Bitzer Z, Reilly SM, Trushin N, Foulds J, Muscat J, et al. Variation in free radical yields from U.S. marketed cigarettes. Chem Res Toxicol. 2017;30(4):1038–45.
7. Zinellu E, Zinellu A, Fois AG, Carru C, Pirina P. Circulating biomarkers of oxidative stress in chronic obstructive pulmonary disease: a systematic review. Resp Res. 2016;17(1):150.
8. Guo CY, Sun L, Chen XP, Zhang DS. Oxidative stress, mitochondrial damage and neurodegenerative diseases. Neural Regen Res. 2013;8(21):2003–14.
9. Lingappan K. NF-kappaB in oxidative stress. Curr Opin Toxicol. 2018;7:81–6.
10. Barnes PJ. Inflammatory mechanisms in patients with chronic obstructive pulmonary disease. J Allergy Clin Immunol. 2016;138(1):16–27.
11. Hsu CC, Tseng LM, Lee HC. Role of mitochondrial dysfunction in cancer progression. Exp Biol Med. 2016;241(12):1281–95.
12. Parris BA, O'Farrell HE, Fong KM, Yang IA. Chronic obstructive pulmonary disease (COPD) and lung cancer: common pathways for pathogenesis. J Thorac Dis. 2019;11:S2155–S72.
13. Xia YF, Shen S, Verma IM. NF-kappa B, an active player in human cancers. Cancer Immunol Res. 2014;2(9):823–30.
14. Willett J, Achenbach S, Pinto FJ, Poppas A, Elkind MSV. The tobacco endgame-eradicating a worsening epidemic: a joint opinion from the American Heart Association, world heart federation, American College of Cardiology, and the European Society of Cardiology. Circulation. 2021;144(1):e1–5.
15. The Health consequences of smoking—50 years of progress: A report of the surgeon general. Atlanta: U.S. Department of Health and Human Services, Centers for Disease Control and Prevention, National Center for Chronic Disease Prevention and Health Promotion, Office on Smoking and Health; 2014.
16. Cornelius ME, Loretan CG, Wang TW, Jamal A, Homa DM. Tobacco product use among adults–United States, 2020. MMWR Morb Mortal Wkly Rep. 2022;71(11):397–405.
17. Andriani H, Kosasih RI, Putri S, Kuo HW. Effects of changes in smoking status on blood pressure among adult males and females in Indonesia: a 15-year population-based cohort study. BMJ Open. 2020;10(4):e038021.
18. Khan SS, Ning HY, Sinha A, Wilkins J, Allen NB, Vu THT, et al. Cigarette smoking and competing risks for fatal and nonfatal cardiovascular disease subtypes across the life course. J Am Heart Assoc. 2021;10(23):e021751.
19. Duncan MS, Freiberg MS, Greevy RA, Kundu S, Vasan RS, Tindle HA. Association of smoking cessation with subsequent risk of cardiovascular disease. JAMA. 2019;322(7):642–50.
20. Jacob L, Freyn M, Kalder M, Dinas K, Kostev K. Impact of tobacco smoking on the risk of developing 25 different cancers in the UK: a retrospective study of 422,010 patients followed for up to 30 years. Oncotarget. 2018;9(25):17420–9.
21. Hou W, Hu S, Li C, Ma H, Wang Q, Meng G, et al. Cigarette smoke induced lung barrier dysfunction, EMT, and tissue remodeling: a possible link between COPD and lung cancer. Biomed Res Int. 2019;2019:2025636.
22. Su X, Wu W, Zhu Z, Lin X, Zeng Y. The effects of epithelial-mesenchymal transitions in COPD induced by cigarette smoke: an update. Respir Res. 2022;23(1):225.
23. Bhatt SP, Kim YI, Harrington KF, Hokanson JE, Lutz SM, Cho MH, et al. Smoking duration alone provides stronger risk estimates of chronic obstructive pulmonary disease than pack-years. Thorax. 2018;73(5):414–21.
24. Chang JT, Meza R, Levy DT, Arenberg D, Jeon J. Prediction of COPD risk accounting for time-varying smoking exposures. PLoS One. 2021;16(3):e0248535.
25. Centers for Disease Control and Prevention (CDC). Chronic obstructive pulmonary disease among adults–United States, 2011. MMWR Morb Mortal Wkly Rep. 2012;61:938–43.
26. Secondhand smoke exposure and cardiovascular effects: Making sense of the evidence. Washington (DC): National Academies Press (US); 2010. Report No.: 9780309138390.
27. The health consequences of involuntary exposure to tobacco smoke: A report of the surgeon general. Atlanta: U.S. Department of Health and Human Services, Centers for Disease

Control and Prevention, National Center for Chronic Disease Prevention and Health Promotion, Office on Smoking and Health 2006.
28. Park SA, Lee DH, Lim HS. Factors influencing exposure to secondhand smoke: passive inhalation in student nurses. Osong Public Health Res Perspect. 2019;10(2):78–84.
29. Lindsay RP, Tsoh JY, Sung HY, Max W. Secondhand smoke exposure and serum cotinine levels among current smokers in the USA. Tob Control. 2016;25(2):224–31.
30. Asomaning K, Miller DP, Liu G, Wain JC, Lynch TJ, Su L, et al. Second hand smoke, age of exposure and lung cancer risk. Lung Cancer. 2008;61(1):13–20.
31. Park YS, Lee CH, Kim YI, Ahn CM, Kim JO, Park JH, et al. Association between secondhand smoke exposure and hypertension in never smokers: a cross-sectional survey using data from Korean National Health and nutritional examination survey V, 2010-2012. BMJ Open. 2018;8(5):e021217.
32. Gopal SH, Mukherjee S, Das SK. Direct and second hand cigarette smoke exposure and development of childhood asthma. J Environ Health Sci. 2016;2(6)
33. Putcha N, Barr RG, Han MK, Woodruff PG, Bleecker ER, Kanner RE, et al. Understanding the impact of second-hand smoke exposure on clinical outcomes in participants with COPD in the SPIROMICS cohort. Thorax. 2016;71:411–20.
34. Korsbaek N, Landt EM, Dahl M. Second-hand smoke exposure associated with risk of respiratory symptoms, asthma, and COPD in 20,421 adults from the general population. J Asthma Allergy. 2021;14:1277–84.
35. Hagstad S, Bjerg A, Ekerljung L, Backman H, Lindberg A, Ronmark E, et al. Passive smoking exposure is associated with increased risk of COPD in never smokers. Chest. 2014;145(6):1298–304.
36. Ahmad FB, Anderson RN. The leading causes of death in the US for 2020. JAMA. 2021;325(18):1829–30.
37. Thomashow BM, Mannino DM, Tal-Singer R, Crapo JD. A rapidly changing understanding of COPD: world COPD day from the COPD foundation. Am J Physiol-Lung C. 2021;321(5):L983–L7.
38. Meza D, Khuder B, Bailey JI, Rosenberg SR, Kalhan R, Reyfman PA. Mortality from COVID-19 in patients with COPD: a US study in the N3C data enclave. Int J Chronic Obstr. 2021;16:2323–6.
39. Croft JB, Wheaton AG, Liu Y, Xu F, Lu H, Matthews KA, et al. Urban-rural county and state differences in chronic obstructive pulmonary disease–United States, 2015. MMWR Morb Mortal Wkly Rep. 2018;67(7):205–11.
40. Zafari Z, Li SK, Eakin MN, Bellanger M, Reed RM. Projecting long-term health and economic burden of COPD in the United States. Chest. 2021;159(4):1400–10.
41. MedlinePlus [Internet]. Bethesda (MD): National Library of Medicine (US); [updated 2021 Oct 5]. COPD [cited 2022 Sep 19]. Available from: https://medlineplus.gov/copd.html.
42. Rajendra KC, Shukla SD, Gautam SS, Hansbro PM, O'Toole RF. The role of environmental exposure to non-cigarette smoke in lung disease. Clin Transl Med. 2018;7(1):39.
43. Diver WR, Jacobs EJ, Gapstur SM. Secondhand smoke exposure in childhood and adulthood in relation to adult mortality among never smokers. Am J Prev Med. 2018;55(3):345–52.
44. Wells AD, Woods A, Hilleman DE, Malesker MA. Alpha-1 antitrypsin replacement in patients with COPD. P T. 2019;44(7):412–5.
45. Torres-Duran M, Lopez-Campos JL, Barrecheguren M, Miravitlles M, Martinez-Delgado B, Castillo S, et al. Alpha-1 antitrypsin deficiency: outstanding questions and future directions. Orphanet J Rare Dis. 2018;13(1):114.
46. Kalfopoulos M, Wetmore K, ElMallah MK. Pathophysiology of alpha-1 antitrypsin lung disease. Methods Mol Biol. 2017;1639:9–19.
47. Rodrigues SO, Cunha C, Soares GMV, Silva PL, Silva AR, Goncalves-de-Albuquerque CF. Mechanisms, pathophysiology and currently proposed treatments of chronic obstructive pulmonary disease. Pharmaceuticals (Basel). 2021;14(10):979.
48. Lin CR, Bahmed K, Kosmider B. Dysregulated cell signaling in pulmonary emphysema. Front Med (Lausanne). 2021;8:762878.

49. Hartman JE, Klooster K, Augustijn SWS, van Geffen WH, Garner JL, Shah PL, et al. Identifying responders and exploring mechanisms of action of the endobronchial coil treatment for emphysema. Respiration. 2021;100(5):443–51.
50. Wang C, Zhou J, Wang J, Li S, Fukunaga A, Yodoi J, et al. Progress in the mechanism and targeted drug therapy for COPD. Signal Transduct Tar. 2020;5(1):248.
51. Lei X, Goel R, Sun D, Bhangu G, Bitzer ZT, Trushin N, et al. Free radical and nicotine yields in mainstream smoke of chinese marketed cigarettes: variation with smoking regimens and cigarette brands. Chem Res Toxicol. 2020;33(7):1791–7.
52. Rahman I, MacNee W. Role of oxidants/antioxidants in smoking-induced lung diseases. Free Radical Bio Med. 1996;21(5):669–81.
53. Rahman I, Adcock IM. Oxidative stress and redox regulation of lung inflammation in COPD. Eur Respir J. 2006;28(1):219–42.
54. Luo J, Mills K, le Cessie S, Noordam R, van Heemst D. Ageing, age-related diseases and oxidative stress: what to do next? Ageing Res Rev. 2020;57:100982.
55. Magallon M, Navarro-Garcia MM, Dasi F. Oxidative stress in COPD. J Clin Med. 2019;8(11):1953.
56. Ayala A, Munoz MF, Arguelles S. Lipid peroxidation: production, metabolism, and signaling mechanisms of malondialdehyde and 4-hydroxy-2-nonenal. Oxidative Med Cell Longev. 2014;2014:360438.
57. Dhakal N, Lamsal M, Baral N, Shrestha S, Dhakal SS, Bhatta N, et al. Oxidative stress and nutritional status in chronic obstructive pulmonary disease. J Clin Diagn Res. 2015;9(2):BC01–4.
58. Jiang XL, Zhong P, Huang CL, He F, Fan XM, Chen XR. The relationship between nutritional status and oxidative stress markers, pulmonary function in patients with stable chronic obstructive pulmonary disease. Zhonghua Jie He He Hu Xi Za Zhi. 2017;40(1):40–5.
59. Singh S, Verma SK, Kumar S, Ahmad MK, Nischal A, Singh SK, et al. Evaluation of oxidative stress and antioxidant status in chronic obstructive pulmonary disease. Scand J Immunol. 2017;85(2):130–7.
60. Breitzig M, Bhimineni C, Lockey R, Kolliputi N. 4-Hydroxy-2-nonenal: a critical target in oxidative stress? Am J Physiol Cell Physiol. 2016;311(4):C537–C43.
61. Liu J, Huang J, Liu H, Chen C, Xu J, Zhong L. Elevated serum 4HNE plus decreased serum thioredoxin: unique feature and implications for acute exacerbation of chronic obstructive pulmonary disease. PLoS One. 2021;16(1):e0245810.
62. Rahman I, van Schadewijk AA, Crowther AJ, Hiemstra PS, Stolk J, MacNee W, et al. 4-Hydroxy-2-nonenal, a specific lipid peroxidation product, is elevated in lungs of patients with chronic obstructive pulmonary disease. Am J Respir Crit Care Med. 2002;166(4):490–5.
63. Graille M, Wild P, Sauvain JJ, Hemmendinger M, Canu IG, Hopf N. Urinary 8-isoprostane as a biomarker for oxidative stress. A systematic review and meta -analysis. Toxicol Lett. 2020;328:19–27.
64. Kinnula VL, Ilumets H, Myllarniemi M, Sovijarvi A, Rytila P. 8-Isoprostane as a marker of oxidative stress in nonsymptomatic cigarette smokers and COPD. Eur Respir J. 2007;29(1):51–5.
65. Ashmawi SSA, Dewdar IA, Mohamed NA, Elhefny AA. Measurement of 8-isoprostane in exhaled breath condensate of patients with chronic obstructive pulmonary disease. Egypt J Chest Dis Tu. 2018;67(3):226–30.
66. Birben E, Sahiner UM, Sackesen C, Erzurum S, Kalayci O. Oxidative stress and antioxidant defense. World Allergy Organ J. 2012;5(1):9–19.
67. Caliri AW, Tommasi S, Besaratinia A. Relationships among smoking, oxidative stress, inflammation, macromolecular damage, and cancer. Mutat Res-Rev Mutat. 2021;787:108365.
68. Tian R, Seim I, Ren W, Xu S, Yang G. Contraction of the ROS scavenging enzyme glutathione S-transferase gene family in cetaceans. G3 (Bethesda). 2019;9(7):2303–15.
69. Chia SB, Elko EA, Aboushousha R, Manuel AM, van de Wetering C, Druso JE, et al. Dysregulation of the glutaredoxin/S-glutathionylation redox axis in lung diseases. Am J Physiol Cell Physiol. 2020;318(2):C304–C27.

70. Hayes JD, Flanagan JU, Jowsey IR. Glutathione transferases. Annu Rev Pharmacol Toxicol. 2005;45:51–88.
71. Benhar M. Roles of mammalian glutathione peroxidase and thioredoxin reductase enzymes in the cellular response to nitrosative stress. Free Radic Biol Med. 2018;127:160–4.
72. Bajpai J, Prakash V, Kant S, Verma AK, Srivastava A, Bajaj DK, et al. Study of oxidative stress biomarkers in chronic obstructive pulmonary disease and their correlation with disease severity in north Indian population cohort. Lung India. 2017;34(4):324–9.
73. Al-Azzawi MA, Al-Rubaeaee AA, Ghoneim AH, AboZaid MMN. The effect of cigarette smoking on the oxidant-antioxidant imbalance in patients with chronic obstructive pulmonary disease. Egypt J Chest Dis Tu. 2019;68(4):462–70.
74. Sotgia S, Paliogiannis P, Sotgiu E, Mellino S, Zinellu E, Fois AG, et al. Systematic review and meta-analysis of the blood glutathione redox state in chronic obstructive pulmonary disease. Antioxidants (Basel). 2020;9(11):1146.
75. Yin H, Xu L, Porter NA. Free radical lipid peroxidation: mechanisms and analysis. Chem Rev. 2011;111(10):5944–72.
76. Ito F, Sono Y, Ito T. Measurement and clinical significance of lipid peroxidation as a biomarker of oxidative stress: oxidative stress in diabetes, atherosclerosis, and chronic inflammation. Antioxidants-Basel. 2019;8(3):72.
77. Mihalas BP, De Iuliis GN, Redgrove KA, McLaughlin EA, Nixon B. The lipid peroxidation product 4-hydroxynonenal contributes to oxidative stress-mediated deterioration of the ageing oocyte. Sci Rep. 2017;7(1):6247.
78. Taso OV, Philippou A, Moustogiannis A, Zevolis E, Koutsilieris M. Lipid peroxidation products and their role in neurodegenerative diseases. Ann Res Hosp. 2019;3:2.
79. Feng ZH, Hu WW, Marnett LJ, Tang MS. Malondialdehyde, a major endogenous lipid peroxidation product, sensitizes human cells to UV- and BPDE-induced killing and mutagenesis through inhibition of nucleotide excision repair. Mutat Res-Fund Mol M. 2006;601(1–2):125–36.
80. Sapkota M, Burnham EL, DeVasure JM, Sweeter JM, Hunter CD, Duryee MJ, et al. Malondialdehyde-acetaldehyde (MAA) protein adducts are found exclusively in the lungs of smokers with alcohol use disorders and are associated with systemic anti-MAA antibodies. Alcohol Clin Exp Res. 2017;41(12):2093–9.
81. Duryee MJ, Clemens DL, Opperman PJ, Thiele GM, Duryee LM, Garvin RP, et al. Malondialdehyde-acetaldehyde modified (MAA) proteins differentially effect the inflammatory response in macrophage, endothelial cells and animal models of cardiovascular disease. Int J Mol Sci. 2021;22(23):12948.
82. Sapkota M, DeVasure JM, Kharbanda KK, Wyatt TA. Malondialdehyde-acetaldehyde (MAA) adducted surfactant protein induced lung inflammation is mediated through scavenger receptor a (SR-A1). Resp Res. 2017;18:36.
83. Bhatia R, Thompson CM, Clement EJ, Ganguly K, Cox JL, Rauth S, et al. Malondialdehyde-acetaldehyde extracellular matrix protein adducts attenuate unfolded protein response during alcohol and smoking-induced pancreatitis. Gastroenterology. 2022;163(4):1064–78.e10.
84. Castro JP, Jung T, Grune T, Siems W. 4-Hydroxynonenal (HNE) modified proteins in metabolic diseases. Free Radical Bio Med. 2017;111:309–15.
85. Erejuwa OO, Sulaiman SA, Ab Wahab MS. Evidence in support of potential applications of lipid peroxidation products in cancer treatment. Oxidative Med Cell Longev. 2013;2013:931251.
86. Breau M, Houssaini A, Lipskaia L, Abid S, Born E, Marcos E, et al. The antioxidant N-acetylcysteine protects from lung emphysema but induces lung adenocarcinoma in mice. Jci Insight. 2019;4(19):e127647.
87. Li CL, Yan Y, Shi QH, Kong YH, Gao LX, Bao HP, et al. Recuperating lung decoction attenuates inflammation and oxidation in cigarette smoke-induced COPD in rats via activation of ERK and Nrf2 pathways. Cell Biochem Funct. 2017;35(5):278–86.

88. Hwang HV, Sandeep N, Paige SL, Ranjbarvaziri S, Hu DQ, Zhao M, et al. 4HNE impairs myocardial bioenergetics in congenital heart disease-induced right ventricular failure. Circulation. 2020;142(17):1667–83.
89. Feng ZH, Hu WW, Tang MS. Trans-4-hydroxy-2-nonenal inhibits nucleotide excision repair in human cells: a possible mechanism for lipid peroxidation-induced carcinogenesis. Proc Natl Acad Sci U S A. 2004;101(23):8598–602.
90. Yasuo M, Droma Y, Kitaguchi Y, Ito M, Imamura H, Kawakubo M, et al. The relationship between acrolein and oxidative stress in COPD: in systemic plasma and in local lung tissue. Int J Chronic Obstr. 2019;14:1527–37.
91. Esterbauer H, Schaur RJ, Zollner H. Chemistry and biochemistry of 4-hydroxynonenal, malonaldehyde and related aldehydes. Free Radical Bio Med. 1991;11(1):81–128.
92. Pizzimenti S, Ciamporcero E, Daga M, Pettazzoni P, Arcaro A, Cetrangolo G, et al. Interaction of aldehydes derived from lipid peroxidation and membrane proteins. Front Physiol. 2013;4:242.
93. Seiner DR, LaButti JN, Gates KS. Kinetics and mechanism of protein tyrosine phosphatase 1B inactivation by acrolein. Chem Res Toxicol. 2007;20(9):1315–20.
94. Cheng G, Guo JH, Carmella SG, Lindgren B, Ikuemonisan J, Jensen J, et al. Increased acrolein-DNA adducts in buccal brushings of e-cigarette users. Carcinogenesis. 2022;43(5):437–44.
95. Liu XY, Zhu MX, Xie JP. Mutagenicity of acrolein and acrolein-induced DNA adducts. Toxicol Mech Methods. 2010;20(1):36–44.
96. Feng ZH, Hu WW, Hu Y, Tang MS. Acrolein is a major cigarette-related lung cancer agent: preferential binding at p53 mutational hotspots and inhibition of DNA repair. Proc Natl Acad Sci U S A. 2006;103(42):15404–9.
97. Murphy E, Ardehali H, Balaban RS, DiLisa F, Dorn GW 2nd, Kitsis RN, et al. Mitochondrial function, biology, and role in disease: a scientific statement from the American Heart Association. Circ Res. 2016;118(12):1960–91.
98. Bhatti JS, Bhatti GK, Reddy PH. Mitochondrial dysfunction and oxidative stress in metabolic disorders–a step towards mitochondria based therapeutic strategies. BBA-Mol Basis Dis. 2017;1863(5):1066–77.
99. Ramaccini D, Montoya-Uribe V, Aan FJ, Modesti L, Potes Y, Wieckowski MR, et al. Mitochondrial function and dysfunction in dilated cardiomyopathy. Front Cell Dev Biol. 2021;8:624216.
100. Ryter SW, Rosas IO, Owen CA, Martinez FJ, Choi ME, Lee CG, et al. Mitochondrial dysfunction as a pathogenic mediator of chronic obstructive pulmonary disease and idiopathic pulmonary fibrosis. Ann Am Thorac Soc. 2018;15:S266–S72.
101. Ma KL, Chen G, Li WH, Kepp O, Zhu YS, Chen Q. Mitophagy, mitochondrial homeostasis, and cell fate. Front Cell Dev Biol. 2020;8:467.
102. Gloire G, Legrand-Poels S, Piette J. NF-kappaB activation by reactive oxygen species: fifteen years later. Biochem Pharmacol. 2006;72(11):1493–505.
103. Liu T, Zhang L, Joo D, Sun SC. NF-kappaB signaling in inflammation. Signal Transduct Target Ther. 2017;2:17023.
104. Tait SWG, Green DR. Mitochondrial regulation of cell death. Cold Spring Perspect Biol. 2013;5(9):a008706.
105. Riley JS, Quarato G, Cloix C, Lopez J, O'Prey J, Pearson M, et al. Mitochondrial inner membrane permeabilisation enables mtDNA release during apoptosis. EMBO J. 2018;37(17):e99238.
106. Slebos DJ, Ryter SW, van der Toorn M, Liu F, Guo F, Baty CJ, et al. Mitochondrial localization and function of heme oxygenase-1 in cigarette smoke-induced cell death. Am J Respir Cell Mol Biol. 2007;36(4):409–17.
107. Mizumura K, Cloonan SM, Nakahira K, Bhashyam AR, Cervo M, Kitada T, et al. Mitophagy-dependent necroptosis contributes to the pathogenesis of COPD. J Clin Invest. 2014;124(9):3987–4003.

108. Cloonan SM, Glass K, Laucho-Contreras ME, Bhashyam AR, Cervo M, Pabon MA, et al. Mitochondrial iron chelation ameliorates cigarette smoke-induced bronchitis and emphysema in mice. Nat Med. 2016;22(2):163–74.
109. Hoffmann RF, Zarrintan S, Brandenburg SM, Kol A, de Bruin HG, Jafari S, et al. Prolonged cigarette smoke exposure alters mitochondrial structure and function in airway epithelial cells. Respir Res. 2013;14:97.
110. Hara H, Araya J, Ito S, Kobayashi K, Takasaka N, Yoshii Y, et al. Mitochondrial fragmentation in cigarette smoke-induced bronchial epithelial cell senescence. Am J Physiol Lung Cell Mol Physiol. 2013;305(10):L737–46.
111. Haji G, Wiegman CH, Michaeloudes C, Patel MS, Curtis K, Bhavsar P, et al. Mitochondrial dysfunction in airways and quadriceps muscle of patients with chronic obstructive pulmonary disease. Respir Res. 2020;21(1):262.
112. Giordano L, Gregory AD, Perez Verdaguer M, Ware SA, Harvey H, DeVallance E, et al. Extracellular release of mitochondrial DNA: triggered by cigarette smoke and detected in COPD. Cell. 2022;11(3):369.
113. Rovina N, Koutsoukou A, Koulouris NG. Inflammation and immune response in COPD: where do we stand? Mediat Inflamm. 2013;2013:413735.
114. Bartoli ML, Costa F, Malagrino L, Nieri D, Antonelli S, Decusatis G, et al. Sputum inflammatory cells in COPD patients classified according to GOLD 2011 guidelines. Eur Respir J. 2016;47(3):978–80.
115. Mendy A, Forno E, Niyonsenga T, Gasana J. Blood biomarkers as predictors of long-term mortality in COPD. Clin Respir J. 2018;12(5):1891–9.
116. Ostridge K, Williams N, Kim V, Bennett M, Harden S, Welch L, et al. Relationship between pulmonary matrix metalloproteinases and quantitative CT markers of small airways disease and emphysema in COPD. Thorax. 2016;71(2):126–32.
117. Verma AK, Pandey AK, Singh A, Kant S, Mahdi AA, Prakash V, et al. Increased serum levels of matrix-metalloproteinase-9, cyclo-oxygenase-2 and prostaglandin E-2 in patients with chronic obstructive pulmonary disease (COPD). Indian J Clin Bioche. 2022;37(2):169–77.
118. Hao W, Li M, Zhang Y, Zhang C, Xue Y. Expressions of MMP-12, TIMP-4, and neutrophil elastase in PBMCs and exhaled breath condensate in patients with COPD and their relationships with disease severity and acute exacerbations. J Immunol Res. 2019;2019:7142438.
119. Voynow JA, Shinbashi M. Neutrophil elastase and chronic lung disease. Biomol Ther. 2021;11(8):1065.
120. Voynow JA, Young LR, Wang Y, Horger T, Rose MC, Fischer BM. Neutrophil elastase increases MUC5AC mRNA and protein expression in respiratory epithelial cells. Am J Phys. 1999;276(5):L835–43.
121. Zhao J, Yang T, Qiao W, Ye Y, Zhang J, Luo Q. Human neutrophil elastase mediates MUC5AC hypersecretion via the tumour necrosis factor-alpha converting enzyme-epidermal growth factor receptor signalling pathway in vivo. ORL J Otorhinolaryngol Relat Spec. 2021;83(5):310–8.
122. Jackson PL, Xu X, Wilson L, Weathington NM, Clancy JP, Blalock JE, et al. Human neutrophil elastase-mediated cleavage sites of MMP-9 and TIMP-1: implications to cystic fibrosis proteolytic dysfunction. Mol Med. 2010;16(5–6):159–66.
123. Wilkinson DJ, Falconer AMD, Wright HL, Lin H, Yamamoto K, Cheung K, et al. Matrix metalloproteinase-13 is fully activated by neutrophil elastase and inactivates its serpin inhibitor, alpha-1 antitrypsin: implications for osteoarthritis. FEBS J. 2022;289(1):121–39.
124. Itoh Y, Nagase H. Preferential inactivation of tissue inhibitor of metalloproteinases-1 that is bound to the precursor of matrix metalloproteinase-9 (Progelatinase-B) by human neutrophil elastase. J Biol Chem. 1995;270(28):16518–21.
125. Vlahos R, Bozinovski S. Role of alveolar macrophages in chronic obstructive pulmonary disease. Front Immunol. 2014;5:435.
126. Akata K, van Eeden SF. Lung macrophage functional properties in chronic obstructive pulmonary disease. Int J Mol Sci. 2020;21(3):853.

127. Lee JW, Chun W, Lee HJ, Min JH, Kim SM, Seo JY, et al. The role of macrophages in the development of acute and chronic inflammatory lung diseases. Cell. 2021;10(4):897.
128. Chen XR, Wang DX. Serum MCP-1 and NGAL play an important role in the acute inflammatory event of chronic obstructive pulmonary disease. COPD. 2021;18(4):425–31.
129. Monzon ME, Forteza RM, Casalino-Matsuda SM. MCP-1/CCR2B-dependent loop upregulates MUC5AC and MUC5B in human airway epithelium. Am J Physiol Lung Cell Mol Physiol. 2011;300(2):L204–15.
130. Malaviya R, Laskin JD, Laskin DL. Anti-TNF alpha therapy in inflammatory lung diseases. Pharmacol Therapeut. 2017;180:90–8.
131. Fujita M, Ouchi H, Ikegame S, Harada E, Matsumoto T, Uchino J, et al. Critical role of tumor necrosis factor receptor 1 in the pathogenesis of pulmonary emphysema in mice. Int J Chron Obstruct Pulmon Dis. 2016;11:1705–12.
132. Feng Q, Yu YZ, Meng QH. Blocking tumor necrosis factor-alpha delays progression of chronic obstructive pulmonary disease in rats through inhibiting MAPK signaling pathway and activating SOCS3/TRAF1. Exp Ther Med. 2021;22(5):1311.
133. Chen G, Zhou M, Chen L, Meng ZJ, Xiong XZ, Liu HJ, et al. Cigarette smoke disturbs the survival of CD8(+) Tc/Tregs partially through muscarinic receptors-dependent mechanisms in chronic obstructive pulmonary disease. PLoS One. 2016;11(1):e0147232.
134. Siena L, Gjomarkaj M, Elliot J, Pace E, Bruno A, Baraldo S, et al. Reduced apoptosis of CD8+ T-lymphocytes in the airways of smokers with mild/moderate COPD. Respir Med. 2011;105(10):1491–500.
135. Zhu XH, Gadgil AS, Givelber R, George MP, Stoner MW, Sciurba FC, et al. Peripheral T cell functions correlate with the severity of chronic obstructive pulmonary disease. J Immunol. 2009;182(5):3270–7.
136. Paats MS, Bergen IM, Hoogsteden HC, van der Eerden MM, Hendriks RW. Systemic CD4+ and CD8+ T-cell cytokine profiles correlate with GOLD stage in stable COPD. Eur Respir J. 2012;40(2):330–7.
137. Wang HY, Ying HJ, Wang S, Gu X, Weng YS, Peng WD, et al. Imbalance of peripheral blood Th17 and Treg responses in patients with chronic obstructive pulmonary disease. Clin Respir J. 2015;9(3):330–41.
138. Li XN, Pan X, Qiu D. Imbalances of Th17 and Treg cells and their respective cytokines in COPD patients by disease stage. Int J Clin Exp Med. 2014;7(12):5324–9.
139. Zhang JC, Chen G, Chen L, Meng ZJ, Xiong XZ, Liu HJ, et al. TGF-beta/BAMBI pathway dysfunction contributes to peripheral Th17/Treg imbalance in chronic obstructive pulmonary disease. Sci Rep. 2016;6:31911.
140. Zhao P, Li JS, Tian YE, Mao J, Liu XF, Feng SX, et al. Restoring Th17/Treg balance via modulation of STAT3 and STAT5 activation contributes to the amelioration of chronic obstructive pulmonary disease by Bufei Yishen formula. J Ethnopharmacol. 2018;217:152–62.
141. Xu W, Li R, Sun Y. Increased IFN-γ-producing Th17/Th1 cells and their association with lung function and current smoking status in patients with chronic obstructive pulmonary disease. BMC Pulm Med. 2019;19(1):137.
142. Hogg JC, Chu F, Utokaparch S, Woods R, Elliott WM, Buzatu L, et al. The nature of small-airway obstruction in chronic obstructive pulmonary disease. New Engl J Med. 2004;350(26):2645–53.
143. Bracke KR, Verhamme FM, Seys LJM, Bantsimba-Malanda C, Cunoosamy DM, Herbst R, et al. Role of CXCL13 in cigarette smoke-induced lymphoid follicle formation and chronic obstructive pulmonary disease. Am J Resp Crit Care. 2013;188(3):343–55.
144. Ni L, Dong C. Roles of myeloid and lymphoid cells in the pathogenesis of chronic obstructive pulmonary disease. Front Immunol. 2018;9:1431.
145. Polverino F, Seys LJM, Bracke KR, Owen CA. B cells in chronic obstructive pulmonary disease: moving to center stage. Am J Physiol-Lung C. 2016;311(4):L687–L95.
146. Polverino F, Cosio BG, Pons J, Laucho-Contreras M, Tejera P, Iglesias A, et al. B cell-activating factor an orchestrator of lymphoid follicles in severe chronic obstructive pulmonary disease. Am J Resp Crit Care. 2015;192(6):695–705.

147. Knight DA, Holgate ST. The airway epithelium: structural and functional properties in health and disease. Respirology. 2003;8(4):432–46.
148. Gao W, Li L, Wang Y, Zhang S, Adcock IM, Barnes PJ, et al. Bronchial epithelial cells: the key effector cells in the pathogenesis of chronic obstructive pulmonary disease? Respirology. 2015;20(5):722–9.
149. Fujisawa T, Chang MM, Velichko S, Thai P, Hung LY, Huang F, et al. NF-kappaB mediates IL-1beta- and IL-17A-induced MUC5B expression in airway epithelial cells. Am J Respir Cell Mol Biol. 2011;45(2):246–52.
150. Gray T, Nettesheim P, Loftin C, Koo JS, Bonner J, Peddada S, et al. Interleukin-1beta-induced mucin production in human airway epithelium is mediated by cyclooxygenase-2, prostaglandin E2 receptors, and cyclic AMP-protein kinase a signaling. Mol Pharmacol. 2004;66(2):337–46.
151. Takeyama K, Jung B, Shim JJ, Burgel PR, Pick TD, Ueki IF, et al. Activation of epidermal growth factor receptors is responsible for mucin synthesis induced by cigarette smoke. Am J Physiol-Lung C. 2001;280(1):L165–L72.
152. Richter A, O'Donnell RA, Powell RM, Sanders MW, Holgate ST, Djukanovic R, et al. Autocrine ligands for the epidermal growth factor receptor mediate interleukin-8 release from bronchial epithelial cells in response to cigarette smoke. Am J Respir Cell Mol Biol. 2002;27(1):85–90.
153. Wu YL, Lin AH, Chen CH, Huang WC, Wang HY, Liu MH, et al. Glucosamine attenuates cigarette smoke-induced lung inflammation by inhibiting ROS-sensitive inflammatory signaling. Free Radic Biol Med. 2014;69:208–18.
154. Manzel LJ, Shi L, O'Shaughnessy PT, Thorne PS, Look DC. Inhibition by cigarette smoke of nuclear factor-kappaB-dependent response to bacteria in the airway. Am J Respir Cell Mol Biol. 2011;44(2):155–65.
155. Xiao K, Liu C, Tu Z, Xu Q, Chen S, Zhang Y, et al. Activation of the NF-kappaB and MAPK signaling pathways contributes to the inflammatory responses, but not cell injury, in IPEC-1 cells challenged with hydrogen peroxide. Oxidative Med Cell Longev. 2020;2020:5803639.
156. Park J, Min JS, Kim B, Chae UB, Yun JW, Choi MS, et al. Mitochondrial ROS govern the LPS-induced pro-inflammatory response in microglia cells by regulating MAPK and NF-kappa B pathways. Neurosci Lett. 2015;584:191–6.
157. Durham AL, Adcock IM. The relationship between COPD and lung cancer. Lung Cancer. 2015;90(2):121–7.
158. Siegel RL, Miller KD, Jemal A. Cancer statistics, 2017. CA Cancer J Clin. 2017;67(1):7–30.
159. Young RP, Hopkins RJ, Christmas T, Black PN, Metcalf P, Gamble GD. COPD prevalence is increased in lung cancer, independent of age, sex and smoking history. Eur Respir J. 2009;34(2):380–6.
160. Wang W, Dou S, Dong WY, Xie MS, Cui LW, Zheng CY, et al. Impact of COPD on prognosis of lung cancer: from a perspective on disease heterogeneity. Int J Chronic Obstr. 2018;13:3767–76.
161. Ahn SV, Lee E, Park B, Jung JH, Park JE, Sheen SS, et al. Cancer development in patients with COPD: a retrospective analysis of the National Health Insurance Service-National Sample Cohort in Korea. BMC Pulm Med. 2020;20(1):170.
162. Park HY, Kang D, Shin SH, Yoo KH, Rhee CK, Suh GY, et al. Chronic obstructive pulmonary disease and lung cancer incidence in never smokers: a cohort study. Thorax. 2020;75(6):506–9.
163. Kuo CH, Wu CY, Lee KY, Lin SM, Chung FT, Lo YL, et al. Chronic obstructive pulmonary disease in stage I non-small cell lung cancer that underwent anatomic resection: the role of a recurrence promoter. COPD. 2014;11(4):407–13.
164. Husebo GR, Nielsen R, Hardie J, Bakke PS, Lerner L, D'Alessandro-Gabazza C, et al. Risk factors for lung cancer in COPD–results from the Bergen COPD cohort study. Respir Med. 2019;152:81–8.

165. Yang XF, Wisselink HJ, Vliegenthart R, Heuvelmans MA, Groen HJM, Vonder M, et al. Association between chest CT-defined emphysema and lung cancer: a systematic review and meta-analysis. Radiology. 2022;304(2):322–30.
166. Sancar A, Lindsey-Boltz LA, Unsal-Kacmaz K, Linn S. Molecular mechanisms of mammalian DNA repair and the DNA damage checkpoints. Annu Rev Biochem. 2004;73:39–85.
167. Kryston TB, Georgiev AB, Pissis P, Georgakilas AG. Role of oxidative stress and DNA damage in human carcinogenesis. Mutat Res-Fund Mol M. 2011;711(1–2):193–201.
168. Ceylan E, Kocyigit A, Gencer M, Aksoy N, Selek S. Increased DNA damage in patients with chronic obstructive pulmonary disease who had once smoked or been exposed to biomass. Respir Med. 2006;100(7):1270–6.
169. Tzortzaki EG, Dimakou K, Neofytou E, Tsikritsaki K, Samara K, Avgousti M, et al. Oxidative DNA damage and somatic mutations: a link to the molecular pathogenesis of chronic inflammatory airway diseases. Chest. 2012;141(5):1243–50.
170. Aoshiba K, Zhou F, Tsuji T, Nagai A. DNA damage as a molecular link in the pathogenesis of COPD in smokers. Eur Respir J. 2012;39(6):1368–76.
171. Stefanou DT, Kouvela M, Stellas D, Voutetakis K, Papadodima O, Syrigos K, et al. Oxidative stress and deregulated DNA damage response network in lung cancer patients. Biomedicine. 2022;10(6):1248.
172. Caramori G, Papi A. Pathogenic link between chronic obstructive pulmonary disease and squamous cell lung cancer. Expert Rev Respir Med. 2007;1(2):171–5.
173. Mamdani H, Jalal SI. DNA repair in lung cancer: potential not yet reached. Lung Cancer Manag. 2016;5(1):5–8.
174. Sauler M, Lamontagne M, Finnemore E, Herazo-Maya JD, Tedrow J, Zhang XC, et al. The DNA repair transcriptome in severe COPD. Eur Respir J. 2018;52(4):1701994.
175. Scharer OD. Nucleotide excision repair in eukaryotes. Cold Spring Perspect Biol. 2013;5(10):a012609.
176. Hosgood HD, Liu CS, Rothman N, Weinstein SJ, Bonner MR, Shen M, et al. Mitochondrial DNA copy number and lung cancer risk in a prospective cohort study. Carcinogenesis. 2010;31(5):847–9.
177. Wallace DC. Mitochondrial-DNA sequence variation in human-evolution and disease. Proc Natl Acad Sci U S A. 1994;91(19):8739–46.
178. Lee HC, Chang CM, Chi CW. Somatic mutations of mitochondrial DNA in aging and cancer progression. Ageing Res Rev. 2010;9:S47–58.
179. Dasgupta S, Soudry E, Mukhopadhyay N, Shao C, Yee J, Lam S, et al. Mitochondrial DNA mutations in respiratory complex-I in never-smoker lung cancer patients contribute to lung cancer progression and associated with EGFR gene mutation. J Cell Physiol. 2012;227(6):2451–60.
180. Kazdal D, Harms A, Endris V, Penzel R, Kriegsmann M, Eichhorn F, et al. Prevalence of somatic mitochondrial mutations and spatial distribution of mitochondria in non-small cell lung cancer. Brit J Cancer. 2017;117(2):220–6.
181. Ai SSY, Hsu K, Herbert C, Cheng ZJ, Hunt J, Lewis CR, et al. Mitochondrial DNA mutations in exhaled breath condensate of patients with lung cancer. Respir Med. 2013;107(6):911–8.
182. Raghav L, Chang YH, Hsu YC, Li YC, Chen CY, Yang TY, et al. Landscape of mitochondria genome and clinical outcomes in stage 1 lung adenocarcinoma. Cancers. 2020;12(3):755.
183. Bulgakova O, Kausbekova A, Kussainova A, Kalibekov N, Serikbaiuly D, Bersimbaev R. Involvement of circulating cell-free mitochondrial DNA and proinflammatory cytokines in pathogenesis of chronic obstructive pulmonary disease and lung cancer. Asian Pac J Cancer Prev. 2021;22(6):1927–33.
184. Lin CS, Yeh YC, Pan SC, Lu SY, Chen YJ, Chueh WY, Wei YH. Role of mitochondrial DNA copy number alteration in non-small cell lung cancer. Formos J Surg. 2020;53:165–76.
185. Chen J, Zhang L, Yu X, Zhou H, Luo Y, Wang W, et al. Clinical application of plasma mitochondrial DNA content in patients with lung cancer. Oncol Lett. 2018;16(6):7074–81.

186. Lee KM, Giltnane JM, Balko JM, Schwarz LJ, Guerrero-Zotano AL, Hutchinson KE, et al. MYC and MCL1 cooperatively promote chemotherapy-resistant breast cancer stem cells via regulation of mitochondrial oxidative phosphorylation. Cell Metab. 2017;26(4):633–647.e7.
187. Han SY, Jeong YJ, Choi Y, Hwang SK, Bae YS, Chang YC. Mitochondrial dysfunction induces the invasive phenotype, and cell migration and invasion, through the induction of AKT and AMPK pathways in lung cancer cells. Int J Mol Med. 2018;42(3):1644–52.
188. Chuang CH, Dorsch M, Dujardin P, Silas S, Ueffing K, Holken JM, et al. Altered mitochondria functionality defines a metastatic cell state in lung cancer and creates an exploitable vulnerability. Cancer Res. 2021;81(3):567–79.
189. Houghton AM. Mechanistic links between COPD and lung cancer. Nat Rev Cancer. 2013;13(4):233–45.
190. Balkwill F, Mantovani A. Inflammation and cancer: back to Virchow? Lancet. 2001;357(9255):539–45.
191. Chang SH, Mirabolfathinejad SG, Katta H, Cumpian AM, Gong L, Caetano MS, et al. T helper 17 cells play a critical pathogenic role in lung cancer. Proc Natl Acad Sci U S A. 2014;111(15):5664–9.
192. Ritzmann F, Jungnickel C, Vella G, Kamyschnikow A, Herr C, Li D, et al. IL-17C-mediated innate inflammation decreases the response to PD-1 blockade in a model of Kras-driven lung cancer. Sci Rep. 2019;9(1):10353.
193. Moghaddam SJ, Li HH, Cho SN, Dishop MK, Wistuba II, Ji L, et al. Promotion of lung carcinogenesis by chronic obstructive pulmonary disease-like airway inflammation in a K-ras-induced mouse model. Am J Resp Cell Mol. 2009;40(4):443–53.
194. De la Garza MM, Cumpian AM, Daliri S, Castro-Pando S, Umer M, Gong L, et al. COPD-type lung inflammation promotes K-ras mutant lung cancer through epithelial HIF-1alpha mediated tumor angiogenesis and proliferation. Oncotarget. 2018;9(68):32972–83.
195. Shen HM, Tergaonkar V. NF kappa B signaling in carcinogenesis and as a potential molecular target for cancer therapy. Apoptosis. 2009;14(4):348–63.
196. Gu LJ, Wang ZY, Zuo J, Li HM, Zha L. Prognostic significance of NF-kappa B expression in non-small cell lung cancer: a meta-analysis. PLoS One. 2018;13(5):e0198223.
197. Tergaonkar V, Pando M, Vafa O, Wahl G, Verma I. P53 stabilization is decreased upon NF kappa B activation: a role for NF kappa B in acquisition of resistance to chemotherapy. Cancer Cell. 2002;1(5):493–503.
198. Kashatus D, Cogswell P, Baldwin AS. Expression of the Bcl-3 proto-oncogene suppresses p53 activation. Genes Dev. 2006;20(2):225–35.
199. Xia DR, Srinivas H, Ahn YH, Sethi G, Sheng XY, Yung WKA, et al. Mitogen-activated protein kinase kinase-4 promotes cell survival by decreasing PTEN expression through an NF kappa B-dependent pathway. J Biol Chem. 2007;282(6):3507–19.
200. Julien S, Puig I, Caretti E, Bonaventure J, Nelles L, van Roy F, et al. Activation of NF-kappa B by Akt upregulates snail expression and induces epithelium mesenchyme transition. Oncogene. 2007;26(53):7445–56.
201. Escriva M, Peiro S, Herranz N, Villagrasa P, Dave N, Montserrat-Sentis B, et al. Repression of PTEN phosphatase by Snail1 transcriptional factor during gamma radiation-induced apoptosis. Mol Cell Biol. 2008;28(5):1528–40.
202. Zhang S, Yu D. PI(3)king apart PTEN's role in cancer. Clin Cancer Res. 2010;16(17):4325–30.
203. Ho YS, Chen CH, Wang YJ, Pestell RG, Albanese C, Chen RJ, et al. Tobacco-specific carcinogen 4-(methylnitrosamino)-1-(3-pyridyl)-1-butanone (NNK) induces cell proliferation in normal human bronchial epithelial cells through NF kappa B activation and cyclin D1 up-regulation. Toxicol Appl Pharm. 2005;205(2):133–48.
204. Xia Y, Yeddula N, Leblanc M, Ke E, Zhang Y, Oldfield E, et al. Reduced cell proliferation by IKK2 depletion in a mouse lung-cancer model. Nat Cell Biol. 2012;14(3):257–65.
205. Zaynagetdinov R, Stathopoulos GT, Sherrill TP, Cheng DS, McLoed AG, Ausborn JA, et al. Epithelial nuclear factor-kappa B signaling promotes lung carcinogenesis via recruitment of regulatory T lymphocytes. Oncogene. 2012;31(26):3164–76.

206. Takahashi H, Ogata H, Nishigaki R, Broide DH, Karin M. Tobacco smoke promotes lung tumorigenesis by triggering IKK beta- and JNK1-dependent inflammation. Cancer Cell. 2010;17(1):89–97.
207. Carneiro-Lobo TC, Scalabrini LC, Magalhaes LD, Cardeal LB, Rodrigues FS, dos Santos EO, et al. IKK beta targeting reduces KRAS-induced lung cancer angiogenesis in vitro and in vivo: a potential anti-angiogenic therapeutic target. Lung Cancer. 2019;130:169–78.
208. Gu Y, Pais G, Becker V, Korbel C, Ampofo E, Ebert E, et al. Suppression of endothelial miR-22 mediates non-small cell lung cancer cell-induced angiogenesis. Mol Ther-Nucl Acids. 2021;26:849–64.
209. Zhao J, Wang X, Mi Z, Jiang X, Sun L, Zheng B, et al. STAT3/miR-135b/NF-kappa B axis confers aggressiveness and unfavorable prognosis in non-small-cell lung cancer. Cell Death Dis. 2021;12(5):493.
210. Hoang PH, Landi MT. DNA methylation in lung cancer: mechanisms and associations with histological subtypes, molecular alterations, and major epidemiological factors. Cancers. 2022;14(4):961.
211. Benincasa G, DeMeo DL, Glass K, Silverman EK, Napoli C. Epigenetics and pulmonary diseases in the horizon of precision medicine: a review. Eur Respir J. 2021;57(6):2003406.
212. Sundar IK, Mullapudi N, Yao HW, Spivack SD, Rahman I. Lung cancer and its association with chronic obstructive pulmonary disease: update on nexus of epigenetics. Curr Opin Pulm Med. 2011;17(4):279–85.
213. Bruse S, Petersen H, Weissfeld J, Picchi M, Willink R, Do K, et al. Increased methylation of lung cancer-associated genes in sputum DNA of former smokers with chronic mucous hypersecretion. Resp Res. 2014;15(1):2.
214. Sato T, Arai E, Kohno T, Takahashi Y, Miyata S, Tsuta K, et al. Epigenetic clustering of lung adenocarcinomas based on DNA methylation profiles in adjacent lung tissue: its correlation with smoking history and chronic obstructive pulmonary disease. Int J Cancer. 2014;135(2):319–34.
215. Tessema M, Yingling CM, Picchi MA, Wu GD, Liu YS, Weissfeld JL, et al. Epigenetic repression of CCDC37 and MAP 1B links chronic obstructive pulmonary disease to lung cancer. J Thorac Oncol. 2015;10(8):1181–8.
216. Suzuki M, Wada H, Yoshino M, Tian L, Shigematsu H, Suzuki H, et al. Molecular characterization of chronic obstructive pulmonary disease-related non-small cell lung cancer through aberrant methylation and alterations of EGFR signaling. Ann Surg Oncol. 2010;17(3):878–88.
217. Wauters E, Janssens W, Vansteenkiste J, Decaluwe H, Heulens N, Thienpont B, et al. DNA methylation profiling of non-small cell lung cancer reveals a COPD-driven immune-related signature. Thorax. 2015;70(12):1113–22.
218. Fregonese L, Stolk J. Hereditary alpha-1-antitrypsin deficiency and its clinical consequences. Orphanet J Rare Dis. 2008;3:16.
219. Senn O, Russi EW, Imboden M, Probst-Hensch NM. Alpha(1)-antitrypsin deficiency and lung disease: risk modification by occupational and environmental inhalants. Eur Respir J. 2005;26(5):909–17.
220. Zhang LY, Sun XW, Ding YJ, Yan YR, Li CX, Wang Y, et al. SERPINA1 methylation levels are associated with lung cancer development in male patients with chronic obstructive pulmonary disease. Int J Chronic Obstr. 2022;17:2117–25.
221. Zhao JJ, Tao YJ, Zhou Y, Qin NL, Chen C, Tian D, et al. MicroRNA-7: a promising new target in cancer therapy. Cancer Cell Int. 2015;15:103.
222. Rosas-Alonso R, Galera R, Sanchez-Pascuala JJ, Casitas R, Burdiel M, Martinez-Ceron E, et al. Hypermethylation of anti-oncogenic microRNA 7 is increased in emphysema patients. Arch Bronconeumol. 2020;56(8):506–13.
223. Cheng MW, Shen ZT, Hu GY, Luo LG. Prognostic significance of microRNA-7 and its roles in the regulation of cisplatin resistance in lung adenocarcinoma. Cell Physiol Biochem. 2017;42(2):660–72.
224. Ling H, Fabbri M, Calin GA. MicroRNAs and other non-coding RNAs as targets for anticancer drug development. Nat Rev Drug Discov. 2013;12(11):847–65.

225. Keller A, Fehlmann T, Ludwig N, Kahraman M, Laufer T, Backes C, et al. Genome-wide microRNA expression profiles in COPD: early predictors for cancer development. Genom Proteom Bioinf. 2018;16(3):162–71.
226. Mateu-Jimenez M, Curull V, Rodriguez-Fuster A, Aguilo R, Sanchez-Font A, Pijuan L, et al. Profile of epigenetic mechanisms in lung tumors of patients with underlying chronic respiratory conditions. Clin Epigenetics. 2018;10:7.
227. Chen CY, Chen JY, He L, Stiles BL. PTEN: tumor suppressor and metabolic regulator. Front Endocrinol. 2018;9:338.
228. Bickeboller M, Tagscherer KE, Kloor M, Jansen L, Chang-Claude J, Brenner H, et al. Functional characterization of the tumor-suppressor MARCKS in colorectal cancer and its association with survival. Oncogene. 2015;34(9):1150–9.
229. Jancik S, Drabek J, Radzioch D, Hajduch M. Clinical relevance of KRAS in human cancers. J Biomed Biotechnol. 2010;2010:150960.
230. Fathinavid A, Ghobadi MZ, Najafi A, Masoudi-Nejad A. Identification of common microRNA between COPD and non-small cell lung cancer through pathway enrichment analysis. BMC Genomic Data. 2021;22(1):41.
231. O'Farrell HE, Bowman RV, Fong KM, Yang IAA. Plasma extracellular vesicle miRNAs can identify lung cancer, current smoking status, and stable COPD. Int J Mol Sci. 2021;22(11):5803.
232. Huang DD, She J, Limper AH, Yang Y, Yang P. A comprehensive (biological and computational) investigation on the role of microRNA: mRNA regulations performed in chronic obstructive pulmonary disease and lung cancer. In: 2016 IEEE international conference on bioinformatics and biomedicine (BIBM); 2016. p. 1062–7.
233. Gritz ER, Talluri R, Domgue JF, Tami-Maury I, Shete S. Smoking behaviors in survivors of smoking-related and non-smoking-related cancers. JAMA Netw Open. 2020;3(7):e209072.
234. Parker AS, Arnold ML, Diehl ND, Hassan L, Thiel DD. Evaluation of awareness of risk factors for kidney cancer among patients presenting to a urology clinic. Scand J Urol. 2014;48(3):239–44.
235. Sorensen LT, Horby J, Friis E, Pilsgaard B, Jorgensen T. Smoking as a risk factor for wound healing and infection in breast cancer surgery. Eur J Surg Oncol. 2002;28(8):815–20.
236. Chen AM, Chen LM, Vaughan A, Sreeraman R, Farwell DG, Luu Q, et al. Tobacco smoking during radiation therapy for head-and-neck cancer is associated with unfavorable outcome. Int J Radiat Oncol. 2011;79(2):414–9.
237. Khuder SA, Mutgi AB. Effect of smoking cessation on major histologic types of lung cancer. Chest. 2001;120(5):1577–83.
238. Tobacco Control: Reversal of Risk after Quitting Smoking. Vol. 11 in the series IARC Handbooks of Cancer Prevention. Anticancer Res. 2008;28(4A):2029.
239. Parsons A, Daley A, Begh R, Aveyard P. Influence of smoking cessation after diagnosis of early stage lung cancer on prognosis: systematic review of observational studies with meta-analysis. BMJ. 2010;340:b5569.

Sun Exposure and Skin Cancer

6

Kelvin Allenson, Nestor Esnaola, and Eric H. Bernicker

Foreword

Many articles described the rising incidence of melanoma and non-melanoma skin cancer in recent years. Current estimates indicate that the incidence of melanoma is increasing by at least 4% by year [1–3]. There are multiple contributing factors to this increase, one of which is the potential and probable role of climate change. The idea of "climate change" has the potential to be a contentious and politicized subject. Therefore, in order to provide appropriate context for this chapter, it was necessary to explore the evidence to date for climate change, to describe the way in which climate change is studied, and how the predicted changes in climate in the near to intermediate future will impact the diagnosis and management of skin cancer. This brief review is not meant to be exhaustive in detail or scope, but rather to provide a functional understanding of the subject of climate change.

Brief History and Evidence for Climate Change

Human civilization has been tracking weather cycles and it's variances for thousands of years beginning with animal migrations in hunter-gatherer culture, and seasonal patterns in agrarian societies. In more recent times, Europe, the United

K. Allenson · N. Esnaola
Division of Surgical Oncology and Gastrointestinal Surgery, Department of Surgery, Houston Methodist Hospital, Houston, TX, USA

Neal Cancer Center, Houston Methodist Hospital, Houston, TX, USA

E. H. Bernicker (✉)
Neal Cancer Center, Houston Methodist Hospital, Houston, TX, USA

Department of Medical Oncology, Houston Methodist Hospital, Houston, TX, USA
e-mail: EricBernicker@Centura.org

E. H. Bernicker (ed.), *Environmental Oncology*,
https://doi.org/10.1007/978-3-031-33750-5_6

States and portions of Asian underwent the Industrial Revolution whereby the pace of technological and manufacturing advancements accelerated. Fossil fuels and combustion engines shifted modern society into a carbon-based economy [4, 5].

By the early nineteenth century, biological and physical scientists had established the main principles contributing to Earth's climate. Even at that time, weather stations (Blue Hill Meteorological Observatory, Central England Temperature Record) and commercial entities that follow weather occurrences, like maritime shipping had already recognized and noted warming temperatures. Industrialization would continue and similar observations would happen for the next 100 years. In the twentieth century, increasingly detailed and sophisticated scientific reports on Earth climate changes would continue to be published. One data point is that global average temperature has increased 1.8 °F between 1901 and 2016. 1.2 of this 1.8° increased has occurred more recently from 1985 onwards. Natural factors that would contribute to temperature change include volcanic activity, solar irradiation output and natural variability. However, since the 1800s, natural factors alone cannot explain the observed rise in temperature [4].

While there are numerous case studies of climate changes, the topic is perhaps best illustrated by the depletion of atmospheric ozone and the "ozone hole" that developed over Antarctica. By the 1970s, refrigerant chlorofluorocarbons (CFCs) were widely in use. Within the atmosphere, these reactive molecules cause the catalytic breakdown of ozone to oxygen. The ozone layer normally shields the lower Earth atmosphere and Earth itself from harmful ultraviolet radiation (UVR) [3, 6, 7]. With depletion of the ozone layer, increasing amounts of radiation were able to penetrate the atmosphere. In a complex sequence of events to be described later, CFCs serving as greenhouse gases (GHGs) resulted in ozone depletion, increasing levels of UV radiation reaching the Earth's surface, and warming the Earth's temperature. Within this context, the "ozone hole" over Antarctica was publicized and garnered wide-spread recognition. Worse still, it was already known that CFCs would persist in the atmosphere for up to 100 years catalyzing the breakdown of ozone, further propagating the cycle.

Cumulatively, there was increasing physical and biological evidence of environmental changes attributable to human activity, resulting in a body of evidence in support of "climate change." Concern existed that if the use of CFCs continued unabated, there would be continued deterioration of the Earth system at large with resulting deleterious human health consequences [4, 5]. Particularly with the concern of atmospheric ozone depletion due to CFCs, there was now enough scientific evidence, and social and political motivation to result in policy change.

The Montreal Protocol represents one of the first international efforts to address climate change. This treaty was signed by 198 nations in 1987 and went into effect in 1989. The Montreal Protocol functionally served to phase CFCs out of production. Their use in industry was replaced by hydrochlorofluorocarbons (HCFCs) which were less harmful than CFCs but still potent GHGs which propagate the same effect as CFCs. Subsequent amendments to the Montreal Protocol would also phase HCFCs out of production.

Concurrent with the Montreal Protocol and with the topic of climate change having achieved considerable momentum throughout the 1980s, representatives of the international community began to meet regularly. Namely, the Intergovernmental Panel on Climate Change (IPCC) through the United Nations, was created in 1988 and convenes at regular intervals with the final work product of these meetings culminating in the release of climate "Assessment Reports." A trend emerged whereby political representatives would meet following the release of these assessments in order to discuss and formulate international agreements on industrial policy [6, 8, 9].

Since 1988, the IPCC has released six Assessment Reports. Two major international agreements have resulted over the history of the IPCC Assessments. The Kyoto protocol was signed in 1997 and became active in 2005 with the major goal to reduce GHG emissions. In 2014, the IPCC Fifth Assessment Report was released where "representative concentration pathways" (RCPs) were used to model future climate change scenarios. Increasing RCP model number corresponds to a greater GHG concentration and therefore less favorable future climate and environment. These individual RCPs represent GHG concentrations and not emissions of such gasses. From the climate change perspective, the RCP 1.9 model is "Aspirational" and represents the most optimistic model for atmospheric GHG concentration. This model translates to global warming estimate of an increase below 1.5 °C by the year 2100. RCP 8.5 is the most deleterious model, and also considered to be unlikely, but predicts an increase in global temperature up to 5 °C. The findings of this report contributed significantly to the policy changes of the Paris Climate Accords signed in 2016 [10].

The IPCC is currently in the process of releasing its Sixth Assessment Report which is composed of multiple reports which will be released between 2021 and 2023. Within the already released Working Group 1 report, carbon emissions are modeled by Shared Socioeconomic Pathways (SSPs) as opposed the RCPs in the previous report. The SSPs are noteworthy in that they represent projections of both socioeconomic and geopolitical interests of constituent countries. Thus far, no major policy changes have emerged in response to the Sixth Assessment Report [10–12].

Complex Mechanisms of the Earth System Model

Earth's ecosystem is considerably more complex than the summary measure of global temperature, or "global warming," would indicate [4]. What is more, changes in one aspect of the ecosystem has the potential to result in positive feedback loops imparting a greater than incremental change on the system at large. It is rather straightforward to follow the sequence of events that increased GHGs result in the depletion of atmospheric ozone, allowing greater penetrance of solar irradiation, contributing to an overall rise in temperature. However, climate is not accurately depicted by the static measure of temperature alone. While the mean global temperature may increase, certain parts of the world may be warmer or colder, but other factors that contribute to the environmental risk of cutaneous melanoma may also have changed, including the length of seasons, cloud cover, ground level UV radiation, and albedo (the degree to which visible light or other wavelengths is reflected by the local environment) [3, 8, 13].

As previously mentioned, the global average temperature has increased 1.8 °F since 1900, 1.2° of which has occurred since 1985. Climate modeling, as reflected by the IPCC's fifth and sixth Assessment Reports, predicts the average global temperature to increase between 1.0 and 5.7 °C by the year 2100 [5, 11, 12]. International agreements to date have set a goal of limiting the temperature increase to 2 °C [10].

Coinciding with this increase in temperature, as well as with the increase in carbon dioxide and other GHGs, the oceans have absorbed 93% of the excess heat produced by global warming and in serving as a an acid-base buffer, have become more acidic over time. Average sea level has increased 7–8 inches since 1900. Warmer temperature has resulted in more humid conditions and a resultant rise in annual precipitation, estimated at 4% since 1900 [4].

How Climate Change Is Studied

As previously mentioned, observational data describing the environment has existed for the last few hundred years. In the modern era, these observations are augmented by satellite recording and measurement of Earth and atmospheric conditions [5]. While past observations provide important frames of reference, future predictions depend on complex modeling of the various factors impacting the weather and climate [10, 12, 13].

Notably, the main predictive models for short-, intermediate-, and long-term for climate are the results of the IPCC Assessment Reports. In the last two versions of the report, greenhouse gas concentrations were estimated by RCPs, and greenhouse gas emissions predicted by SSPs, respectively. Both the RCPs and SSPs outline a wide a spectrum of future climate scenarios based on current environmental and economic conditions [12]. Inherently, there are many limitations to such models. Economic interests may run counter to environmental interests and emissions goals. The models are quite complex and the degree to which positive feedback loops may shift a climate scenario/model into a qualitatively different state of affairs is difficult to model [4]. The pace of technological advancement is ever-increasing and not able to be accounted for in models. At some point, technologies may emerge whereby humans will be able to extract GHGs from the atmosphere and therefore actively change the atmosphere amongst other aspects of the environment.

For all of the possible limitations, the IPCC climate change models do create scaffolds upon which further research can be conducted. For example, researchers have used change in global temperature to model expected changes in surface UV-radiation [13]. This is germane not just in the further understanding on local changes in weather/temperature by physical scientists, but also has been used to model predicted changes in melanoma and non-melanoma skin cancer risk by epidemiologists and physicians [3, 8, 14, 15].

Climate Change Predictions and Models for the Current Century

The RCPs highlighted by the IPCC Fifth Assessment Report model an increase in average global temperature anywhere between 1 and 5 °C by the year 2100. However, Assessment Reports have been released since 1990, and observation data reflection the current real climate, and not just the projected climate exists. In the last 20 years, global carbon emissions are consistent with RCP8.5, the "worst-case-scenario" model within the Fifth Assessment [10]. However, since 2014, carbon emissions have begun to slow, and in 2016, the trend was approaching zero growth [5].

Ground-level ultraviolet radiation (UVR) is variable, related to GHG concentration, temperature and cloud cover amongst other factors. Even within an RCP model, the predicted change in UVR may change dependent upon, for example, northern vs southern latitude. For the most likely climate scenario, RCP 4.5, UVR is expected to increase by 2% in the tropics,6% in the northern mid-latitudes, and decreased by 5% in the southern med-latitudes [13]. As temperature and UV-exposure represent the primary environmental risk factors for cutaneous melanoma, predicted changes to other aspects of climate will not be further discussed.

The Mechanism of UV Radiation-Related Cutaneous Carcinogenesis

Ultraviolet radiation (UVR) exists on the electromagnetic spectrum, shorter in wavelength than visible light and longer than X-rays. UV-A is observed from 320–400 nm, UV-B from 280–315 nm, and UV-C from 200–280 nm. The ozone layer and atmospheric oxygen absorb all UV-C before it reaches the Earth's surface. Further, 98.7% of all UV light that reaches the Earth surface is UV-A. UV-A penetrates 5 times deeper into the skin than UV-B [16]. UV-A causes indirect cellular DNA damage through the generation of reactive oxygen species, while UV-B causes direct DNA damage through the creation of thymidine DNA dimers. If these DNA dimers are not repaired through nucleotide excision repair, DNA replication becomes error prone, resulting in a carcinogenesis pathway [1].

Multiple protective mechanisms have evolved to preserve the integrity of DNA within skin cells. Photoconversion is the process whereby UV radiation is absorbed in the skin and dissipated as heat. DNA itself is a component of the photoconversion pathway, absorbing and dissipating energy. UV energy that is not dissipated as heat, results in reactive oxygen species that causes DNA damage, and which must subsequently be repaired if the fidelity of DNA replication is to be preserved [17].

The primary photoconverter is skin pigment. Melanosomes are specialized organelles that store and concentrate pigment. Melanosomes within keratinocytes form a UV-protective barrier around the nuclei and genetic material of the cell. Eumelanin is a brown-black pigment and is more prevalent in darker skinned individuals. This pigment evolved over time in populations that lived closer to the equator, and overall is 99.9% effective in dissipating UVR as heat. Pheomelanin is red-yellow in color and more prevalent in fairer skinned individuals. While pheomelanin is not as effective at UV-photoprotection, it is more effective at cutaneous Vitamin D absorption, an evolutionary adaptation which is well suited to fairer skinned individuals living further from the equatorial latitudes at which locations the UV-exposure is less [16, 17].

Lastly, temperature is potentiator of skin cancer development. This finding initially emerged in mouse-based lab experiments. While this mechanism has been explored in mouse models, no such study is feasible in humans, but has been explored in epidemiologic studies. Specifically, the incidence of skin cancers for populations at similar latitude but at varying elevations/temperatures have been compared. 80% of the variance in skin cancer incidence is explained by UV-exposure. The remainder of the variance is composed in part from differences in ambient temperature, as well as ethnic background and sun-exposure behavior [3, 14, 15].

Non-UVR Risk Factors for Skin Cancer

Melanoma and non-melanoma skin cancer risk is multifactorial but can be divided into two groups; genetic susceptibility and environmental exposure [16].

About 10% of patients with a melanoma diagnosis have a family history of melanoma. However, only about 1% of patients with melanoma are believed to have a hereditary syndrome. While these syndromes do not contribute significantly to the overall incidence, they are noteworthy for the high personal risk experienced by those affected. Familial atypical multiple mole melanoma (FAMMM) syndrome is characterized by a personal history of multiple nevi, atypical nevi, and a family history of melanoma. These patients can have germline mutations in CDKN2A. Genetic penetrance is variable and lifetime melanoma *incidence* is as high as 28%. BAP1 tumor syndrome is associated with risk of both cutaneous (13%) and uveal (36%) melanoma. Xeroderma pigmentosum is a rare autosomal recessive disorder. The genetic mutations responsible for the disease function in the repair of UV-induced DNA damage. The disorder has 100% penetrance. The risk of melanoma is increased 2000-fold, while NMSC is increased 10,000-fold [1].

The risk of melanoma has been well described related to personal phenotypic features [18]. Light hair, particularly red-haired individuals are at risk of melanoma with relative risk of 3.6. Increasing number of both typical and atypical nevi also confer an increased risk of melanoma. A parent or sibling with melanoma increases the individual relative risk up to 4.4 and 4.4 respectively. Lastly, a personal history of melanoma carries an 8% risk of a diagnosis of a second primary melanoma, while diagnosis of a NMSC carries a relative risk of 4.2.

The Impact of Climate Change on Melanoma and Non-Melanoma Skin Cancer

The risk of melanoma has increased significantly over time with increased ground-level UV radiation as the major contributing factor. In the United States in 1935, the lifetime risk of melanoma was 1:500. By 2021, the risk was 1:27. Currently, the incidence of new melanoma diagnosis, year-over-year in the United States is increasing between 3 and 7% [1]. While some of this rising incidence may be explained by increased and effective skin cancer screening, the majority of cases are attributable to increased UVR exposure in the context of climate change at large [19].

The experience in Australia is illustrative. Stratospheric ozone depletion is known to be highest over Antarctica during the twentieth century and into the current century. This resulted in an up to 80% higher surface UV-B radiation level. In neighboring Australia, clear sky UVR has increased between 2 and 6% per year since 1990. In subsequent years, the incidence of melanoma and non-melanoma skin cancer was noted to rise [19]. Models estimate that for each 1% decrease in ozone layer thickness, the incidence of melanoma is expected to increase 1–2%, while BCC and SCC will increase up 2.7% and 4.6% respectively [3, 14, 17]. Under current conditions, the ozone depletion over Antarctica is not expected to recover until 2060. Furthermore, combining ozone depletion with the synergistic effect of an expected 2 °C increase in global temperature, the global incidence of skin cancer is expected to increase 11% by the year 2050 [3].

Similar results have been modeled by climate researchers and validate the Australian experience. Van der Leun found that each °C increase in daily maximum temperature translated to an effective 2% increase in the effective dose/exposure to UVR. This translated into a 5.5% and 2.9% increase in the incidence of SCC and BCC per increase in °C [14]. In a similar study by Piacentini, the incidence of SCC and BCC by the year 2100 from the least to greatest risk RCP models ranged from 5.8–21.4%, and 2.8–9.9%, respectively [15]. UVR was modeled by Watanabe, and ground-level UVR was found to increase by as much as 6% within the IPCC RCP 4.5 model by the year 2100 [13].

While climate change in the form of increasing global temperature and UVR is expected to get worse before it gets better, the effectiveness of prevention, again highlighted by Australia, is extremely encouraging. An estimated 60–70% of melanoma is attributable to UV-exposure. Fair skinned populations in high-sun exposure latitudes, namely, Australia, New Zealand, and the United States are particularly at risk. While UV exposure is the primary risk factor for the development of melanoma, it is also the most modifiable. The Australian Sun Smart Program highlights the effectiveness of screening and secondary prevention of melanoma. The Sun Smart Program was a public service campaign across Australian which sought to increase melanoma awareness as well as reinforce UV-protective strategies such as appropriate use of sunblock. Within 3 months of its roll-out, public service announcements resulted in a 167% increase in new melanoma diagnoses. Particularly in this era, pre-dating the introduction of effective immunotherapy, early diagnosis if not secondary prevention emphasizes a major benefit of the program. Since its inception in 1988, the program is estimated to have prevented 50,000 skin cancers and over 1400 deaths. Similarly, public education on the ABCD risk factors for melanoma diagnosis, again before the immunotherapy era, increased the detection of suspicious lesions from 57 to 90%, with an estimated decreased mortality of 63% [16].

Importantly, given that melanoma arises in the setting of many years of cumulative sun-exposure, it can take years for the benefit of such public awareness campaigns to be realized. For example, nearly 20 years after its introduction, the Sun Smart program is in large part responsible for the peak in melanoma incidence in the year 2005; incidence is now declining. Within the United States, incidence is estimated to peak by 2026. Europe will not peak until 2030 at the earliest [16].

While fair-skinned populations remain at significant risk of the harmful effects of UV-exposure and skin cancer, a systematic review by Lopes examined the role of UV-exposure and melanoma in "skin of color." While greater than 11,000 articles were screened, only 13 could be included and all included studies were of low to moderate quality. This review found that UV exposure was not related to melanoma risk in patients with darker skin color. However, the review qualifies that patients are generally characterized by ethnicity within clinical studies, whereas for melanoma or skin cancer-specific studies, risk would more appropriately be stratified according to Fitzpatrick skin types. Higher quality research is needed within more darkly pigmented patient populations in order to appropriately characterize the risk that increasing future UVR levels and temperature will pose [20].

While there is a wealth of data on climate change, and how these changes are expected to impact and increase the burden of cutaneous melanoma, there is a paucity of research on immunotherapy in the context of climate change. Immunotherapy was FDA-approved for the treatment of cutaneous melanoma beginning in 2011. Multiple, newer immunotherapies have emerged over the last decade. Whereas before the immunotherapy era, the efficacy of systemic therapies for melanoma was poor, immunotherapy has revolutionized the treatment of advanced melanoma, with a mortality decrease of 30%. Importantly, long-term data for many lines of therapy are not yet mature. Mechanistically, immunotherapy is effective in the treatment of melanoma because the primary exposure of UV radiation results in a high burden of genetic mutations within the melanoma, known as tumor mutational burden (TMB). In a study by Dousset, TMB was found to be higher within melanomas arising within chronically sun-exposed areas compared to intermittently exposed areas compared to non-exposed areas. Furthermore, a putative UV-exposure mutational signature was directly related to the TMB. Both the UV molecular signature and TMB as potential biomarkers predicted response to anti-PD1 immunotherapy. Therefore, while we may be subject to increased temperature and UV-exposure and therefore risk of melanoma in the setting of future climate change, we are quite fortunate to now have effective systemic therapies that mechanistically address the underlying risk factors contributing to the genesis of the disease [16, 21].

Summary

A wealth of data exists in support of the topic of climate change. Much of the current and predicted "change" is due to greenhouse gases produced by industrial practices that have already been phased out. Cleaner technologies are already in development and more recently, there is a trend towards carbon-neutral economies and industries. Public policy related to climate change has the potential to be politically contentious. Models predicting future climate change scenarios are increasingly complex, and are imperfect, but have been accurate thus far in predicting our climate.

Damage to the environment has already been done and the Earth ecosystem is already warmer and with greater exposure to UV-radiation. The incidence of melanoma and non-melanoma skin cancer is increasing but large-scale prevention efforts have proven to be effective in mitigating skin cancer risk [22]. Melanoma is very treatable and curable in the early clinical stages, and effective systemic immunotherapies give cause for optimism for disease management even in the setting of an increasing burden of disease.

References

1. Šerman N, Vranić S, Glibo M, Šerman L, Mokos ZB. Genetic risk factors in melanoma etiopathogenesis and the role of genetic counseling: a concise review. Bosn J Basic Med Sci. 2022;22:673–82. https://doi.org/10.17305/bjbms.2021.7378.
2. Cust AE, Mishra K, Berwick M. Melanoma-role of the environment and genetics. Photochem Photobiol Sci. 2018;17:1853–60. https://doi.org/10.1039/c7pp00411g.
3. Parker ER. The influence of climate change on skin cancer incidence – a review of the evidence. Int J Women's Dermatol. 2021;7:17–27. https://doi.org/10.1016/j.ijwd.2020.07.003.
4. Hayhoe K, Wuebbels DJ, Easterling DR et al. Our changing climate. In: Impacts, risks, and adaptation in the United States: Fourth National Climate Assessment, Volume II. U.S. global change research program. Washington; p. 72–144. https://doi.org/10.7930/NCA2.2018.CH2
5. Chen D, Rojas M, Samset BH, et al. Framing, context, and methods. In: Climate change 2021: the physical science basis. Contribution of working group I to the sixth assessment report of the intergovernmental panel on climate change. Cambridge: Cambridge University Press; 2021. p. 147–286. https://doi.org/10.1017/9781009157896.003
6. Bharath AK, Turner RJ. Impact of climate change on skin cancer. J R Soc Med. 2009;102:215–8. https://doi.org/10.1258/jrsm.2009.080261.
7. Cordero RR, Feron S, Damiani A, Redondas A, Carrasco J, Sepúlveda E, et al. Persistent extreme ultraviolet irradiance in Antarctica despite the ozone recovery onset. Sci Rep. 2022;12:1–10. https://doi.org/10.1038/s41598-022-05449-8.
8. Umar SA, Tasduq SA. Ozone layer depletion and emerging public health concerns - an update on epidemiological perspective of the ambivalent effects of ultraviolet radiation exposure. Front Oncol. 2022;12:1–11. https://doi.org/10.3389/fonc.2022.866733.
9. Barnes PW, Robson TM, Neale PJ, Williamson CE, Zepp RG, Madronich S, et al. Environmental effects of stratospheric ozone depletion, UV radiation, and interactions with climate change: UNEP environmental effects assessment panel, update 2021. Photochem Photobiol Sci. 2022;21:275–301. https://doi.org/10.1007/s43630-022-00176-5.
10. Tollefson J. How hot will earth get by 2100? Nature. 2020;580:444–6.
11. Kirtman B, Power SB, Adedoyin JA et al. Near-term climage change: projections and predictability. In: Climate change 2013: The physical science basis. Contribution of working group I to the fifth assessment report of the intergovernmental panel on climate change. Cambridge: Cambridge University Press; 2013. p. 953–1028.
12. Lee J-Y, Marotzke J, Bala G, Cao L, Corti S, Dunne JP, et al. Future global climate: scenario-based projections and near-term information. In: Climate change 2021: The physical science basis. Contribution of working group I to the sixth assessment report of the intergovernmental panel on climate change. New York: Cambridge University Press; 2021. p. 553–672. https://doi.org/10.1017/9781009157896.006.
13. Watanabe S, Sudo K, Nagashima T, Takemura T, Kawase H, Nozawa T. Future projections of surface UV-B in a changing climate. J Geophys Res Atmos. 2011;116:1–11. https://doi.org/10.1029/2011JD015749.

14. Van Der Leun JC, Piacentini RD, De Gruijl FR. Climate change and human skin cancer. Photochem Photobiol Sci. 2008;7:730–3. https://doi.org/10.1039/b719302e.
15. Piacentini RD, Della Ceca LS, Ipiña A. Climate change and its relationship with non-melanoma skin cancers. Photochem Photobiol Sci. 2018;17:1913–7. https://doi.org/10.1039/c7pp00405b.
16. Dzwierzynski WW. Melanoma risk factors and prevention. Clin Plastic Surg. 2021;48:543–50. https://doi.org/10.1016/j.cps.2021.05.001.
17. Bolick NL, Geller AC. Epidemiology of melanoma. Hematol Oncol Clin North Am. 2021;35:57–72. https://doi.org/10.1016/j.hoc.2020.08.011.
18. O'Neill CH, Scoggins CR. Melanoma. J Surg Oncol. 2019;120:873–81.
19. Volkovova K, Bilanicova D, Bartonova A, Letaiová S, Dusinska M. Associations between environmental factors and incidence of cutaneous melanoma. Rev Environ Heal A Glob Access Sci Source. 2012;11:1–13. https://doi.org/10.1186/1476-069X-11-S1-S12.
20. Lopes FCPS, Sleiman MG, Sebastian K, Bogucka R, Jacobs EA, Adamson AS. UV exposure and the risk of cutaneous melanoma in skin of color: a systematic review. JAMA Dermatol. 2021;157:213–9. https://doi.org/10.1001/jamadermatol.2020.4616.
21. Dousset L, Poizeau F, Robert C, et al. Positive association between location of melanoma, ultraviolet signature, tumor mutational burden, and response to anti-PD-1 therapy. JCO Precis Oncologia. 2021;5:1821–9.
22. Dzwierzynski WW. Melanoma risk factors and prevention. Clin Plastic Surg. 2021;48:435–550.

Part III

Industry and Cancer

The Climate Crisis and Cancer

7

Joan H. Schiller and Jasmine Kamboj

Introduction

Climate change is a real threat to humanity and our planet, and unfortunately the field of healthcare is not excluded from its impact. The impact of climate change on health prompted the editors of over 260 health journals to call the health effects of climate change as "catastrophic" [1, 2]. Cancer care is a rapidly evolving field and oncologists are committed to providing the best outcomes for their patients, be it by prevention, utilization of novel therapeutics, cancer treatments, lifestyle modifications, or protecting their environment. Climate Activism is an area which needs research, education, and engagement by oncologists, so as to best protect and treat their patients.

Climate Change Review

97% or more of climate scientists suggest climate change is real, is caused by greenhouse gases, and that human activity is causing almost all of the increase [3].

Greenhouse gases trap solar radiation that would otherwise have been reflected into space after hitting the earth's surface. This radiation is re-radiated back to earth

J. H. Schiller (✉)
University of Virginia, Charlottesville, VA, USA

UT Southwestern Simmons Comprehensive Cancer Center,
Dallas, TX, USA

Oncologists United for Climate and Health, Vienna, Virginia, USA

J. Kamboj
Mayo Clinic, Mayo Clinic Health System, Rochester, MN, USA

E. H. Bernicker (ed.), *Environmental Oncology*,
https://doi.org/10.1007/978-3-031-33750-5_7

as infrared radiation, thus warming the planet by 1.1 °C (2 °F) since preindustrial times [4]. This global warming -- otherwise known as climate change – has a number of consequences, such as higher temperatures; more extreme weather events; thawing permafrost and glaciers leading to rising sea levels: warming seas damaging coral reefs; impacts on water and food quality, supply and production; environmental degradation; and changes in the life cycles, habitat, and migration of insects, birds, and animal life, leading to more pandemics and spread of diseases [5].

Largely overlooked, however, is that climate change and air pollution also influence the prevention, prognosis, diagnosis, treatment, and access to care of cancer patients. The carbon footprint caused by healthcare and the field of oncology is an area that warrants further research.

Air Pollution

Air pollution and climate change are caused by the burning of fossil fuels. Air pollution also has major impacts on human health like resulting in respiratory and cardiac diseases, including asthma, increasing allergies, respiratory infections, and an increase incidence of pregnancy complications [6]. In 2018, it was responsible for approximately 8.7 million deaths worldwide - one in five deaths globally [7] – and has been categorized as a group 1 carcinogen (carcinogenic to humans) by the International Agency for Research on Cancer (IARC) [8]. Air pollution accounts for 14% lung cancer worldwide [7].

Air Pollution and Lung Cancer

Burning of fossil fuels not only emits greenhouse gases and other carcinogenic elements such as nitrogen dioxide and sulfur dioxide, but particulate matter (PM) as well. Fine particles (PM2.5, particles smaller than 2.5 microns, or about 1/30th the width of a human hair) are particularly harmful and have been categorized as a group 1 carcinogen by the IARC. In addition to causing local inflammation and oxidative stress, these particles can penetrate deep into the lung and terminal bronchioles, and from there translocate into the circulation, causing systemic inflammation. These pollutants can be found in ambient air pollution (emitted into the outdoor environment) as well as those primarily emitted in the indoor environment, largely through the burning of methane gas (also known as natural gas) in leaky gas-powered stoves, even when those stoves are not in use. It is estimated that annual methane emissions from all gas stoves in U.S. homes have a climate impact comparable to the annual carbon dioxide emissions of 500,000 cars [9]. In addition to methane emissions, when methane burns, other health damaging air pollutants such as benzene and nitrogen oxides (NOx) are released into home air. It is estimated that families who don't use their range hoods or who have poor ventilation can surpass the 1-hour national standard of NO2 within a few minutes of stove usage, particularly in smaller kitchens [9].

PM2.5 can also come from natural sources, such as wildfires. PM 2.5 from wildfires is up to 10 times more harmful on human health than PM 2.5 from other sources since it is mostly carbonaceous (with 5–20% elemental carbon and at least 50% organic carbon) and has more oxidative potential than ambient urban particulate due to the presence of more polar organic compounds [10]. The estimated mortality and morbidity attributable to smoke plumes in the US is not just in the west – smoke can travel thousands of miles. Although the majority of large landscape fires occur in the western US, one study found that the majority of mortality (74%) and asthma morbidity (about 75%) attributable to smoke PM 2.5 occurs outside the West [11].

Wildfire exposure has been associated with slightly increased incidence of lung cancer and brain tumors. One population based observational cohort study found that people exposed to a wildfire within 50 km of residential locations in the past 10 years had a 4•9% relatively higher incidence [HR] 1•049, 95% CI 1•028–1•071) of lung cancer than unexposed populations, and a 10% relatively higher incidence (adjusted HR 1•100, 1•026–1•179) of brain tumors [12]. In 2022, IARC classified occupational exposure as a firefighter as carcinogenic to humans [13].

The research around the link between air pollution, PM2.5 and lung cancerLung cancer is robust. Numerous epidemiological studies have shown that people exposed to long term air pollution are more likely to die of lung cancer than those who do not. For example, one study looked at 352,053 patients with newly diagnosed lung cancer living in California between 1988- 2009, and after adjusting for histology and other potential confounders, found that HR associated with 1 standard deviation increases in NO2, O3, PM 10, PM 2.5 for patients with localized stage at diagnosis were 1.30 (95% CI: 1.28–1.32), 1.04 (95% CI: 1.02–1.05), 1.26 (95% CI: 1.25–1.28), and 1.38 (95% CI: 1.35–1.41), respectively [14]. After controlling for mortality controlling for age, sex, race, smoking status, body mass, income, education, marital status, rural versus urban, region, and survey year, a significant adverse association of ambient PM2.5 and lung cancer mortality was observed among over 635,000 US National Health Interview Survey participants (HR: 10 mcg/m^3 1.13; 95% CI, 1.00–1.26 [15]. A meta-analysis involving numerous cohort studies over the past 25 years, found the estimated hazard ratio, adjusted for age, sex, and smoking status, was 1·13 (95% CI 1·07–1·20) per 10 μg/m^3 elevation in PM2·5 [16].

Air pollution and particulate matter also affect people who have never smoked. For example, Turner and colleagues performed a Cox proportional hazard regression model adjusting for numerous variables--- smoking, passive smoking, occupational exposures (asbestos, coal dust, diesel engine exhaust etc.), an occupational “dirtiness” index, radon exposure, among others-- and found a dose-response relationship between PM2.5 concentration and lung cancer mortality, with each 10 mg/m increase in PM2.5 concentrations associated with a 15–27% increase in lung cancer mortality [17]. A 2020 study reported in Thorax that patients with COPD who have never smoked are more likely to get lung cancer compared to never-smokers without COPD (HR 2.67, 95% CI [2.09, 3.40]) [18]. An analysis of the data from the National Health Interview Survey by Coleman and colleagues writing in Cancer Causes and Control found lung cancer mortality was adversely associated with increases in PM2.5, in both the overall population as well as in a cohort of over 340,000 never smokers. The risk of all cancer mortality was adversely associated with PM 2.5 (per 10 μg/m^3 increase) in the full cohort (hazard ratio [HR] 1.15, 95% confidence interval [CI] 1.08–1.22) and the never-smokers' cohort (HR 1.19, 95% CI 1.06–1.33) [15].

A study reported in Environmental Health Perspectives also showed that exposure to air pollution increases the incidence and mortality from lung cancer, with lung cancer risk associated with PM2.5 exposure being greatest for former smokers (HR 1.44; 95% CI [1.04, 2.01]), followed by never-smokers (HR 1.18; 95% CI [1.00, 1.39]), and then current smokers (HR 1.06; 95% CI [0.97, 1.15]).

Air pollution also affects patients who already have lung cancer. Air pollution exposures after the diagnosis of lung cancer shortens survival. For example, a 2016 study looking at 352,053 patients with newly diagnosed lung cancer between 1988 - 2009 in California found the median survival for patients with early-stage lung cancer at diagnosis was 2.4 years for those with high PM2.5 exposure ($\geq$ 16 μg/m^3) and 5.7 years for those with low PM 2.5 exposure (< 10 μg/m^3). [14] Adjusting for potential confounders, the HR associated with 1 standard deviation increases in NO2, O3, PM 10, PM 2.5 for patients with localized stage at diagnosis were 1.30 (95% CI: 1.28–1.32), 1.04 (95% CI: 1.02–1.05), 1.26 (95% CI: 1.25–1.28), and 1.38 (95% CI: 1.35–1.41), respectively [14].

Climate Change and Other Cancers

Melanoma

International Agency for Research on Cancer (IARC) of the World Health Organization (WHO) and the National Toxicology Program (NTP) of the United States have deemed ultraviolet radiation to be a human carcinogen [19].

Photocarcinogenesis results when skin cancer develops from the impact of ultraviolet radiation exposure. Non melanoma skin cancer (NMSC) incidence has almost doubled in the Nordic countries and has seen a four-fold rise in Australia since 1960. Cutaneous malignant melanoma (CMM) has seen a three-fold rise in the United States and a 4–5% rise annually, globally. The year 2020 witnessed almost 63,000 deaths worldwide due to non-melanoma skin cancer [20, 21].

Synergism between UVR and heat leads to generation of heat shock proteins resulting in greater survival of DNA-damaged cells. Heat exposure also results in reduction of p53 mediated cell cycle arrest and apoptosis in UVR-exposed cells [22, 23]. A 2 °C (3.6 °F) increase in temperature can lead to an increase of the incidence of skin cancer by 11% globally by 2050 [24].

Stratospheric Ozone Depletion (SOD) can cause increased UVR penetration, and therefore increased incidence of skin cancer. A 1% decrease in ozone layer can result in increase in the incidence of melanoma by 1–2%, squamous cell carcinoma by 2–4.6% and basal cell carcinoma by up to 2.7% [25].

Ozone depleting substances can have long life in the stratosphere, resulting in delayed recovery of ozone depletion over Antarctica. And further, as of 2011, a large hole in the stratospheric ozone layer has developed in the Northern Hemisphere over the Arctic. Such factors combined, can cause up to 60% increase in ultraviolet exposure [26, 27]. The 2022 index for the Antarctic has projected recovery of ozone layer over the region to occur sometime around 2070.

Hydrofluorocarbons have replaced ozone depleting substances; however, they are potent greenhouse gases and have global warming potentials that range from hundreds to thousands of times that of carbon dioxide.

Ambient AP has also been linked with an increased risk of non-melanoma skin cancer, and cutaneous malignant melanoma [28–30] Particles of PM2.5 can penetrate the epidermis and may pass transdermally [31]. According to a study by Vierkotter and colleagues, PM2.5 from traffic emissions can cause skin proliferation and melanogenesis resulting in up to 20% increase in pigmented facial lesions [32]. Black carbon comprises a large fraction of PM2.5 and is a known carcinogen shown to increase risk of bladder, mesothelioma, and melanoma [29, 32, 33].

Breast Cancer

Breast cancer is the most common cancer in the women in the United States. Only up to 10% of breast cancer is deemed to be due to genetic causes leaving almost 90% of breast cancer due to sporadic multifactorial reasons. Needless to say, environmental factors have significant role to play in this disease.

Recent research done by Johnson et al. reflects the rising incidence of invasive breast cancer in the adolescent and young adult (AYA) population, since 2004, with almost 12,000 AYA women having been diagnosed with breast cancer in the United States. This is concerning as two thirds of such diagnoses are regional or distant disease. Additionally, younger women are more likely to present with unfavorable tumor biology, resulting in poorer outcomes. These trends are not related to any genetic changes since this change has occurred over the last 30 years or so, which is too short of a period for a genetic shift [34] and hence warrant further research into environmental etiologies that could be responsible for such a change.

Most studies on environmental factors and cancer, have relied on air pollution exposure evaluation at a single time point in adulthood. Due to long latency of breast cancer, more recent exposures may not have been incorporated in such evaluations [35]. Evidence suggests breast tissue is most susceptible to carcinogenesis during periods like adolescence and pregnancy [36]. Studies of early life exposure have been limited by the lack of available historical criteria air pollution exposures monitoring during critical time periods, although one study reported an increased incidence of breast cancer among women who reported a median/barrier dividing either their primary childhood residential road or the nearest cross-street and among women whose nearest cross-street had the highest traffic, ≥3 lanes, and/or a median/barrier, possible indicators of a higher traffic load [37–39].

The National Institute of Environmental Health Sciences' (NIEHS) Sister Study includes more than 50,000 sisters of women with breast cancer from the U.S. and Puerto Rico. This landmark observational study, evaluating environmental exposures that may affect a woman's chance of developing breast cancer, found that women who live in areas of air pollution with higher levels of lead, mercury, and cadmium have a greater chance of developing postmenopausal breast cancer [40].

NIEHS researchers, studying how environmental exposures can interact with genetic factors to affect breast cancer risk, found that women living in areas with higher levels of air pollution, particularly PM 2.5 and other traffic related pollutants, may have a greater chance of developing breast cancer. This result is based on a large study of about 1900 demographically diverse women. The subgroup in the study who had the highest air pollution exposure were non-Hispanic Black women living in midwestern Urban areas [40, 41].

In a meta-analysis, Gabet et al. found that a 10 microgram/meter cube increase in nitrogen dioxide (NO2), a marker of traffic related air pollution, is associated with a 3% higher risk of breast cancer [42].

Effects of Climate Change on Patients During Pandemics

As climate change leads to changes in temperature and rainfall, it is likely to change the geographical range of vectors and their hosts, leading to more opportunities for people to become exposed to novel vector-borne diseases, leading to an increasing number of pandemics. For example, climate change has shifted the global distribution of bats, increasing their range to southern Chinese Yunnan province and neighboring regions in Myanmar and Laos [43]. COVID-19 is hypothesized to have jumped from bats to humans at a wet market in Wuhan, China.

Although we cannot be certain that the COVID-19 pandemic was due to climate change, it illustrates some of the problems we will face as pandemics become more common and more frequent. For example, in the COVID pandemic cancer patients have worse outcomes, including an increase in rate of infection, complications, and deaths [44]. A large, population based cohort study involving 13,594 patients found that the 30-day in-hospital mortality was higher in patients with solid cancer compared with patients without cancer (31.7% vs 20.0%, respectively; (adjusted OR 1.34; 95% CI 1.13 to 1.58) [45]. Another population-based study reported that people with lung cancer are nearly three times as likely to die of COVID-19 as people without cancer [46, 47]. Early results of the TERAVOLT (Thoracic cancERs international coVid19 cOLlaboraTion) global registry study revealed that of the first 200 patients with COVID-19 and thoracic cancers, 152 (76.0%) patients were hospitalized and 66 (33.3%) died, regardless of cancer type or therapy [48]. Other cancer-related consequences of the COVID-19 pandemic include a delay or decrease in cancer screening, and changes or delays in cancer treatments [49].

Severe Weather Events and Delivery of Care

A systemic review of 85 articles looking at the effects of disaster events on oncology services, survival outcomes, and psychological issues identified major areas of disruption to cancer care: damage to infrastructure, workforce management, loss of medical records and tissue samples, the need for the evacuation of patients and staff, disruption to communication services, and a lack of medications [50]. We discuss the effects on infrastructure, treatment, medical records, medications, and tissue samples below

Infrastructure

Importantly, extreme weather events also have major impacts on the ability of cancer patients to access and receive care. Infrastructure damage due to flooding can result in loss of power, electricity, water, and transportation, resulting in the loss of radiotherapy equipment, clinical facilities, medical records, and access to staff, as evidenced by Hurricane Katrina in 2005 [50]. It is estimated that the prolonged power outage and lack of a disaster management plan were partly responsible for the death of 3052 Puerto Ricans who experienced extended interruptions in access to medical care following Hurricane Maria in 2005 [51], and prompted a compilation of recommendations from radiation oncologists who were in Puerto Rico during the disaster, and from a panel of American Society for Radiation Oncology (ASTRO) [52] [53]. Recommendations included the development of an emergency operations plan tailored to specific circumstances, prospective coordination with other radiation oncology clinics before a disaster, ongoing communications with emergency management organizations, and routine practice of alternate methods to disseminate information among providers and patients.

Other consequences of extreme weather events include the effects on the supply chain. For example, interruption of drug supply chains and consequent shortages, as occurred in 2018 when Hurricane Maria struck Puerto Rico the closure of factories that supplied the bulk of small volume IVs in the continental US [50].

Medical Records and Medications

Damage to communication systems and medical record losses substantially disrupt oncology care [50]. The loss of paper and electronic medical records from either water or collapsed infrastructure were the most common health-care issues reported across all natural disaster events [50]. This is particularly important because in a natural disaster, medications are often lost or left behind and many patients are unaware of their treatment plan and medications. In one study, approximately 32–48% of evacuees did not save their prescription medication and those that did only had enough for a few days [54]. In another, following Hurricane Katrina, the percentage of those without medication ranged between 32% and 48%, and 42.7% had run out of their medications even when they brought them [13, 55].

Delays in Treatment

Multiple studies have documented the impact of extreme events such as hurricanes and flooding on cancer care delivery [50, 56]. However, the effects of treatment interruption on cancer outcomes are often difficult to measure since they may not be apparent for months or years. A recent study published in the Journal of the American Medical Association found that a poorer survival was associated with patients with non-small cell lung cancer receiving definitive radiation therapy during hurricane disasters compared to a matched cohort of patients who underwent treatment in the absence of a hurricane disaster [57].

Effects on Screening

Disruptions to cancer care can lead to a decrease in cancer screening, resulting in a decrease in the number of patients presenting with early-stage cancers. During the first months of the pandemic, many countries suspended cancer screening or classified it as low priority [49]. In the US, a study involving 278,778 patients, including 258,598 (92.8%) from before the pandemic and 20,180 (7.2%) during a three-month period of time during the pandemic, found that the mean weekly number of newly identified patients fell 46.4% (from 4310 to 2310) for 6 cancers. Declines were seen in all six cancer types studied, (breast, colorectal, lung, pancreatic, gastric, and esophageal) ranging from 24.7% for pancreatic cancer (from 271 to 204; $P = .01$) to 51.8% for breast cancer (from 2208 to 1064; $P < .001$) [58]. This delay in diagnosis is expected to result in more patients presenting with advanced disease. Estimates range from 10,000 excess breast or colorectal deaths in the US to 33,890 excess deaths in the US for all cancers [59, 60].

Effects on Research and Clinical Trials

One of the most notable impacts on cancer research occurred when Tropical Storm Allison made landfall in June of 2001, dropping more than 40 inches of rain, and flooding the basement and first floor of McGovern Medical School at UTHealth, where research animals and biological samples were located. Hundreds of thousands of research animals and tumor samples were lost -- more than 4700 research animals, along with filing cabinets of research findings and records, and 25 years of research data, including, for many scientists, their life's work. The gross anatomy lab and storage, as well as the university's cyclotron facility, were also rendered inoperable [61].

Health Care's Contribution to Climate Change

Health care also has an" outsized" impact on the environment [62] The health care industry is among the most carbon-intensive service sectors in the industrialized world [62]. In the US, the health care sector is responsible for about 8.5% of the nation's greenhouse gas emissions: roughly equivalent to the emissions from driving 12 million cars, powering three million homes, or burning 240,000 railcars filled with coal [63]. These emissions come from energy used for heating, cooling, lighting and water; transportation and commuting, anesthetic gases [64], waste disposal, medical devices and equipment, pharmaceuticals, and food procurement, among others.

Cancer care also has an environmental impact. The operating room, for example, accounts for 70% of a hospital's waste and in 2014, the release of hydrofluorocarbon and chlorofluorocarbon anesthetic gases stood at the equivalent of three million tons of CO_2 [65]. A typical palliative course of fractionated external beam radiation therapy usually involves five treatments per week for two to four weeks., requiring daily trips to the hospital [66]. Hypofractionated stereotactic radiation therapy involves 1-5 treatments and would cut down on the emissions associated due to reduced patient/staff travel, total operating energy, and resource utilization [66].

Chemotherapy can harm the environment, either through improper disposal of unused medicines or contamination from manufacturing facilities [67]. It can enter the aquatic environment via human excretion and wastewater treatment facilities [68]. Life cycle analysis are needed to further characterize the activities (Scope 3 emissions) associated with the manufacture and distribution of drugs. Scope 3 emissions (emissions associated with activities not directly related to care delivery, including the supply chain (e.g., food, pharmaceuticals, chemicals, medical devices, and supplies), typically represent about 82% of health care emissions [62].

As office visits fell during the COVID pandemic, telehealth visits soaed, reducing the environmental impact associated with traveling to receive treatment and improving patients' access to cancer care, as having to travel long distances is often associated with missed visits and reduced compliance [69].

Another often unrecognized source of environmental is the impact that large professional society meetings have on the environment. Teleconferencing is associated with reduced travel emissions, as well as increased accessibility for individuals with financial, personal, professional, medical, and/or geographic constraints [66]. One study estimated that if the American Radium Society's Annual meeting in 2021 had transitioned to a virtual platform, 469.4 metric tons of CO_2, the equivalent of the annual emissions of 102.4 passenger vehicles, would have been saved from transportation alone [70]. Other organizations are attempting to minimize the carbon footprint of their International Conferences. The American Cancer Society initiated a series of reforms to decrease their carbon footprint of their 2022 Annual Conference in San Francisco by reducing their use of paper (e.g. making the final program available only as a PDF, decreasing the use of hotel drop bags) and distributing complimentary reusable water bottle [71].

Disparities

As a result of structural racism in the US, Black communities have disproportionately higher exposure to adverse neighborhood-level factors, many of which are linked to lung cancer risk, such as air pollution exposure, traffic proximity, low neighborhood socioeconomic status. For example, data from the Black Women's Health Study, which researched a cohort of approximately 60,000 black women across the United States between a period of 1995–2018, suggested worse neighborhood concentrated disadvantage was associated with increased lung cancer risk in Black women who never smoked leading to disparities [72].

The National Longitudinal Mortality Study analyzed individual-level socioeconomic and racial/ethnic patterns in mortality, incidence, and survival rates from all-cancers combined and major cancers from 1950 to 2014 [73]. Individuals in more deprived areas had higher mortality and incidence rates; particularly for lung, colorectal, cervical, stomach, and liver cancer. Mortality was highest amongst Blacks. Deprived neighborhoods had poorer cancer specific survival [73].

Another study highlighted that fine particulate matter-attributable health risks are unevenly and inequitably distributed across Washington DC. Higher PM2.5 attributable disease burden for lung cancer and other illnesses, were found in neighborhoods with larger proportions of people of color. Hence, pointing towards inequity due to heterogeneity of distribution and exposure to air pollution [74].

A report published in Geophysical Research Letters showed that low income and minority neighborhoods in the US are exposed to 28% more nitrogen dioxide pollution [75]. That is in turn related to the fact that such neighborhoods are located closer to tracking routes and major highways. In this study, satellite captured data tracked N02 emissions over a span of two years in 52 major cities in the US. Newark, Phoenix, Los Angeles had the highest N02 related disparities, surpassing up to 40% amongst neighborhoods. This study not only reinforces the disparities related to pollution but also calls for a need to utilize satellite data and integrate it with demographics to produce meaningful conclusions and strategies to reduce climate related disparities.

The Hope

The most important way to reduce climate change and air pollution is to reduce our dependency on burning of fossil fuels if we have any hope of meeting goals laid out in the 2015 Paris Climate Accords. Fortunately, a number of new technologies are fast starting to replace coal and natural gas, including solar and wind power, heat pumps, hydrogen, nuclear and carbon-removal, enhanced energy storage, direct CO2 capturing, and transition to electric vehicles.

A number of professional health organizations are working to mobilize the health care community. The Medical Society Consortium on Climate and Health, for example, has been hosting their annual meetings since 2018 and now enjoys membership by over 40 societies including American Society of Clinical Oncology

(ASCO), American Cancer Society (ASCO), the International Association for the Study of Lung Cancer (IASLC) and Society of Gynecologic Oncology (SGCO). Affiliates include "Oncologists United for Climate and Health," "Association of American Cancer Institutes," "Massachusetts and Wisconsin Medical Societies" and many more. Thereby the Consortium represents 70% of all U.S. physicians.

Their Mission & Consensus Statement reads as follows, *"The mission of the Consortium is to mobilize and amplify the voices of U.S. doctors, in partnership with public health experts and fellow health professionals, to successfully advocate for equitable climate solutions that protect and promote the health of all people"* [76]. *F*ormation and success of such a Consortium and others underscores the rapidly expanding reach of this cause and it is immensely reassuring to see a robust representation by the physicians and oncologists.

Sustainable building design is another area that is gaining momentum. Many hospitals are setting carbon neutrality goals for the next 10 to 20 years. American Society for Health Care Engineering (ASHE) is providing vision and guidance to healthcare systems for their carbon neutrality and sustainability targets. Health care facilities receive the Energy to Care Award for saving 10% of their energy use in a single year and 15% over two years. ASHE had 9 awardees in 2006 when they started this program and a whopping 255 awardees in the year 2020.

Green buildings will also lower our carbon footprint. Leadership in Energy and Environmental Design (LEED), by the U.S. Green Building Council (USGBC), is the most widely used green building rating system. To achieve LEED certification a project collects points by engaging in carbon neutral practices under energy, water, waste, transportation, materials, health, and indoor environmental quality. Top tier health care facilities have embraced this approach. For example, The Ohio State University Comprehensive Cancer Center's Arthur G. James Cancer Hospital and Richard J. Solove Research Institute is a LEED Gold building [76]. Memorial Sloan Kettering Cancer Center has utilized the concepts of limiting carbon emissions, energy and water use, hazardous, regulated, and solid waste generation, and supply chain impacts with focus on healthy environment and communities.

We also have the capacity of running hybrid or remote conferencing models for the annual conferences, and telemedicine models for running our practices and research. Such practices can reduce enormous carbon footprint. The United Nations Environment Programme has issued a guide for planning large scale events with sustainability, and they have elaborately discussed ways under the categories of, raising awareness, carbon neutrality, transportation, venue and accommodation, catering, material & merchandise and waste management [77].

Conclusions

We have sufficient evidence to highlight the potential ramifications of climate change on the field of oncology and our patients. Studies show that climate change impacts cancer care, particularly on the underprivileged, minorities and those in lower socioeconomic communities. Acknowledging the problem is the first step towards finding its solution. The field of oncology itself has a significant amount of carbon footprint, and it is time that we recognize the climate picture fully, take responsibility and act now.

References

1. Atwoli L, Baqui AH, Benfield T, et al. Call for emergency action to limit global temperature increases, restore biodiversity, and protect health. N Engl J Med. 2021;385:1134–7.
2. Atwoli L, Baqui AH, Benfield T, et al. Call for emergency action to limit global temperature increases, restore biodiversity and protect health. Frontline Gastroenterol. 2022;13:346–8.
3. Cook J, et al. Quantifying the consensus on anthropogenic global warming in the scientific literature. Environ Res Lett; 2013.
4. Watts N, Amann M, Arnell N, et al. The 2020 report of The Lancet Countdown on health and climate change: responding to converging crises. Lancet. 2021;397:129–70.
5. Prevention CCfDCa: Climate Effects on Health; 2022.
6. Perera F, Nadeau K. Climate change, fossil-fuel pollution, and children's health. N Engl J Med. 2022;386:2303–14.
7. Vohra K, Vodonos A, Schwartz J, et al. Global mortality from outdoor fine particle pollution generated by fossil fuel combustion: results from GEOS-Chem. Environ Res. 2021;195:110754.
8. Loomis D, Grosse Y, Lauby-Secretan B, et al. The carcinogenicity of outdoor air pollution. Lancet Oncol. 2013;14:1262–3.
9. Lebel E, Finnegan C, O'uyang Z, et al. Methane and NOx emissions from natural gas stoves, cooktops, and ovens in residential homes. Environ Sci Technol. 2022;56:2529–39.
10. Aguilera R, Corringham T, Gershunov A, et al. Wildfire smoke impacts respiratory health more than fine particles from other sources: observational evidence from Southern California. Nat Commun. 2021;12:1493.
11. O'Dell K, Bilsback K, Ford B, et al. Estimated mortality and morbidity attributable to smoke plumes in the United States: not just a western US problem. Geohealth. 2021;5:e2021GH000457.
12. Korsiak J, Pinault L, Christidis T, et al. Long-term exposure to wildfires and cancer incidence in Canada: a population-based observational cohort study. Lancet Planet Health. 2022;6:e400–9.
13. Demers PA, DeMarini DM, Fent KW, et al. Carcinogenicity of occupational exposure as a firefighter. Lancet Oncol. 2022;23:985–6.
14. Eckel SP, Cockburn M, Shu YH, et al. Air pollution affects lung cancer survival. Thorax. 2016;71:891–8.
15. Coleman NC, Burnett RT, Higbee JD, et al. Cancer mortality risk, fine particulate air pollution, and smoking in a large, representative cohort of US adults. Cancer Causes Control. 2020;31:767–76.
16. Pope CA 3rd, Coleman N, Pond ZA, et al. Fine particulate air pollution and human mortality: 25+ years of cohort studies. Environ Res. 2020;183:108924.
17. Turner MC, Krewski D, Pope CA 3rd, et al. Long-term ambient fine particulate matter air pollution and lung cancer in a large cohort of never-smokers. Am J Respir Crit Care Med. 2011;184:1374–81.
18. Park HY, Kang D, Shin SH, et al. Chronic obstructive pulmonary disease and lung cancer incidence in never smokers: a cohort study. Thorax. 2020;75:506–9.
19. Ultraviolet (UV) radiation. Environ Health, 2000?
20. Ferlay J, Colombet M, Soerjomataram I, et al. Estimating the global cancer incidence and mortality in 2018: GLOBOCAN sources and methods. Int J Cancer. 2019;144:1941–53.
21. Sung H, Ferlay J, Siegel RL, et al. Global Cancer Statistics 2020: GLOBOCAN estimates of incidence and mortality worldwide for 36 cancers in 185 countries. CA Cancer J Clin. 2021;71:209–49.
22. Calapre L, Gray ES, Ziman M. Heat stress: a risk factor for skin carcinogenesis. Cancer Lett. 2013;337:35–40.

23. Calapre L, Gray ES, Kurdykowski S, et al. Heat-mediated reduction of apoptosis in UVB-damaged keratinocytes in vitro and in human skin ex vivo. BMC Dermatol. 2016;16:6.
24. van der Leun JC, de Gruijl FR. Climate change and skin cancer. Photochem Photobiol Sci. 2002;1:324–6.
25. López Figueroa F. Climate change and the thinning of the ozone layer: implications for dermatology. Actas Dermo-Sifiliográficas (English Edition). 2011;102:311–5.
26. Climate N. Path to recovery of ozone layer passes a significant milestone. NOAA Climate.org; 2021.
27. Parker ER. The influence of climate change on skin cancer incidence – a review of the evidence. Int J Women's Dermatology. 2021;7:17–27.
28. Datzmann T, Markevych I, Trautmann F, et al. Outdoor air pollution, green space, and cancer incidence in saxony: a semi-individual cohort study. BMC Public Health. 2018;18:715.
29. Kim KE, Cho D, Park HJ. Air pollution and skin diseases: adverse effects of airborne particulate matter on various skin diseases. Life Sci. 2016;152:126–34.
30. Vogeley C, Esser C, Tüting T, et al. Role of the aryl hydrocarbon receptor in environmentally induced skin aging and skin carcinogenesis. Int J Mol Sci. 2019;20:6005.
31. Parrado C, Mercado-Saenz S, Perez-Davo A, et al. Environmental stressors on skin aging. Mechanistic insights. Front Pharmacol. 2019;10:759.
32. Vierkötter A, Schikowski T, Ranft U, et al. Airborne particle exposure and extrinsic skin aging. J Invest Dermatol. 2010;130:2719–26.
33. Puntoni R, Ceppi M, Gennaro V, et al. Occupational exposure to carbon black and risk of cancer. Cancer Causes Control. 2004;15:511–6.
34. J C: Why are young adult women developing later-stage, more aggressive breast cancer than older women? ASCO Post; 2022.
35. Lynch J, Smith GD. A life course approach to chronic disease epidemiology. Annu Rev Public Health. 2005;26:1–35.
36. Terry MB, Michels KB, Brody JG, et al. Environmental exposures during windows of susceptibility for breast cancer: a framework for prevention research. Breast Cancer Res. 2019;21:96.
37. Bonner MR, Han D, Nie J, et al. Breast cancer risk and exposure in early life to polycyclic aromatic hydrocarbons using total suspended particulates as a proxy measure. Cancer Epidemiol Biomark Prev. 2005;14:53–60.
38. Nie J, Beyea J, Bonner MR, et al. Exposure to traffic emissions throughout life and risk of breast cancer: the Western New York Exposures and Breast Cancer (WEB) study. Cancer Causes Control. 2007;18:947–55.
39. Shmuel S, White AJ, Sandler DP. Residential exposure to vehicular traffic-related air pollution during childhood and breast cancer risk. Environ Res. 2017;159:257–63.
40. White AJ, O'Brien KM, Niehoff NM, et al. Metallic air pollutants and breast cancer risk in a Nationwide Cohort Study. Epidemiology. 2019;30:20–8.
41. Program NT: 15th report on carcinogens; 2021.
42. Gabet S, Lemarchand C, Guenel P, et al. Breast Cancer Risk in Association with Atmospheric Pollution Exposure: a meta-analysis of effect estimates followed by a health impact assessment. Environ Health Perspect. 2021;129:57012.
43. Beyer RM, Manica A, Mora C. Shifts in global bat diversity suggest a possible role of climate change in the emergence of SARS-CoV-1 and SARS-CoV-2. Sci Total Environ. 2021;767:145413.
44. Kuderer NM, Choueiri TK, Shah DP, et al. Clinical impact of COVID-19 on patients with cancer (CCC19): a cohort study. Lancet. 2020;395:1907.
45. de Azambuja E, Brandão M, Wildiers H, et al. Impact of solid cancer on inhospital mortality overall and among different subgroups of patients with COVID-19: a nationwide, population-based analysis. ESMO Open. 2020;5.
46. Dai M, Liu D, Liu M, et al. Patients with cancer appear more vulnerable to SARS-COV-2: a multi-center study during the COVID-19 outbreak. Cancer Discov; 2020.
47. Yu J, Ouyang W, Chua MLK, et al. SARS-CoV-2 transmission in patients with cancer at a Tertiary Care Hospital in Wuhan, China. JAMA Oncol; 2020.

48. Garassino MC, Whisenant JG, Huang LC, et al. COVID-19 in patients with thoracic malignancies (TERAVOLT): first results of an international, registry-based, cohort study. Lancet Oncol. 2020;21:914.
49. Richards M, Anderson M, Carter P, et al. The impact of the COVID-19 pandemic on cancer care. Nat Cancer. 2020;1:565–7.
50. Man RX, Lack DA, Wyatt CE, et al. The effect of natural disasters on cancer care: a systematic review. Lancet Oncol. 2018;19:e482–99.
51. Rodríguez-Madera SL, Varas-Díaz N, Padilla M, et al. The impact of Hurricane Maria on Puerto Rico's health system: post-disaster perceptions and experiences of health care providers and administrators. Glob Health Res Policy. 2021;6:44.
52. Gay HA, Santiago R, Gil B, et al. Lessons learned from Hurricane Maria in Puerto Rico: practical measures to mitigate the impact of a Catastrophic Natural Disaster on Radiation Oncology Patients. Pract Radiat Oncol. 2019;9:305–21.
53. Lopez-Araujo J, Burnett OL 3rd. Letter from Puerto Rico: the state of radiation oncology after Maria's landfall. Int J Radiat Oncol Biol Phys. 2017;99:1071–2.
54. Tomio J, Sato H, Mizumura H. Interruption of medication among outpatients with chronic conditions after a flood. Prehosp Disaster Med. 2010;25:42–50.
55. Ochi S, Hodgson S, Landeg O, et al. Disaster-driven evacuation and medication loss: a systematic literature review. PLoS Curr. 2014;6.
56. Grew D, Vatner R, DeWyngaert K, et al. The impact of superstorm Sandy on the care of radiation oncology patients. Int J Radiat Oncol Biol Phys. 2013;87:S490–1.
57. Nogueira LM, Sahar L, Efstathiou JA, et al. Association between declared hurricane disasters and survival of patients with lung cancer undergoing radiation treatment. JAMA. 2019;322:269–71.
58. Kaufman H, Chen Z, Niles J, et al. Changes in the number of US patients with newly identified cancer before and during the coronavirus disease 2019 (COVID-19) Pandemic. JAMA Netw Open. 2020;3:e2017267.
59. Sharpless NE. COVID-19 and Cancer. Science. 2020;368:1290.
60. Lai A, Pasea L, Banerjee A. Estimating excess mortality in people with cancer and multimorbidity in the COVID-19 emergency. ResearchGate; 2020.
61. Harper F. Remembering Tropical Storm Allison 20 years later: a lesson in resiliency. UT Health Houston News; 2021.
62. Eckelman MJ. Health care pollution and public health damage in the United States: an update. Health Aff. 2020;39:2071–9.
63. Quinlan D. Health care and climate change: an opportunity for tansformative leadership. Health Care without Harm; 2014.
64. Wittenberg A. Hospitals take on greening the planet. Greenwire; 2022.
65. Agbafe V, Berlin N, O'ffodile A. Prescriptions for mitigating climate change–related externalities in cancer care: a surgeon's perspective. J Clin Oncol. 2022;40:1976–9.
66. Lichter KE, Anderson J, Sim AJ, et al. Transitioning to environmentally sustainable, climate-smart radiation oncology care. Int J Radiat Oncol Biol Phys. 2022;113:915–24.
67. Journal P. Chemotherapy is harming the environment: it is time for regulators, manufacturers and pharmacy to take action. Pharm J. 2021;306.
68. Wormington A, Martynluk C. Antineoplastic agents: environmental prevalence and adverse outcomes in aquatic organisms. Environ Toxicol Chem. 2020;39:967.
69. Nogueira LM, Yabroff KR, Bernstein A. Climate change and cancer. CA Cancer J Clin. 2020;70:239.
70. Lichter KE, Drew T, Demeulenaere S, et al. Environmental outcomes associated with transition from in-person to a virtual oncology conference during the COVID-19 pandemic. JAMA Oncol. 2022;8:1351–2.
71. News AC. Go green with ATS. ATS Conference News; 2022.
72. Erhunmwunsee L, Wing SE, Zou X, et al. Neighborhood disadvantage and lung cancer risk in a national cohort of never smoking Black women. Lung Cancer. 2022;173:21–7.

73. Singh GK, Jemal A. Socioeconomic and racial/ethnic disparities in cancer mortality, incidence, and survival in the United States, 1950–2014: Over six decades of changing patterns and widening inequalities. J Environ Public Health. 2017;2017:2819372.
74. Castillo MD, Kinney PL, Southerland V, et al. Estimating intra-urban inequities in PM2.5-attributable health impacts: a case study for Washington, DC. GeoHealth. 2021;5:e2021GH000431.
75. Demetillo MAG, Harkins C, McDonald BC, et al. Space-based observational constraints on NO2 air pollution inequality from diesel traffic in major US cities. Geophys Res Lett. 2021;48:e2021GL094333.
76. Consortium MS, 2021 Annu Rep; 2021.
77. Program UNE: SUSTAINABLE EVENTS GUIDE; 2012.

Pesticides and Cancer

8

Taehyun Roh, Anisha Aggarwal, Nishat Tasnim Hasan, Alka Upadhyay, and Nusrat Fahmida Trisha

Pesticides are chemicals used to control weeds, insects, fungi, rodents, and other pests in various settings, including agriculture. While necessary for agricultural productivity, pesticides have been associated with detrimental effects on the environment and human health. One of the potential health risks of pesticide exposure is the development of cancer. There is growing evidence linking pesticide exposure to various types of cancer, with evidence suggesting that any classes of pesticides can be a risk factor. The impact of exposure to pesticides on cancer incidence has been a subject of interest for many years, and a significant body of research has accumulated in this field. This chapter provides an overview of the relationship between pesticide exposure and cancer, discussing the potential mechanisms of carcinogenicity and identifying specific pesticides and cancer types that have been linked in epidemiological studies. The goal of this review is to provide a comprehensive understanding of the association between pesticide exposure and cancer and to reduce the risks associated with pesticide exposure.

Sales and Usage of Pesticides

According to the US EPA, global annual pesticide use is approximately 6 billion pounds [1] (Fig. 8.1). Between 2008 and 2012, herbicides were used the most, accounting for 50% of annual use, followed by fumigants, insecticides, and fungicides. In the US, around 1.1 billion pounds of pesticides are used each year. This

T. Roh (✉) · N. T. Hasan · A. Upadhyay · N. F. Trisha
Department of Epidemiology and Biostatistics, School of Public Health, Texas A&M University, College Station, TX, USA
e-mail: taehyunroh@tamu.edu

A. Aggarwal
Department of Health Behavior, School of Public Health, Texas A&M University, College Station, TX, USA

E. H. Bernicker (ed.), *Environmental Oncology*,
https://doi.org/10.1007/978-3-031-33750-5_8

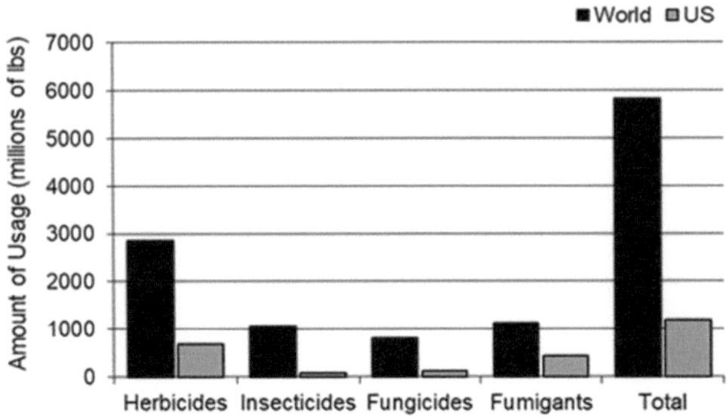

Fig. 8.1 Amounts of pesticide usage by pesticide type in the US and globally (reproduced from [1])

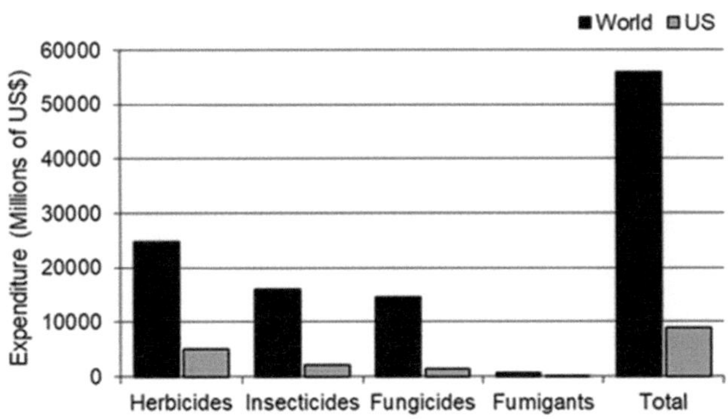

Fig. 8.2 Expenditure of pesticide by pesticide type in the US and globally (reproduced from [1])

accounts for 23% of all global pesticide use in pounds, 25% of the total pounds of herbicides, 43% of total pounds of fumigants, 12% of fungicides, and 6% of insecticides.

In 2012, global spending on pesticides was approximately 56 billion dollars [1] (Fig. 8.2). Herbicides consistently made up the largest share of overall expenditure between 2008 and 2012, accounting for 45% annually, followed by insecticides, fungicides, and other pesticides. In the US, about 9 billion dollars were spent on pesticides in 2012, which represents 16–18% of the world's total expenditure between 2008 and 2012. In 2012, US spending constituted 21% of expenditure on herbicides, 14% on insecticides, 10% on fungicides, and 23% on fumigants.

Glyphosate, atrazine, metolachlor, dichloropropene, and 2,4-dichlorophenoxyacetic acid (2,4-D) are the most widely used pesticides in agriculture, whereas 2,4-D, glyphosate, methylchlorophenoxypropionic acid (MCPP), pendimethalin, and carbaryl are the most frequently used pesticides in residential settings [1] (Table 8.1).

Table 8.1 Top ten most widely used pesticides for agricultural and residential purposes in the US in 2012 (amount in million pounds)

	Agricultural			Residential		
Rank	Active ingredient	Type	Amount	Active ingredient	Type	Amount
1	Glyphosate	H	270–290	2,4-D	H	270–290
2	Atrazine	H	64–74	Glyphosate	H	64–74
3	Metolachlor-S	H	34–44	MCPP	H	34–44
4	Dichloropropene	Fum	32–42	Pendimethalin	Fum	32–42
5	2,4-D	H	30–40	Carbaryl	H	30–40
6	Metam	Fum	30–40	Acephate	Fum	30–40
7	Acetochlor	H	28–38	Pyrethroids	H	28–38
8	Metam Potassium	Fum	16–26	Dicamba	Fum	16–26
9	Chloropicrin	Fum	8–18	MCPA	Fum	8–18
10	Chlorothalonil	F	6–16	Malathion	F	6–16

Note: H, herbicide; Fum, fumigant; F, fungicide
Source: Atwood and Paisley-Jones et al. [1]

Routes of Pesticide Exposure

Exposure to pesticides mainly occurs through dermal, oral, and respiratory routes [2]. Humans are exposed to pesticides directly or indirectly. Direct exposure happens during occupational, agricultural, and residential activities that involve the manufacturing, transportation, mixing, and application of pesticides [3]. Indirect exposure can occur through environmental media such as soil, air, drinking water, and diet [4]. Pesticides can be transferred from the treated area to other environmental media, such as soil, air, and water, through adsorption, spray drift, volatilization, runoff, and leaching. People living close to agricultural lands might be exposed to pesticides through spray drift and volatilization [5]. High levels of pesticides were found in carpet and other indoor dust samples in homes in agricultural regions [6, 7]. Multiple studies have shown that the levels of pesticides and their metabolites in biological samples, such as urine and blood, increased with the proximity of residences to agricultural land [8, 9]. The annual report from the US Department of Agriculture Pesticide Data Program indicated that 76% of agricultural products had detectable levels of pesticide residues in 2021 [10]. These pesticide residues can be ingested by consumers through the consumption of fruits, vegetables, grains, and other agricultural products. Pesticides also contaminate surface and ground water through runoff and leaching from target sites. A recent study by the US Geological Survey (USGS) found that pesticide compounds were detected in 41% of public wells [11]. Private well users are at higher risk because private wells are not regulated, and owners are responsible for the quality of water. The USGS tested 2000 wells between 1991 and 2004 in the US and found that around 90% of domestic wells had detectable levels of pesticides [12]. A recent USGS study conducted in the San Joaquin Valley in California revealed that a fumigant, 1,2-dichloropropane and its metabolites, and herbicides such as atrazine, simazine, hexazinone, metolachlor, and prometon were detected in domestic or small-system wells [13].

Current Carcinogenicity Classifications of Pesticides

Identification of carcinogens is essential for determining the causes of cancer, primary prevention, and surveillance of exposed workers and the general population. Multiple national and international authorities examine the evidence to determine whether a certain chemical has the potential to cause cancer. These authorities include the World Health Organization's International Agency for Research on Cancer (IARC), the US Environmental Protection Agency (EPA), and the California EPA [14–16].

WHO IARC

The IARC, a part of the World Health Organization (WHO), has formed an advisory committee of expert groups to perform comprehensive reviews on the carcinogenicity of chemicals and has been issuing Monographs since 1971. IARC assesses carcinogenic hazards based on the qualitative assessment of the evidence in animals and humans. The IARC Working Group divides agents, mixtures, and exposures into four classes: Group 1, Group 2A, Group 2B, and Group 3. The list and classifications are periodically updated as the IARC Monograph Program is ongoing. The classification is based on the strength of evidence for carcinogenicity in humans and experimental animals, as well as mechanistic information. IARC's evaluation aims to identify and inform carcinogenicity hazards of chemicals, rather than providing recommendations on regulatory or legislative decisions. The classification of carcinogens by IARC is as follows (Table 8.2):

- Group 1 - Carcinogenic to humans.
- Group 2A - Probably carcinogenic to humans.
- Group 2B - Possibly carcinogenic to humans.
- Group 3 - Not classifiable as to its carcinogenicity to humans.

US EPA

Since the issuance of the first set of principles in 1976, the EPA has updated its guidance to evaluate the carcinogenic potential in humans. In the 1986 guidelines, chemicals were classified into six categories based on their carcinogenic potential in humans (A, B1, B2, C, D, and E). The categories were as follows:

- Group A - Human carcinogen.
- Group B - Probable human carcinogen.
- Group C - Possible human carcinogen.
- Group D - Not classifiable as to human carcinogenicity.
- Group E - Evidence of non-carcinogenicity for humans.

In 1996, the EPA released the "Proposed Guidelines for Carcinogen Risk Assessment." These guidelines emphasized the importance of understanding the mode of action of chemicals in inducing cancers and other biological considerations

Table 8.2 WHO IARC carcinogenicity classification of pesticide

Class	Evidence in human	Evidence in animal	Example pesticides
Group 1 Human carcinogen	Sufficient	Sufficient OR Less than sufficient	Arsenic and arsenic compounds Ethylene oxide Lindane TCDD
Group 2A Probable human carcinogen	Limited	Sufficient	Captafol; Diazinon; DDT Ethylene dibromide Glyphosate Malathion Nonarsenical insecticides
Group 2B Possible human carcinogen	Limited	Less than sufficient	Chlorophenoxy herbicides
	Inadequate	Sufficient	2,4-D; Aramite; Chlordane; Chlordecone; Chlorothalonil; 1,2-Dibromo-3-chloropropane (DBCP) para-Dichlorobenzene Dichlorvos Hexachlorobenzene Hexachlorocyclohexanes Mirex; Nitrofen Sodium ortho-phenylphenate Toxaphene (polychlorinated camphenes) Parathion; Tetrachlorvinphos
Group 3 Not classifiable as human carcinogen	Inadequate	Less than sufficient	Permethrin; Trifluralin; Others

Source: [17]

to decrease the uncertainty in describing the potential for harm. The interim final guidelines, published in 1999, addressed the risk to children and other concerns regarding the quantity and quality of data on a chemical. The EPA's final Guidelines for Carcinogen Risk Assessment, released in 2005, included the revised categories as follows (Table 8.3):

- Carcinogenic to human.
- Likely to be carcinogenic to humans.
- Suggestive evidence of carcinogenic potential.
- Inadequate information to assess carcinogenic potential.
- Not likely to be carcinogenic to humans.

California EPA

In California, the Office of Environmental Health Hazard Assessment (OEHHA) within the California EPA releases a list of chemicals known to cause cancer, birth defects, or other reproductive harm to warn people living in California, as required by the Safe Drinking Water and Toxic Enforcement Act of 1986, also known as Proposition 65. The Carcinogen Identification Committee (CIC), which

Table 8.3 US EPA carcinogenicity classification of pesticide

Class 1986	2005	Evidence in human	Evidence in animal	Example pesticides
Group A Human carcinogen	Carcinogenic to humans	Sufficient		Arsenic compounds
Group B Probable human carcinogen	Likely to be carcinogenic to humans	Inadequate OR Limited	Sufficient	Alachlor; Aviglycine Benalaxyl; Carbaryl Epoxiconazole; Pirimicarb
Group C Possible human carcinogen	Suggestive evidence of carcinogenic potential	Limited	Less than Sufficient	Dichlorvos; Lindane Malathion; Parathion Pendimethalin; Permethrin Tetrachlorvinphos Trifluralin
Group D Not classifiable as human carcinogen	Inadequate data to assess carcinogenic potential	Inadequate	Inadequate	Azafenidin; Cyhexatin Dazomet; Ethephon
Group E Non-carcinogenicity for humans	Not likely to be carcinogenic to humans	Data robust for declining hazard		Acequinocyl; Bensulide Chlorpyrifos; Diazinon Glyphosate; Metolachlor

Source: [18]

meets at least once a year, serves as the state's qualified experts to evaluate evidence of carcinogenicity based on a comprehensive literature review. The list of Proposition 65 is updated based on the CIC's decisions. The carcinogens identified by IARC should also be included on Proposition 65's list of substances known to cause cancer. Proposition 65 is enforced by the California Attorney General's Office, and products sold in California containing any of the compounds on the list must carry warning labels. The list includes more than 100 pesticides, such as alachlor, carbaryl, dieldrin, dichlorodiphenyltrichloroethane (DDT), glyphosate, lindane, malathion, parathion, tetrachlorvinphos, among others.

Evidence of Cancer Risks of Pesticide Exposure by Cancer Type

Lung Cancer

Globally, the incidence rate of lung cancer is 22.4 per 100,000 people, and the mortality rate is 18.0 per 100,000 [19]. In the United States, as of 2019, the incidence rate was 52.0 new cases per 100,000 people, and the mortality rate was 35.0 deaths per 100,000 people [20]. Lung cancer is primarily caused by tobacco smoking, with 80–90% of all cases being attributable to smoking [21]. Other risk factors for lung cancer include air pollution, radon, arsenic, and pesticides, in addition to smoking [22].

Alavanja and Bonner reported significant associations between the development of lung cancer and various insecticides (such as chlorpyrifos, diazinon, and dieldrin) and herbicides (such as metolachlor, pendimethalin, and dicamba) after conducting a systematic review [23]. A meta-analysis of cohort studies showed that chlorophenol and related phenoxyacetic acids increased the risk of lung cancer mortality [24]. In a French study, pea and field bean growers who used pesticides were highly affected by small cell lung carcinoma (Hazard Ratio (HR) 2.38, 95% Confidence Interval (CI) 1.07–5.28) [25]. A cohort study based on data from the Agricultural Health Study (AHS) found that longer lifetime exposure days to diazinon, an organophosphate insecticide, significantly increased the risk of developing lung cancer [26]. Another AHS-based study conducted from 1999–2011 found a distinct increase in the risk of lung cancer among ever-users of acetochlor (Relative Risk (RR) 1.74, 95% CI 1.07–2.84) [27, 28]. This study also reported that acetochlor-atrazine co-exposure potentiated lung carcinogenicity (RR 2.33, 95% CI 1.30–4.17). Bonner et al. provided evidence of a link between exposure to pendimethalin, dieldrin, parathion, and chlorimuron ethyl and the increased risk of lung cancer [29]. A Korean cohort study showed that occupational pesticide exposure is closely related to lung cancer, with a strong impact of the duration of pesticide use (20 years or more) (HR 2.43, 95% CI 1.05–5.65) [30]. Additionally, their propensity score model added more depth and reliability to the analysis, showing an extremely significant risk (HR 2.84, 95% CI 1.81–4.46). Spanish study found that people who lived in areas with high pesticide use had a higher risk of developing lung cancer than those who lived in areas with low pesticide use (Odds Ratio (OR) 1.21, 95% CI 1.15–1.28) [31].

Previous research has suggested that exposure to certain types of pesticides, such as organochlorine, organophosphate, and carbamate pesticides, may inhibit the activity of acetylcholinesterase, leading to elevated levels of acetylcholine in the body and increased proliferation of lung cancer cells [32–34]. In addition, exposure to pesticides has been shown to increase levels of reactive oxygen species in lung tissue, which can damage DNA and contribute to the development of cancer [35]. Studies have also found that certain organophosphate pesticides, such as monocrotophos and chlorpyrifos, can induce oxidative stress and DNA damage, and alter signal transduction pathways, which may promote the growth of lung cancer cells [36].

Breast Cancer

Breast cancer is a disease in which cells in the breast proliferate abnormally and continue to grow abnormally even after the stimuli are removed. There were 2,088,849 breast cancer incidences and 626,679 deaths in 2018 globally [37]. The International Agency for Research on Cancer (IARC) classifies certain organophosphates (glyphosate and malathion) as possibly carcinogenic to humans [17].

The Copenhagen City Heart Study showed a dose-response relationship between dieldrin exposure and the risk of breast cancer from a 17-year follow-up study (OR

2.05, 95% CI 1.17–3.57) [38]. Ferro et al. also reported the risk of developing breast cancer elevated by exposure to multiple types of pesticides in the classes of organochlorine, organophosphates, and phenoxy herbicides [39]. Based on the data from Chile between 1995 and 2005, women with more malathion exposure were 5.7 times more likely to be diagnosed with breast cancer, and 30.5% had metastases [40]. In another case-control study in Tunisia [41], a positive dose-response relationship was found between the risk of breast cancer and exposure to p,p′-DDE and heptachlor. Cohn et al. compared the effect of DDT on breast cancer between premenopausal and postmenopausal women [42]. Results showed that DDT was associated with early postmenopausal breast cancer risk. However, it was only linked to the risk of premenopausal breast cancer for women first exposed during infancy through puberty. Tayour et al. reported that exposure to chlorpyrifos was associated with a three times higher risk of breast cancer among postmenopausal women in California Central Valley (OR 3.22, 95% CI 1.38–7.53) [43].

In Ethiopia, a case-control study showed higher concentrations of organochlorine pesticides in the serum of breast cancer patients than controls, with p,p'-DDT, and gamma-chlordane being significant predictors (OR 2.03, 95% CI 1.04–3.97 and OR 3.12, 95% CI 1.19–8.20, respectively) [44]. Franke et al. reported a significant association between urinary glyphosate metabolite, aminomethyl phosphonic acid (AMPA) levels, and the risk of breast cancer [45]. Another biomonitoring study in Egypt revealed that breast cancer tissues contained organochlorines such as hexachlorobenzene, chlordane, methoxychlor, and DDT at higher levels than normal tissues [46]. In India, subjects with detectable serum beta-hexachlorocyclohexane showed an increased risk of breast cancer among young women (OR 1.18, 95%CI 1.05–1.34) [47].

These pesticides can interact with the estrogen receptor, which can mimic the effect of estrogen on breast tissue [48]. An animal study on rats showed that another widely used pesticide, chlorpyrifos affects mammary gland structure and hormonal balance with chronic exposure. Laboratory studies also showed that malathion and chlorpyrifos have cancer-promoting properties and estrogenic potential [49]. A proteomics study showed that women exposed to pesticides had downregulation of immune response, coagulation, and estrogen-mediated events, affecting the levels of tumor necrosis factor-alpha and interleukin-1 beta [50]. Cardona and Rudel reviewed the pesticides with EPA Reregistration Eligibility Decisions (REDs) documents. They found that 35 pesticides, including alachlor, atrazine, and chlordane, affect the mammary gland by causing mammary tumors or other effects [51].

Prostate Cancer

Prostate cancer is the most common malignancy among aging men. The US has an incidence rate of about 650.7 per 100,000 men over the age of 65 having prostate cancer, compared to 3.9 per 100,000 men under age 50 [20]. The etiology for developing prostate cancer is multifactorial. These include the age of the individual, environmental exposure to metals, personal/family history of cancer, and tobacco smoking [52]. Multiple studies suggested that pesticides and their derivatives have androgenic and estrogenic effects, disrupting the normal physiology of endogenous hormones and promoting cancers [53].

Settimi et al. evaluated 124 Italian cases of prostate cancer and 659 controls and found an association between ever being employed in agriculture and a 40% increased risk of developing prostate cancer (OR 1.4, 95% CI 0.9–2.0) [54]. Specifically, an increased risk was found when they were exposed to acaricides and organochlorine insecticides (OR 2.5, 95% CI 1.4–4.2), in particular dicofol (OR 2.8, 95% CI 1.5–5.0) and DDT (OR 2.1, 95% CI 1.2–3.8). Multigner et al. report a significantly increased risk of developing prostate cancer among individuals with increased plasma concentration of chlordecone (OR 1.77, 95% CI 1.21–2.58) [55]. Another case-control study reported a significant association with exposure to DDT, lindane, and simazine [56]. Abhishek et al. conducted a case-control study among 102 prostate cancer patients and found higher levels of malathion, fenvalerate, beta-hexachlorocyclohexane (β-HCH), and chlorpyrifos were found in the prostate cancer patients [57]. Especially a significant correlation between malathion and β-HCH was found with a higher grade of prostate carcinoma.

The Agricultural Health Study data revealed statistically significant associations with organophosphate insecticides (malathion, fonofos, terbufos, and dimethoate) among pesticide applicators [58, 59]. In addition, they found an organochlorine insecticide, aldrin, was found to be associated with an increased risk of developing aggressive prostate carcinoma (RR 1.49, 95% CI 1.03–2.18). Cockburn et al. also found an increased risk of prostate carcinoma among individuals exposed to a group of organochlorines (OR 1.64, 95% CI 1.02–2.63) and methyl bromide (OR 1.62, 95% CI 1.02–2.59) in California Central Valley (2005–2006) [60]. Recent systemic reviews presented that among pesticide applicators, significantly positive associations were found between increased risks of prostate cancers and exposure to pesticides, including fonofos, methyl bromide, chlorpyrifos, terbufos, butylate, malathion, atrazine, lindane, and DDT [61, 62].

Methyl bromide was reported to have a genotoxic potential. Studies in rats and mice showed that the methyl group in methyl bromide covalently binds to DNA, creating DNA adducts [63, 64]. During DNA replication, O6-methylguanine can pair with thymine and cytosine, resulting in a G:C to A:T transition mutation. These mutations represent the early stages of prostate carcinogenesis [65]. Some organochlorines may induce overexpression of erbB-2, an oncogene implicated in prostate cancer, causing prostate cancer cell proliferation [60, 66].

Kidney Cancer

According to GLOBOCAN data, there are about 403,000 new cases of kidney cancer every year, accounting for 2.2% of all cancer cases [67]. In the US, the number of new cases of kidney cancer is 73,750, and the number of deaths is 14,830, accounting for 2.4% of all cancer-related deaths [20]. Modifiable risk factors include obesity, diet, lack of physical activity, smoking, alcohol drinking, and environmental and occupational exposure [68, 69]. According to the IARC, 2,3,7,8-tetrachlorodibenzo-p-dioxin (TCDD) is a human carcinogen [17].

Agricultural workers who directly handle pesticides, fertilizers, soil treatment, and other chemicals are at potential risk of developing kidney cancer [70]. A recent study based on the Agricultural Health Study (AHS) found that people who used 2,4,5-T, chlorpyrifos, chlordane, atrazine, cyanazine, and paraquat had a significantly higher risk of renal cell carcinoma (RCC) than people who never used these chemicals [70]. On the other hand, some herbicides, such as glyphosate, pendimethalin, and dicamba, were not statistically significant [70]. Another AHS-based cohort study found that the risks of kidney cancer were higher in the top tertile of diazinon exposure (lifetime exposure days >108.8 exposure days; RR 2.56, 95% CI 1.01–6.47) [26]. A Canadian case-control study found that exposure to herbicides, pesticides, arsenic, benzidine, creosote, cadmium, asbestos, vinyl chloride, and other chemicals at home or work increased the risk of kidney cancer [71]. Recent studies have investigated the effects of pesticides on Wilms' tumors, the most common type of kidney cancer in children. A previous study indicated that pesticides such as pyrethroids and malathion cross the fetal placental barrier, resulting in fetal exposure [72]. Moreover, combined exposure to insecticides during pregnancy has increased the risk (OR 2.6, 95% CI 1.2–5.7) of developing Wilm's tumor in offspring compared to exposure to individual pesticides (OR 1.6, 95% CI 1.1–2.4) [72].

The proposed mechanisms of developing kidney cancer include binding and activation of the aryl hydrocarbon (AhR) receptor, endocrine disruption, activation of the hypothalamic–pituitary–adrenal axis, nephrotoxicity, DNA damage, and mutation [70]. For example, an experimental in vivo study showed that atrazine causes oxidative stress and epigenetic changes, leading to renal carcinoma in rats [73, 74]. A meta-analysis of case-control studies from the United States, Asia, and Europe concluded that polymorphisms in the glutathione S-transferases genes (GSTM1 and GSTT1) are responsible for RCC [75].

Bladder Cancer

Urinary bladder cancer is the tenth most common carcinoma worldwide [76]. According to the GLOBACON data, 3% of all global cancer diagnoses, especially those prevalent in developed nations, are urinary bladder cancer [76]. The disease incidence is four times more among men than women, with 90% of all bladder cancer diagnoses made among individuals above 55 [77]. Previous literature reports an increased relative risk of developing bladder cancer with environmental and occupational exposure to chemicals such as polycyclic aromatic hydrocarbons, aromatic amines, and pesticides [78].

Letasiova et al. reported that environmental exposure to 2-chloroaniline and aromatic amines (2-naphthylamine, 4-aminobiphenyl, and benzidine) was a notable risk for the development of bladder cancer [79]. These compounds are included in fungicides, dyes, rubber industries, and motor vehicle exhausts [79]. Sharma et al. reported that bladder cancer patients in Delhi, India had significantly higher levels of organochlorines such as HCH and DDT in their blood [80]. Boada et al. also observed that the risk of bladder cancer was associated with exposure to organochlorine pesticides including atrazine, aldrin, DDT, lindane, and dieldrin in the Canary Islands [81]. A cohort study in Iowa and North Carolina, based on the Agricultural Health Study, showed pesticide applicators had higher risks of developing bladder cancer with exposure to an aromatic amine and imidazolinone herbicides, imazethapyr and imazaquin [82]. They also found an increased risk of cancer from exposures to organochlorine insecticides and chlorophenoxy herbicides.

Liang et al. treated Wistar rats with an organophosphate pesticide, chlorpyrifos, for 90 days and found a time- and concentration-dependent cytotoxicity and a more condensed shape of bladder T24 cells [83]. Matic et al. reported that the risk of bladder cancer from pesticide exposure is attributable to the genetic polymorphism in glutathione S-transferase (GST), which plays a role in the detoxification of pesticides and other exposures [84]. Sharma et al. also found significant interactions between GST genotypes and β-HCH, presenting the gene-environment interaction as a potential mechanism of urinary bladder cancer [80].

Pancreatic Cancer

Pancreatic cancer is the 14th most common carcinoma in developed parts of the world, affecting every 1 in 60 of the population over their lifetime in this cohort [85]. The incidence of pancreatic adenocarcinoma has more than doubled worldwide between 1990 and 2017 [86]. Previous research reported that medical history of type 2 diabetes, chronic pancreatitis, hepatitis B and C, obesity, and tobacco smoking are potential risk factors for developing pancreatic adenocarcinoma [87, 88]. Environmental pesticide exposure has also been suggested as a risk factor [87].

Previous epidemiological studies consistently report a positive association between exposure to organochlorine pesticides and pancreatic cancer. Porta et al. conducted an extensive prospective case-control study evaluating 513 pancreatic cancer cases and 1020 matched controls and found associations with exposure to p, p'-DDT, β-hexachlorocyclohexane, trans-nonachlor, the sum of 6 organochlorine pesticides [89]. Clary et al. reported higher mortality of pancreatic carcinoma in areas with zip codes with a higher 1,3-dichloropropene application rate among California's agricultural counties [90]. Similar results were observed in a study utilizing the USGS National Water Quality Assessment Projects that significantly increased death rates were observed in the area with high levels of 1,3-dichloropropene in water [91]. However, Fritsci et al. reported no association between occupational exposure to N-nitrosamine compounds via pesticides, especially herbicides, and the risk of developing pancreatic cancer (OR 0.85, 95% CI 0.51–1.42) [92].

Pesticides may trigger metabolic disorders such as diabetes mellitus and obesity, leading to disturbances of hepatic energy storage, bowel absorption, energy and adipose tissue, and skeletal muscle regulation [93]. Pesticides directly cause inflammation of the pancreas via acinar cell injury in the pancreatic cells [87]. Pesticides also disrupt the endocrine metabolism and cell-protective activities, induce insulin resistance and impair mitochondrial functions, increasing cell death and oxidative stress [87]. For example, ROS are essential for oncogenic KRAS-mediated tumorigenicity in the pancreas [94].

Non-Hodgkin Lymphoma

Non-Hodgkin lymphoma (NHL) accounts for 4% of all new cancer cases in the US [95]. The risk factors associated with non-Hodgkin lymphoma include age older than 60, males, family history, radiation, and exposure to pesticides and herbicides [96]. Non-Hodgkin lymphoma arises from B cell, T cell, or natural killer cell when the proto-oncogenes are activated, or tumor suppressor genes are inactivated by chromosomal translocation or mutation/deletion [97, 98].

There has been evidence showing positive associations between pesticides and non-Hodgkin lymphoma. A case-control study based in Iowa and Minnesota found that male subjects involved in farming were at an elevated risk of non-Hodgkin's lymphoma (OR 1.2, 95% CI 1.0–1.5) [99]. In Germany, a case-control study reported that the odds of lymphoma in children exposed to pesticides are 2.6 times higher than in those not exposed to pesticides (OR 2.6, 95% CI 1.2–5.7) [100]. Similar results were found in an Australian study, showing that exposure to any pesticide can increase the risk of non-Hodgkin's lymphoma in both adults and children [101]. A case-control study conducted in Canada examined the effects of individual pesticides and showed a significant association of non-Hodgkin lymphoma with exposures to phenoxy herbicides, dicamba, carbamate, organophosphorus insecticides, amide fungicides, and the fumigant carbon tetrachloride [102]. Yildirim et al. showed that the risk of non-Hodgkin lymphoma was elevated with increasing pesticide usage [103]. The relative risk increased by 2.42-fold between 1995 and 2005 and 2.77-fold between 1995 and 2010 (correlation coefficient of 0.5). A 2014 meta-analysis of 33 studies found a positive association between NHL and exposure to pesticides overall, as well as specific classes of pesticides such as organochlorines and phenoxy herbicides [104]. A recent study specified that carbamate exposure has the strongest association with non-Hodgkin lymphoma was the worst exposure (HR 3.87, 95% CI 1.66–9.01) [105].

Leukemia

Leukemia is a cancer of blood-forming tissue like bone marrow that causes hyperproliferation of abnormal blood cells [106]. There are four main types of leukemia: acute lymphocytic leukemia, chronic lymphocytic leukemia, acute myeloid leukemia, and chronic myeloid leukemia [107]. Myeloid leukemia is the most common in adults, and acute leukemia is the most common type in children [108]. In 2018, leukemia ranked 15th for frequently diagnosed cancers and was the 11th for cancer deaths worldwide [109]. The risk factors associated with leukemia are smoking, radiation exposure, and carcinogenic chemicals such as benzene and Agent Orange [110].

Foucault et al. conducted a meta-analysis and found that occupational pesticide exposures significantly increased the risk of developing acute myeloid leukemia (AML). Insecticide exposure, in particular, increased the risk of myelodysplastic syndromes by 71% (OR 1.71, 95% CI 1.22–2.40) [111]. Brown et al. also found that farmers who were exposed to insecticides like crotoxyphos, pyrethrins, methoxychlor, and dichlorvos were much more likely to develop all types of leukemia [112]. Parental and residential exposures to pesticides also increased the risk of leukemia in children. Nguyen et al. found an association between the risk of developing childhood leukemia and the distance between residences and nurseries. The highest odds (OR 3.09, 95% CI 1.14–8.34) were found for those living within 75 meters of the nursery [113]. A study in California found that children exposed to pesticides were four times more likely to develop acute myelocytic leukemia and two times more likely to develop acute lymphocytic leukemia [114]. In this study, the pesticides, prosulfocarb, bentazone, and thifensulfuron-methyl were found to raise the risk. Park et al. also observed that general pesticide exposure elevated the risk for childhood leukemia and specified the individual pesticides with significant associations, such as diuron, propanil, urea, 2,6-dinitroaniline, and phosmet [115]. A recent meta-analysis revealed that children whose mothers were exposed to pesticides during pregnancy were more likely to develop acute leukemia (OR 2.18, 95% CI 1.44–3.29) [116].

Brain Cancer

There has been an increase in the incidence of brain cancer globally, with large variability of up to a fivefold difference between the highest incidence rates in northern Europe and the lowest rates in Africa [117]. The annual incidence rate of brain cancer is 7–19.1 per 100,000, while the mortality rate is 3.4 per 100,000 globally [118]. In children, brain cancers are the most common type of solid tumors, and the incidence rates of brain tumors in children range from 1.15 to 5.14 per 100,000 worldwide [119]. The risk factors include radiation, genetic abnormality, and environmental and occupational exposures [120].

Miranda-Filho et al. presented that brain cancer mortality increased with pesticide use in an agricultural region, Serrano [121]. In contrast, the metropolitan region showed a decreasing trend from an ecological study conducted in Brazil. A case-control study with 341 cases and 527 controls observed between January 1995 to January 1997 in Iowa, Michigan, Minnesota, and Wisconsin showed no significant associations between pesticide exposures and intracranial glioma [122]. However, the risk of intracranial glioma is three times greater in the exposed group compared to the non-exposed group. Another study in Nebraska reported a significantly increased risk of primary glioma associated with men aged 55 years or more who lived or worked on a farm (OR 3.9, 95% CI 1.8–8.6) [123]. In Southwestern France, the risk of developing all brain tumors was 1.29 times greater among people exposed to pesticides, and the risk of developing glioma was 1.47 times greater [124]. A retrospective study revealed that 90% of brain cancer patients in India were exposed to chlorpyriphos, dimethoate, mancozeb, and captan for more than ten years [125]. Shim et al. examined this association in children living in Florida, New Jersey, New York, and Pennsylvania and found parental residential exposure to herbicides increased the risk of developing childhood brain cancer 1.9 times higher (95% CI 1.2–3.0) [126].

Myeloma

Multiple myeloma (MM) is a malignancy of terminally differentiated B-cells in the bone marrow [127]. About 2% of all new cancer cases and deaths in the US are attributable to MM [20]. Globally, the age-standardized incidence and mortality of multiple myeloma were 1.78 and 1.14 per 100,000, respectively, in 2020 [128].

Exposure to pesticides and agricultural activities, such as farming, has continuously been related to MM. Viel et al. observed the mortality trends conducted between 1984 and 1986 and found that MM both have higher mortality rates among farmers and farm laborers with MM (SMR 1.59, 95% CI 1.32–1.89) [129]. The risk of myeloma increased when exposure was prolonged for more than a decade. In Iowa, positive associations were found with exposure to carbaryl, chlordane, malathion, DDT, and glyphosate [130]. Orsi et al. reported a link between occupational pesticide exposure and MM [131]. Specifically, exposure to pyrethrin, carbamates, picoline, and urea increased the risk of developing MM up to 3 times. Two studies conducted in Canada observed associations between MM and exposure to captan, carbaryl, and mecoprop [132, 133]. A pooled analysis based on men in North America revealed a significant positive association between ever having used carbaryl, captan, and DDT and the risk of MM (OR 2.02, 95% CI 1.28–3.21, OR 1.98, 95% CI 1.04–3.77, and OR 1.44, 95% CI 1.05–1.97, respectively) [134].

Previous studies have proposed several mechanisms. For example, 2,4-D damages DNA, rendering cells more susceptible to oxidative stress and increasing divisions of lymphocytes [135]. Another study revealed that captan affects the control of apoptosis and stimulates the regeneration of cells in the bone marrow, resulting in increased genetic errors and neoplastic transformation [136]. Pesticide exposure also contributes to the development of multiple myeloma by inducing chromosomal abnormalities and mutations in oncogenes and tumor suppressor genes [137].

Evidence of Cancer Risks of Pesticide Exposure by Pesticide Type

Organochlorines

Organochlorines are chlorinated hydrocarbon compounds used worldwide as synthetic pesticides, including DDT, lindane, chlordane, dieldrin, toxaphene, and hexachloride. Organochlorine pesticides are used for both agricultural and residential purposes. In the early days, it was used to control typhus and malaria but it is now banned in most developed countries [138]. The burden of exposure was increased due to their bioaccumulation in the food chain and persistence in the environment [139].

DDT

Dichlorodiphenyltrichloroethane (DDT) is an organochlorine pesticide and insecticide used in agriculture. DDT was widely used in agriculture in various forms between the 1950s and 1980s [140]. DDT is an organic pollutant that stays around for a long time. The half-lives vary according to the environmental conditions. For example, it sticks easily to soils and sediments, and the half-life is up to 30 years [141]. DDT also possesses hydrophobic properties, which makes it dangerous for marine life and aquatic ecosystems [142]. Due to its persistence in the environment and animal tissues, DDT was banned in many countries in the 1970s and 1980s [143]. For example, DDT was banned in 1972 in the US, and China stopped its production in 2007. DDT is still used to control mosquitoes that spread malaria, visceral leishmaniasis, and typhoid in some countries outside of the US [144]. DDT is still manufactured on a large scale in India, which is the largest consumer of DDT [145].

Garabrant et al. found that exposure to DDT increased the risk of developing pancreatic cancer by 4.8 times (95% CI 1.3–17.6) [146]. Exposure for longer than 47 months increased the risk much larger by 7.4 times compared to the population who are not exposed. Another case-control study also showed that exposure to organochlorines such as DDT and DDD was associated with excess pancreatic cancer risks [147]. Two nested case-control studies on Californian women showed that DDT exposure might increase the risk of breast cancer among postmenopausal women [42, 148]. However, the risk of breast cancer among premenopausal women only increased from the exposure during early life between infancy and puberty. A case-control study based on the Servicemen's Testicular Tumor Environmental and Endocrine Determinants Study reported a positive association between plasma DDT levels and the risk of testicular germ cell cancer [149]. Persson et al. conducted a nested case-control study based on Haimen Cohort in China and showed a dose-response relationship between serum DDT level and the risk of liver cancer [150]. The association between DDT exposure and non-Hodgkin lymphoma was also reported from a cohort in Iowa and North Carolina [151].

A proposed mechanism is that DDT and its metabolites affect mitochondria after both short- and long-term exposure, which are essential for maintaining the integrity and functioning of cells such as hepatocytes [152]. Previous studies also showed that DDT induced oxidative stress in human peripheral blood mononuclear cells and stimulated cancer cell proliferation [153]. In addition, its binding to estrogen receptor increases the risk of breast cancer [154]. The U.S. National Toxicology Program (NTP) and the WHO classify DDT as "moderately toxic." The International Association for Research on Cancer classified DDT as a possible carcinogen for humans.

Chlordane and Lindane

Lindane and chlordane are widely used in Asia because they are inexpensive and effective insect and rodent killers [155]. Lindane, a y-isomer of hexachlorocyclohexane (HCH), is no longer used in agriculture [156]. Although the use of lindane in agriculture has been prohibited in developed countries, it is still used in creams, lotions, and shampoos to treat scabies and head lice [157]. While the limited agricultural use of lindane effectively lowered exposure, pharmaceutical exposure is still of concern. These chemicals are found in the immediate home environment as household dust in the carpets, protected from degradation by moisture, microorganisms, and sunlight [158]. These indoor sources majorly contribute to exposure to organochlorine pesticides among children. Lindane has a high persistence half-life of 15 months, and chlordane also has a high persistence half-life of 10 years [159]. The characteristics of organochlorine chemicals are high lipid solubility, high persistence, and low aqueous solubility, leading to a potential to bioaccumulate in humans [159, 160]. The International Agency for Research on Cancer and the US EPA have classified lindane as carcinogenic for humans and a possible human carcinogen, respectively.

Occupational exposure to chlordane has been associated with an increased risk of developing adult acute leukemia. Parada et al. evaluated the blood chlordane level of 633 women and found a positive association between chlordane exposure and 5-year all-cause and breast cancer-specific death among cases with invasive breast cancer [161]. Lindane is significantly associated with an increased risk of developing non-Hodgkin's lymphoma. Kachuri et al. reported an increased risk of developing non-Hodgkin's lymphoma among individuals exposed to lindane (OR 1.60, 95% CI 1.20–2.10) and chlordane (OR 1.59, 95% CI 1.17–2.16) [162]. Higher incidence rates of thyroid cancer in certain states in the U.S. (northeastern, including Rhode Island, New York, Delaware, and New Jersey) were observed with higher use of lindane. Lerro et al. reported an increased risk of developing thyroid cancer among male pesticide applicators who worked with lindane (HR 1.74, 95% CI 1.06–2.84) [163]. Mortazavi et al. reported a significantly higher serum level of organochlorines in patients with bladder carcinoma [164]. As it is excreted in the urine, there is an increased chance of developing bladder cancer in individuals exposed, attributable to the contact of lindane metabolites with the urinary bladder's epithelial cells [165].

Triazine

Triazines have been used for more than 50 years and continue to be used today as herbicides to kill weeds and other plants [166]. Although banned in Italy and other European countries, triazine herbicides like atrazine are still very commonly used today in the United States; atrazine was the second most widely used pesticide in 2016 [167]. Atrazine is the most common triazine pesticide reported in the literature. Through the Atrazine Monitoring Program (AMP), the EPA regularly monitors environmental exposure and contamination in water and soil [168]. Previous epidemiological studies report that triazines act as endocrine disruptors [51], but their contribution to cancer development is still controversial.

Atrazine

Atrazine is a major pesticide found in drinking water supplies in many geographical areas of the US [169]. Occupational and environmental exposure via public water systems to atrazine has been suggested as the two sources of contact with atrazine among humans [170].

Andreotti et al. evaluated the Agricultural Health Study data for associations between RCC incidence and intensity-weighted lifetime days (IWDs) of 38 pesticides and found an exposure-response association between RCC and atrazine [70]. When adjusted for covariates such as the history of hypertension, pack-years of cigarette smoking, alcohol consumption, family history of cancers, and previous use of other correlated pesticides, the associations for atrazine remained unchanged. Rull et al. report an association between the residential proximity to the agricultural application of triazines and an increased risk of acute lymphocytic leukemia [171]. Malagoli et al. consistently found an increased risk of developing childhood leukemias among individuals living close to arable crop fields [172]. The children residing in these areas had 2.04 times higher odds (95% CI 0.50–8.35) of developing leukemia among children under 5. Hopenhayn-Rich et al. investigated the association between atrazine exposure and a 5-year incidence of breast and ovarian cancers in Kentucky and found no positive associations between them [169]. McLeroy et al. reported that no statistically significant association was found between adult exposure to atrazine from drinking water and the risk of developing breast cancer [173]. However, attributable to the small sample in this category, the possible risk of women exposed to atrazine levels at or above the permitted levels of $\geq$3 ppb (OR 1.3, 95% CI 0.3–6.5) cannot be ruled out.

Despite IARC's categorization of atrazine as "not classifiable as to its carcinogenicity to humans", experimental studies on rats suggested that atrazine may cause oxidative stress, DNA damage, and kidney dysfunction and damage [70]. Cardona et al. reported that all triazines are active in the H295R steroidogenesis assay, increasing E1, E2, and P4 [51]. Tinfo et al. reported a similar mechanism that atrazine increases E2 production in ovarian H295R and granulosa cells [174]. These effects may represent a mechanism of action that may increase systemic and local hormone levels in humans and increase the risk of developing breast cancer. A risk assessment for atrazine exposure via raw bovine milk and polluted groundwater reported that 11% of bovine milk samples and 44.4% of groundwater samples from Argentina had atrazine exceeding the internationally established safe level, 0.1 μg/L [175]. The estimated risk of developing cancer was higher for children from both groundwater and bovine milk compared to adults [175].

Organophosphates

Organophosphate is widely used as a main ingredient of herbicides, pesticides, and insecticides. The most commonly used organophosphates include parathion, malathion, chlorpyrifos, diazinon, dichlorvos, glyphosate, and glufosinate [176]. Environmental exposure to organophosphates occurs through dermal, respiratory, and oral exposure to such pesticides.

Glyphosate

Glyphosate is a broad-spectrum herbicide registered for use on multiple food and non-food crops by US EPA [177]. Globally, the use of glyphosate has risen almost 15-fold, and it is the most heavily used herbicide in the world [178]. US EPA had classified glyphosate as a group E carcinogen, meaning "evidence of non-carcinogenicity for humans" [18]. Regulatory agencies such as the European Chemical Agency (ECHA) and the United Nations Food and Agriculture Organization (FAO) have also determined that glyphosate is unlikely to be a carcinogen [179]. However, in March 2015, the International Agency for Research on Cancer (IARC) classified glyphosate as a Group 2A carcinogen (probably carcinogenic to humans) [17].

The increased use of glyphosate in the United States has increased the incidence and mortality rate of multiple diseases, including several cancers [180]. Several epidemiologic cohort and case-control studies on glyphosate and cancer have found that exposure to glyphosate is associated causally with the risk of developing several types of cancer in humans, including breast cancer, leukemia, non-Hodgkin lymphoma, colorectal cancer, pancreatic cancer, glioma, and prostate cancer [177, 181].

One of the critical studies that IARC evaluated to arrive at such a conclusion was the United States Agricultural Health Study (AHS). The AHS is a prospective study that collected data on cancer and pesticide use from more than 89,000 pesticide applicators and their spouses in Iowa and North Carolina from 1993 [182]. The AHS study was used to evaluate associations between glyphosate and breast cancer risk among farmers' wives [183], colon/rectum cancer [184], pancreatic cancer [185], and cutaneous melanoma [186]. Another recent study based on the AHS highlighted a positive association between glyphosate and the risk of AML in the highest exposure quartile compared to never users (RR 2.44, 95% CI 0.94–6.32) [187].

Non-Hodgkin lymphoma (NHL) was also linked to glyphosate in a case-control study among males conducted in Iowa, Minnesota, Nebraska, and Kansa, which found a statistically significant odds ratio for ever-use glyphosate (OR 2.1, 95% CI 1.1–4.0) [188]. The North American Pooled Project (NAPP) has pooled data from case-control studies in the US and Canada. Their analyses also elucidated that subjects who had ever used glyphosate had an excess of NHL overall (OR 1.43, 95% CI 1.11–1.83) [189]. A Swedish study also evaluated NHL by examining the effect of the latency period, where the OR for 1–10 years of latency period was 1.11 (95% CI 0.24–5.08), and for >10 years of the latency period, the OR was 2.26 (95% CI 1.16–4.40) [190]. Statistically significant positive associations were also observed for small lymphocytic lymphoma/chronic lymphocytic leukemia (SLL/CLL) (OR 3.35, 95% CI 1.42–7.89) and for unspecified non-Hodgkin lymphoma (OR 5.63, 95% CI 1.44–22.0) in that study. Another case-control study on glioma found a positive association with glyphosate among the participants in Nebraska (OR 1.5, 95% CI 0.7–3.1) [123]. A recent study examined the association between urinary excretion of a glyphosate metabolite, aminomethylphosphonic acid (AMPA), and breast cancer risk and found a 4.5-fold higher risk of developing breast cancer in the highest vs. lowest quintile of AMPA among the postmenopausal women in Hawaii [45].

Even though there is limited epidemiological evidence linking glyphosate and cancer, studies on animals show that it is biologically plausible. Glyphosate is one of the endocrine-disrupting chemicals (EDCs), showing estrogenic effects on mammary tumor cells in in vitro studies [191]. Glyphosate was also found to be toxic to microbes. Samsel and Seneff found that long-term exposure to glyphosate causes an imbalance of microbes in the gut, which leads to a chronic inflammatory syndrome of the intestine ([192]; [180]). Such chronic inflammatory conditions are known cancer risk factors [192]. Several unpublished studies on rats and mice that Monsanto submitted to the Environmental Protection Agency (EPA) showed a link between glyphosate and the growth of tumors in biological systems [193]. The proposed mechanisms include direct DNA damage in sensitive cells, changes in glycine levels, inhibition of succinate dehydrogenase, chelation of manganese, conversion of glyphosate into more cancer-causing compounds like N-nitrosoglyphosate and glyoxylate, and changes in fructose metabolism.

Malathion and Parathion

Malathion and parathion are the most commonly used organophosphate insecticides in the US. Malathion is widely used in agriculture (e.g., wheat and corn), home and garden applications, mosquito control, Mediterranean fruit fly eradication, and as a treatment for head lice in the US [194]. The International Agency for Research on Cancer (IARC) classified malathion as a probable carcinogen for humans [17]. On the other hand, parathion is used in agricultural and veterinary applications globally, although it is banned in the US because of its high toxicity [195]. Due to its limited solubility in water and strong soil binding, parathion has a low leachability and high persistence [196]. IARC classifies parathion as a possible human carcinogen, and US EPA considers it to have suggestive evidence of carcinogenic potential.

Malathion and parathion have not been established as potent carcinogens for humans. However, previous studies have investigated their roles in causing cancer in farmers and their spouses through direct and indirect contact. Two case-control studies in Kansas and Nebraska also support the association of Hodgkin and non-Hodgkin lymphoma with malathion exposure [197, 198]. In both studies, the risk of developing lymphomas rose proportionately with the increased duration of exposure to malathion compared to never users. Many recent studies have shown that women are more likely to develop breast cancer if they are directly exposed to pesticides or if their farmer spouses are. For example, a case-control study in Chile from 1995–2005 shows women with prolonged exposure to aerial malathion spray were 5.7 times more likely to have breast cancer with a higher rate of metastases (30.5%) than the control group [40]. In the US, a study based on the Agricultural Health Study cohort spouses revealed significant associations of breast cancer with parathion among exposed women at 5-year follow-up (HR 1.9, 95% CI 1.0–3.4) [199]. Moreover, female spouses are highly prone to contracting thyroid cancer from malathion (RR 2.04, 95% CI 1.14–3.63) [27, 28].

There have been reports of other cancers induced by malathion and parathion. For example, those who had contact with malathion as an active ingredient in the pesticide had a significantly higher risk of prostate cancer than non-users (OR 1.34, 95% CI 1.01–1.78) [56]. In addition, dose-response relationships between these two pesticides and the risks of prostate cancer were observed based on the lifetime cumulative exposure. Another prospective cohort study also correlated with an increased risk ratio of prostate cancer in the farming population handling malathion (RR 1.43, 95% CI 1.08–1.88) [58].

Malathion and parathion induce cellular proliferation, oxidative stress, and immunotoxicity, increasing the risks of developing cancers [194]. An in vitro study on human liver carcinoma (HepG2) cells revealed oxidative stress and molecular changes in DNA in the development of hepatocellular carcinoma. Even in vivo studies on animal models revealed the possibility of malathion-induced lung cancer [200]. Recently, an in vitro study examined the simultaneous effects of parathion, malathion, and estrogen exposure on the normal MCF-10F human breast epithelial cell line and found carcinogenic symptoms caused by cellular and nuclear alterations [201].

Chlorophenoxy

Chlorophenoxy herbicides are used as herbicides to control broadleaf weeds and regulate plant growth worldwide [202]. This class includes MCPA, 2,4-D, and dicamba, ranked among the top 10 most commonly used residential pesticides [1]. The IARC has classified phenoxy acid herbicides as possibly carcinogenic to humans (category 2B) [17].

2,4-D

2,4-D is the most common chlorophenoxy herbicide. 2,4-D is an herbicide that has been in use to kill broadleaf plants that are acting as weeds [203]. It acts as a plant growth hormone at low doses, but becomes herbicidal at high doses [204].

A case-control study revealed that the risk of non-Hodgkin lymphoma was increased with exposure to 2,4-D in applicators without personal protection equipment (OR 4.4, 95% CI 1.1–29.1) [205]. Burns et al. study looked at a cohort of 1316 workers who produced 2,4-D over a long period, and a non-significant positive association was found with non-Hodgkin lymphoma [206]. In the AHS-based cohort study with 54,344 men, Koutros et al. presented a positive association between 2,4-D use and urinary bladder cancer [82]. A meta-analysis of observational studies from the US, Canada, and Europe revealed that exposure to 2,4-D significantly increased the risk of non-Hodgkin lymphoma (RR 1.73, 95% CI 1.10–2.72) [207]. Further analysis focusing on the US showed a stronger association (RR 1.96, 95% CI 1.03–3.76).

Several possible mechanisms of carcinogenicity have been proposed, such as oxidative stress, immunosuppression, chronic inflammation, and altered cell proliferation followed by cell death. Animal and cell studies showed that 2,4-D produced free radicals and induced oxidative stress ([208]; [209]). 2,4-D significantly decreased the number of bone-marrow plasma cells, indicating loss of humoral immunity. It also increased lipid peroxidation in vivo and in vitro experiments, leading to chronic inflammation ([210]; [211]).

Dicamba

Dicamba is a selective benzoic acid herbicide used to control broadleaf weeds and woody plants in food and feed crops [212]. Dicamba was the 18th most commonly used agricultural pesticide and the eighth most used in the home and garden sector in 2012 in the US [1].

Previous studies based on the Agricultural Health Study (AHS) cohort showed that exposure to dicamba increased the risk of lung cancer and colon cancer [213, 214]. A Canadian case-control study reported a positive association between dicamba use and prostate cancer [56]. Other studies showed that occupational use of dicamba was associated with non-Hodgkin lymphoma [133] and multiple myeloma [215]. Lerro et al. conducted a cohort study based on the AHS study and found those in the highest quartile of dicamba exposure had increased risks of liver and intrahepatic bile duct cancer (RR 1.80, 95% CI 1.26–2.56) and chronic lymphocytic leukemia (RR 1.20, 95% CI 0.96–1.50) among applicators [212].

Experimental studies showed that the genotoxicity of dicamba including the sister chromatid exchanges and alterations in the cell cycle and cell proliferation in lymphocytes [216]. The DNA damage caused by dicamba resulted from oxidative stress and inflammation. In rats, dicamba altered the activity of enzymes in peroxisomes and increased peroxisome proliferator, increasing the risk of liver cancer [217]. Dicamba also induced monoclonal B-cell lymphocytosis, which is a potential marker of chronic lymphocytic leukemia ([218]; [212]).

Pyrethroid

Pyrethroids are synthetic insecticides that are based on the chemical structure of naturally occurring pyrethrins, which are extracted from certain species of chrysanthemum flowers [219]. Pyrethroids are known for their effectiveness against a broad range of pests and are considered less toxic to mammals than other classes of insecticides [220]. They are widely used for insect control, including as a public health resource to prevent mosquito-borne diseases such as malaria and dengue, as well as to treat army uniforms and mosquito nets [221].

Permethrin

Rusiecki et al. showed that exposure to permethrin elevated the risk of developing leukemia and lymphohematopoietic carcinoma among 49,092 fumigators [222]. A hospital-based case-control study found an association between acute lymphoblastic leukemia and shampoos containing pyrethroids and other insecticides used to treat pediculosis. In Brazil, a case-control study reported a significant association between acute leukemia and in-utero exposure to permethrin and the risk of developing AML (OR 7.28, 95% CI 2.60–20.4) and ALL (OR 2.47, 95% CI 1.17–5.25) among children <2 years old [223]. A systematic review showed that exposure to permethrin elevated the risk of liver and lung cancers based on a systematic review [224]. Shrestha et al. report an elevated risk of mortality from bladder cancer (HR 1.28, 95%CI 0.74–2.22), multiple myeloma (HR 1.26, 95% CI 0.77–2.07), and intrahepatic and liver cancers (HR 1.92, 95% CI 1.03–3.58) among individuals aged >66 years [225].

There has been evidence to suggest the biological plausibility between exposure to pyrethroids and the development of cancers. Previous studies showed that permethrin caused the break in the MLL gene in the BV173 cell line, and cypermethrin induced oncogenic chromosomal translocations, suggesting a relationship between in-utero exposure to pyrethroid and the development of leukemia ([226]; [227]). Multiple in vitro and in vivo studies suggest the genotoxic effects of pyrethroids. Genotoxicity assays such as comet assays, chromosomal aberrations micronucleus, and sister chromatid exchange suggested that exposure to pyrethroids induces DNA damage and aneuploidy as chromosomal aberrations in the peripheral blood lymphocytes and other systems [109]. Pyrethroids also inhibit the enzymatic activity of acetylcholinesterase, increasing the risk of lung cancer ([228]; [34]).

Carbamates

Carbamates are widely used to control a wide range of pests, including insects, nematodes, and weeds across various settings [229]. Despite their relatively low potential for bioaccumulation and toxicity in mammals, carbamate pesticides are still considered to be hazardous to both the environment and human health [230]. Some of the most frequently used carbamate pesticides include carbaryl, carbofuran, methomyl, and aldicarb.

Carbaryl

Carbaryl is a carbamate insecticide commonly used worldwide in agriculture to control a variety of pests, including insects, slugs, and snails [231]. It is also used in public health programs to control disease-carrying insects such as flea and lice [232]. Carbaryl is the second most commonly detected insecticide in water, and it has been found in about half of urban streams [196].

Several epidemiological studies have suggested that there may be a link between carbaryl exposure and the development of certain types of cancer. A study in Florida showed that exposure to carbaryl was related with an increased risk of lung cancer among pest control operators [233]. Engel et al. found that exposure to carbaryl is associated with a higher risk of breast cancer in women, especially among those who are postmenopausal and have a family history of breast cancer [183]. The AHS-based studies found that exposure to carbaryl is linked to an elevated risks of melanoma and leukemia, non-Hodgkin's lymphoma, and melanoma [99, 112, 234]. Another US study found that carbaryl exposure was associated with an increased risk of multiple myeloma [235]. A study in Brazil found that individuals exposed to carbaryl had an increased risk of non-Hodgkin's lymphoma [236], while a US study found an association between cumulative carbaryl exposure and prostate cancer [237].

It is possible that carbaryl exposure may affect hormone signaling pathways and contribute to the development of hormone-sensitive cancers such as breast cancer [183]. Xia et al. suggested that carbaryl may act as a genotoxic agent, causing DNA breaks and chromosomal aberrations [238]. Another study found that exposure to carbaryl induced oxidative stress and inflammation, and promoted the development and progression of liver cancer [239].

Disparities in Pesticide Exposure and Cancer Risk

Pesticide exposure and related cancers are a significant public health concern, and there are several disparities associated with them. Workers in the agricultural industry, including farmworkers and pesticide applicators, have a higher risk of exposure to pesticides than the general population [240, 241]. This puts them at a higher risk of developing pesticide-induced cancers. People who live in or near agricultural areas where pesticides are used are also at a higher risk of exposure [242]. This is particularly for low-income and marginalized communities, who may not have the resources to relocate or advocate for more protective regulations [243]. People of color, particularly African Americans and Latinos, are disproportionately impacted by pesticide exposure and pesticide-induced cancers [244, 245]. This is due in part to systemic racism and the history of discriminatory policies in the agricultural industry. Women are also at a higher risk of pesticide exposure, particularly those who work in agriculture or live in agricultural areas [246]. This is because women often have different roles in the agricultural industry and may have different exposure patterns. Some pesticides are more toxic than others and have been linked to higher rates of cancer. For example, the herbicide glyphosate has been linked to an increased risk of non-Hodgkin lymphoma [247]. Overall, addressing the disparities associated with pesticide exposure and pesticide-induced cancers will require a multi-faceted approach that includes regulatory changes, increased access to protective equipment, education and training for workers, and increased public awareness about the risks associated with pesticides.

Conclusion

Exposure to pesticides is a recognized environmental risk factor associated with the development of various types of cancer. Evidence suggests that exposure to different classes of pesticides, including insecticides, herbicides, fungicides, and fumigants, can increase the risk of cancer. Specific chemicals within several classes of pesticides, such as organochlorines, organophosphates, triazine, chlorophenoxy, pyrethroids, and carbamates, have been linked to cancer development. However, some pesticides are not considered carcinogenic to humans due to the lack of sufficient human evidence, although in vitro and in vivo studies have suggested underlying carcinogenesis mechanisms for some of these pesticides. Some epidemiological studies have been inconclusive in establishing a significant relationship between pesticide exposure and cancer risks, likely due to methodological limitations such as small sample sizes and difficulties in assessing exposure. Current risk assessments also fail to address chronic low-dose exposure and exposure to pesticide mixtures fully. Furthermore, exposure to pesticides disproportionately affects people living in rural areas, low-income urban communities, and developing countries, leading to disparities in cancer risks. Therefore, multidisciplinary studies, including epidemiological and toxicological research, are required to evaluate the carcinogenicity of pesticides and to identify the health impacts of pesticides on vulnerable populations. Intervention to reduce exposure to pesticides can contribute to reducing the risks of developing cancer and improving health disparities.

References

1. Atwood D, Paisley-Jones C. Pesticides industry sales and usage: 2008–2012 market estimates. US Environmental Protection Agency; 2017.
2. Tudi M, Li H, Li H, et al. Exposure routes and health risks associated with pesticide application. Toxics. 2022;10(6):335.
3. Gangemi S, Miozzi E, Teodoro M, et al. Occupational exposure to pesticides as a possible risk factor for the development of chronic diseases in humans. Mol Med Rep. 2016;14(5):4475–88.
4. Wang Y, Abd El-Aty AM, Wang S, et al. Competitive fluorescent immunosensor based on catalytic hairpin self-assembly for multiresidue detection of organophosphate pesticides in agricultural products. Food Chem. 2023;413:5607.
5. De Troeyer K, Casas L, Bijnens EM, et al. Higher proportion of agricultural land use around the residence is associated with higher urinary concentrations of AMPA, a glyphosate metabolite. Int J Hyg Environ Health. 2022;246:114039.
6. Gunier RB, Ward MH, Airola M, et al. Determinants of agricultural pesticide concentrations in carpet dust. Environ Health Perspect. 2011;119(7):970–6.
7. Harnly ME, Bradman A, Nishioka M, et al. Pesticides in dust from homes in an agricultural area. Environ Sci Technol. 2009;43(23):8767–74.
8. Bradman A, Castorina R, Boyd Barr D, et al. Determinants of organophosphorus pesticide urinary metabolite levels in young children living in an agricultural community. Int J Environ Res Public Health. 2011;8(4):1061–83.
9. Cecchi A, Alvarez G, Quidel N, et al. Residential proximity to pesticide applications in Argentine Patagonia: impact on pregnancy and newborn parameters. Environ Sci Pollut Res. 2021;28(40):56565–79.
10. US Department of Agriculture. 2022. Pesticide Data Program Annual Summary Calendar Year 2021.
11. Bexfield LM, Belitz K, Lindsey BD, et al. Pesticides and pesticide degradates in groundwater used for public supply across the United States: occurrence and human-health context. Environ Sci Technol. 2020;55(1):362–72.
12. Desimone L, Hamilton PA, Gilliom RJ. 2009. Quality of water from domestic wells in principal aquifers of the United States, 1991–2004. National Water Quality Assessment Program, Circular 1332. US Geological Survey. https://pubs.usgs.gov/circ/circ1332/includes/circ1332.pdf. Accessed 13 Dec 2022.
13. Levy ZF, Balkan M, Shelton JL. 2023. Quality of groundwater used for domestic supply in the Modesto, Turlock, and Merced Subbasins of the San Joaquin Valley, California. US

Geological Survey. https://pubs.usgs.gov/of/2022/1116/ofr20221116.pdf. Accessed 11 Feb 2023.
14. California Environmental Health Protection Agency. 2023. About Proposition 65. https://oehha.ca.gov/proposition-65/about-proposition-65. Accessed 07 Feb 2023.
15. International Agency for Research on Cancer. 2019. IARC Monographs on the identification of carcinogenic hazards to humans – preamble. https://monographs.iarc.who.int/wp-content/uploads/2019/07/Preamble-2019.pdf. Accessed 07 Feb 2023.
16. US Environmental Protection Agency. 2022. Evaluating pesticides for carcinogenic potential. https://www.epa.gov/pesticide-science-and-assessing-pesticide-risks/evaluating-pesticides-carcinogenic-potential. Accessed 08 Feb 2023.
17. International Agency for Research on Cancer. 2023. Agents classified by the IARC monographs, vols 1–129. http://monographs.iarc.fr/ENG/Classification/index.php. Accessed 07 Feb 2023.
18. US Environmental Protection Agency. 2022. Chemicals evaluated for carcinogenic potential.
19. Sharma R. Mapping of global, regional and national incidence, mortality and mortality-to-incidence ratio of lung cancer in 2020 and 2050. Int J Clin Oncol. 2022;27(4):665–75.
20. National Cancer Institute. 2023. Cancer stat facts: lung and bronchus cancer, surveillance, epidemiology, and end results program. https://seer.cancer.gov/statfacts/html/lungb.html. Accessed 08 Feb 2023.
21. Szalontai K, Gémes N, Furák J, et al. Chronic obstructive pulmonary disease: epidemiology, biomarkers, and paving the way to lung cancer. J Clin Med. 2021;10(13):2889.
22. Boffetta P. Human cancer from environmental pollutants: the epidemiological evidence. Mutat Res Genet Toxicol Environ Mutagen. 2006;608(2):157–62.
23. Alavanja MC, Bonner MR. Occupational pesticide exposures and cancer risk: a review. J Toxicol Environ Health B Crit Rev. 2012;15(4):238–63.
24. Zendehdel R, Tayefeh-Rahimian R, Kabir A. Chronic exposure to chlorophenol related compounds in the pesticide production workplace and lung cancer: a meta-analysis. Asian Pac J Cancer Prev. 2014;15(13):5149–53.
25. Boulanger M, Tual S, Lemarchand C, et al. Lung cancer risk and occupational exposures in crop farming: results from the AGRIculture and CANcer (AGRICAN) cohort. Occup Environ Med. 2018;75(11):776–85.
26. Jones RR, Barone-Adesi F, Koutros S, et al. Incidence of solid tumours among pesticide applicators exposed to the organophosphate insecticide diazinon in the Agricultural Health Study: an updated analysis. Occup Environ Med. 2015;72(7):496–503.
27. Lerro CC, Koutros S, Andreotti G, et al. Organophosphate insecticide use and cancer incidence among spouses of pesticide applicators in the Agricultural Health Study. Occup Environ Med. 2015a;72(10):736–44.
28. Lerro CC, Koutros S, Andreotti G, et al. Use of acetochlor and cancer incidence in the Agricultural Health Study. Int J Cancer. 2015b;137(5):1167–75.
29. Bonner MR, Freeman LE, Hoppin JA, et al. Occupational exposure to pesticides and the incidence of lung cancer in the agricultural health study. Environ Health Perspect. 2017;125(4):544–51.
30. Kim B, Park EY, Kim J, et al. Occupational exposure to pesticides and lung cancer risk: a propensity score analyses. Cancer Res Treat. 2022;54(1):130–9.
31. Parron T, Requena M, Hernandez AF, et al. Environmental exposure to pesticides and cancer risk in multiple human organ systems. Toxicol Lett. 2014;230(2):157–65.
32. Liu CT, Yang CC, Chien WC, et al. Association between long-term usage of acetylcholinesterase inhibitors and lung cancer in the elderly: a nationwide cohort study. Sci Rep. 2022;12(1):3531.
33. Majidi M, Delirrad M, Banagozar Mohammadi A, et al. Cholinesterase level in erythrocyte or serum: which is more predictive of the clinical outcome in patients with acute organophosphate poisoning? Iran J Toxicol. 2018;12(5):23–6.
34. Xi HJ, Wu RP, Liu JJ, et al. Role of acetylcholinesterase in lung cancer. Thorac Cancer. 2015;6(4):390–8.

35. Ziech D, Franco R, Georgakilas AG, et al. The role of reactive oxygen species and oxidative stress in environmental carcinogenesis and biomarker development. Chem Biol Interact. 2010;188(2):334–9.
36. Thakur S, Dhiman M, Mantha AK. APE1 modulates cellular responses to organophosphate pesticide-induced oxidative damage in non-small cell lung carcinoma A549 cells. Mol Cell Biochem. 2018;441:201–16.
37. Bray F, Ferlay J, Soerjomataram I, et al. Global cancer statistics 2018: GLOBOCAN estimates of incidence and mortality worldwide for 36 cancers in 185 countries. CA Cancer J Clin. 2018;68(6):394–424.
38. Høyer AP, Jørgensen T, Rank F, et al. Organochlorine exposures influence on breast cancer risk and survival according to estrogen receptor status: a Danish cohort-nested case-control study. BMC Cancer. 2001;1(1):1–8.
39. Ferro R, Parvathaneni A, Patel S, et al. Pesticides and breast cancer. Adv Breast Cancer Res. 2012;1(3):30–5.
40. Cabello G, Valenzuela-Estrada M, Siques P, et al. Relation of breast cancer and malathion aerial spraying in Arica, Chile. Int J Morphol. 2013;31(2):640–5.
41. Arrebola JP, Belhassen H, Artacho-Cordón F, et al. Risk of female breast cancer and serum concentrations of organochlorine pesticides and polychlorinated biphenyls: a case–control study in Tunisia. Sci Total Environ. 2015;520:106–13.
42. Cohn BA, Cirillo PM, Terry MB. DDT and breast cancer: prospective study of induction time and susceptibility windows. J Natl Cancer Inst. 2019;111(8):803–10.
43. Tayour C, Ritz B, Langholz B, et al. A case–control study of breast cancer risk and ambient exposure to pesticides. Environ Epidemiol. 2019;3(5)
44. Mekonen S, Ibrahim M, Astatkie H, et al. Exposure to organochlorine pesticides as a predictor to breast cancer: A case-control study among Ethiopian women. PLoS One. 2021;16(9):e0257704.
45. Franke AA, Li X, Shvetsov YB, et al. Pilot study on the urinary excretion of the glyphosate metabolite aminomethylphosphonic acid and breast cancer risk: the Multiethnic Cohort study. Environ Pollut. 2021;277:116848.
46. Eldakroory SA, Morsi DE, Abdel-Rahman RH, et al. Correlation between toxic organochlorine pesticides and breast cancer. Hum Exp Toxicol. 2017;36(12):1326–34.
47. Kaur N, Swain SK, Banerjee BD, et al. Organochlorine pesticide exposure as a risk factor for breast cancer in young Indian women: a case–control study. South Asian J Cancer. 2019;8(4):212–4.
48. Wallace DR. Environmental pesticides and heavy metals - role in breast cancer. In: Larramendy ML, Solonesk S, editors. Toxicity and hazard of agrochemicals. London: IntechOpen; 2015. p. p39–70.
49. Yang KJ, Lee J, Park HL. Organophosphate pesticide exposure and breast cancer risk: a rapid review of human, animal, and cell-based studies. Int J Environ Res Public Health. 2020;17(14):5030.
50. Pizzatti L, Kawassaki ACB, Fadel B, et al. Toxicoproteomics disclose pesticides as downregulators of TNF-α, IL-1β and estrogen receptor pathways in breast cancer women chronically exposed. Front Oncol. 2020;10:1698.
51. Cardona B, Rudel RA. US EPA's regulatory pesticide evaluations need clearer guidelines for considering mammary gland tumors and other mammary gland effects. Mol Cell Endocrinol. 2020;518:110927.
52. Green BL, Davis JL, Rivers D, et al. Cancer health disparities. In: Alberts DS, Hess LM, editors. Fundamentals of cancer prevention. 4th ed. Cham: Springer; 2019. p. p199–246.
53. Landau-Ossondo M, Rabia N, Jos-Pelage J, et al. Why pesticides could be a common cause of prostate and breast cancers in the French Caribbean Island, Martinique. An overview on key mechanisms of pesticide-induced cancer. Biomed Pharmacother. 2009;63(6):383–95.
54. Settimi L, Masina A, Andrion A, et al. Prostate cancer and exposure to pesticides in agricultural settings. Int J Cancer. 2003;104(4):458–61.

55. Multigner L, Ndong JR, Giusti A, et al. Chlordecone exposure and risk of prostate cancer. J Clin Oncol. 2010;28(21):3457–62.
56. Band PR, Abanto Z, Bert J, et al. Prostate cancer risk and exposure to pesticides in British Columbia farmers. Prostate. 2011;71(2):168–83.
57. Abhishek A, Ansari NG, Singh V, et al. Genetic susceptibility of CYP1A1 gene and risk of pesticide exposure in prostate cancer. Cancer Biomark. 2020;29(4):429–40.
58. Koutros S, Beane Freeman LE, Lubin JH, et al. Risk of total and aggressive prostate cancer and pesticide use in the Agricultural Health Study. Am J Epidemiol. 2013;177(1):59–74.
59. Pardo LA, Beane Freeman LE, Lerro CC, et al. Pesticide exposure and risk of aggressive prostate cancer among private pesticide applicators. Environ Health. 2020;19(1):1–12.
60. Cockburn M, Mills P, Zhang X, et al. Prostate cancer and ambient pesticide exposure in agriculturally intensive areas in California. Am J Epidemiol. 2011;173(11):1280–8.
61. Lewis-Mikhael AM, Bueno-Cavanillas A, Guiron TO, et al. Occupational exposure to pesticides and prostate cancer: a systematic review and meta-analysis. Occup Environ Med. 2016;73(2):134–44.
62. Silva JF, Mattos IE, Luz LL, et al. Exposure to pesticides and prostate cancer: systematic review of the literature. Rev Environ Health. 2016;31(3):311–27.
63. Djalali-Behzad G, Hussain S, Osterman-Golkar S, et al. Estimation of genetic risks of alkylating agents: VI. Exposure of mice and bacteria to methyl bromide. Mutat Res. 1981;84(1):1–9.
64. Pletsa V, Steenwinkel MJ, van Delft JH, et al. Methyl bromide causes DNA methylation in rats and mice but fails to induce somatic mutations in λlacZ transgenic mice. Cancer Lett. 1998;135(1):21–7.
65. Budnik LT, Kloth S, Velasco-Garrido M, et al. Prostate cancer and toxicity from critical use exemptions of methyl bromide: environmental protection helps protect against human health risks. Environ Health. 2012;11(5):1–13.
66. Tessier DM, Matsumura F. Increased ErbB-2 tyrosine kinase activity, MAPK phosphorylation, and cell proliferation in the prostate cancer cell line LNCaP following treatment by select pesticides. Toxicol Sci. 2001;60(1):38–43.
67. Roberto M, Panebianco M, Aschelter AM, et al. The value of the multidisciplinary team in metastatic renal cell carcinoma: paving the way for precision medicine in toxicities management. Front Oncol. 2023;12:1026978.
68. Chow WH, Dong LM, Devesa SS. Epidemiology and risk factors for kidney cancer. Nat Rev Urol. 2010;7(5):245–57.
69. Padala SA, Barsouk A, Thandra KC, et al. Epidemiology of renal cell carcinoma. World J Oncol. 2020;11(3):79–87.
70. Andreotti G, Beane Freeman LE, Shearer JJ, et al. Occupational pesticide use and risk of renal cell carcinoma in the agricultural health study. Environ Health Perspect. 2020;128(6):067011.
71. Hu J, Mao Y, White K, et al. Renal cell carcinoma and occupational exposure to chemicals in Canada. Occup Med (Lond). 2002;52(3):157–64.
72. Rios P, Bauer H, Schleiermacher G, et al. Environmental exposures related to parental habits in the perinatal period and the risk of Wilms' tumor in children. Cancer Epidemiol. 2020;66:101706.
73. Liu W, Du Y, Liu J, et al. Effects of atrazine on the oxidative damage of kidney in Wister rats. Int J Clin Exp Med. 2014;7(10):3235.
74. Sánchez OF, Lin L, Bryan CJ, et al. Profiling epigenetic changes in human cell line induced by atrazine exposure. Environ Pollut. 2020;258:113712.
75. Abid A, Ajaz S, Khan AR, Zehra F, et al. Analysis of the glutathione S-transferase genes polymorphisms in the risk and prognosis of renal cell carcinomas. Case-control and meta-analysis. Urol Oncol. 2016;34(9):419.e1–2.
76. Saginala K, Barsouk A, Aluru JS, et al. Epidemiology of bladder cancer. Med Sci. 2020;8(1):15.
77. Lin W, Pan X, Zhang C, et al. Impact of age at diagnosis of bladder cancer on survival: a surveillance, epidemiology, and end results-based study 2004–2015. Cancer Control. 2023;30:10732748231152322.

78. Kiriluk KJ, Prasad SM, Patel AR, et al. Bladder cancer risk from occupational and environmental exposures. Urol Oncol. 2012;30(2):199–211.
79. Letašiová S, Medveďová A, Šovčíková A, et al. Bladder cancer, a review of the environmental risk factors. Environ Health. 2012;11(1):1–5.
80. Sharma T, Jain S, Verma A, et al. Gene environment interaction in urinary bladder cancer with special reference to organochlorine pesticide: a case control study. Cancer Biomark. 2013;13(4):243–51.
81. Boada LD, Henríquez-Hernández LA, Zumbado M, et al. Organochlorine pesticides exposure and bladder cancer: evaluation from a gene-environment perspective in a hospital-based case-control study in the Canary Islands (Spain). J Agromedicine. 2016;21(1):34–42.
82. Koutros S, Silverman DT, Alavanja MC, et al. Occupational exposure to pesticides and bladder cancer risk. Int J Epidemiol. 2016;45(3):792–805.
83. Liang YJ, Long DX, Xu MY, et al. Body fluids from the rat exposed to chlorpyrifos induce cytotoxicity against the corresponding tissue– derived cells in vitro. BMC Pharmacol Toxicol. 2021;22:1–8.
84. Matic MG, Coric VM, Savic-Radojevic AR, et al. Does occupational exposure to solvents and pesticides in association with glutathione S-transferase A1, M1, P1, and T1 polymorphisms increase the risk of bladder cancer? The Belgrade case-control study. PLoS One. 2014;9(6):e99448.
85. Ferlay J, Soerjomataram I, Dikshit R, et al. Cancer incidence and mortality worldwide: sources, methods and major patterns in GLOBOCAN 2012. Int J Cancer. 2015;136(5):E359–86.
86. Pourshams A, Sepanlou SG, Ikuta KS, et al. The global, regional, and national burden of pancreatic cancer and its attributable risk factors in 195 countries and territories, 1990–2017: a systematic analysis for the Global Burden of Disease Study 2017. Lancet Gastroenterol Hepatol. 2019;4(12):934–47.
87. Brugel M, Carlier C, Reyes-Castellanos G, et al. Pesticides and pancreatic adenocarcinoma: a transversal epidemiological, environmental and mechanistic narrative review. Dig Liver Dis. 2022;54(12):1605–13.
88. Weinstein B, da Silva A, Carpenter DO. Exocrine pancreatic cancer and living near to waste sites containing hazardous organic chemicals, New York State, USA–an 18-year population-based study. Int J Occup Med Environ Health. 2022;35(4):459–71.
89. Porta M, Gasull M, Pumarega J, et al. Plasma concentrations of persistent organic pollutants and pancreatic cancer risk. Int J Epidemiol. 2022;51(2):479–90.
90. Clary T, Ritz B. Pancreatic cancer mortality and organochlorine pesticide exposure in California, 1989–1996. Am J Ind Med. 2003;43(3):306–13.
91. McGwin G Jr, Griffin RL. An ecologic study of the association between 1, 3-dichloropropene and pancreatic cancer. Cancers. 2022;15(1):150.
92. Fritschi L, Benke G, Risch HA, et al. Occupational exposure to N-nitrosamines and pesticides and risk of pancreatic cancer. Occup Environ Med. 2015;72(9):678–83.
93. He B, Ni Y, Jin Y, et al. Pesticides-induced energy metabolic disorders. Sci Total Environ. 2020;729:139033.
94. Liou GY, Döppler H, DelGiorno KE, et al. Mutant KRas-induced mitochondrial oxidative stress in acinar cells upregulates EGFR signaling to drive formation of pancreatic precancerous lesions. Cell Rep. 2016;14(10):2325–36.
95. Thandra KC, Barsouk A, Saginala K, et al. Epidemiology of non-Hodgkin's lymphoma. Med Sci. 2021;9(1):5.
96. Singh R, Shaik S, Negi BS, et al. Non-Hodgkin's lymphoma: a review. J Family Med Prim Care. 2020;9(4):1834.
97. Kim CJ, Freedman DM, Curtis RE, et al. Risk of non-Hodgkin lymphoma after radiotherapy for solid cancers. Leuk Lymphoma. 2013;54(8):1691–7.
98. Sapkota S, Shaikh H. Non-Hodgkin lymphoma. In: StatPearls [Internet]. Treasure Island, FL: StatPearls Publishing; 2022. https://www.ncbi.nlm.nih.gov/books/NBK559328/.
99. Cantor KP, Blair A, Everett G, et al. Pesticides and other agricultural risk factors for non-Hodgkin's lymphoma among men in Iowa and Minnesota. Cancer Res. 1992;52(9):2447–55.

100. Meinert R, Schüz J, Kaletsch U, et al. Leukemia and non-Hodgkin's lymphoma in childhood and exposure to pesticides: results of a register-based case-control study in Germany. Am J Epidemiol. 2000;151(7):639–46.
101. Fritschi L, Benke G, Hughes AM, et al. Occupational exposure to pesticides and risk of non-Hodgkin's lymphoma. Am J Epidemiol. 2005;162(9):849–57.
102. McDuffie HH, Pahwa P, McLaughlin JR, et al. Non-Hodgkin's lymphoma and specific pesticide exposures in men: cross-Canada study of pesticides and health. Cancer Epidemiol Biomark Prev. 2001;10(11):1155–63.
103. Yildirim M, Karakilinc H, Yildiz M, et al. Non-Hodgkin lymphoma and pesticide exposure in Turkey. Asian Pac J Cancer Prev. 2013;14(6):3461–3.
104. Schinasi L, Leon ME. Non-Hodgkin lymphoma and occupational exposure to agricultural pesticide chemical groups and active ingredients: a systematic review and meta-analysis. Int J Environ Res Public Health. 2014;11(4):4449–527.
105. Poh C, McPherson JD, Tuscano J, et al. Environmental pesticide exposure and non-Hodgkin lymphoma survival: a population-based study. BMC Med. 2022;20(1):165.
106. Faivdullah L, Azahar F, Htike ZZ, et al. Leukemia detection from blood smears. J Med Biol Eng. 2015;4(6):488–91.
107. Sielken RL, Valdez-Flores C. A comprehensive review of occupational and general population cancer risk: 1, 3-Butadiene exposure–response modeling for all leukemia, acute myelogenous leukemia, chronic lymphocytic leukemia, chronic myelogenous leukemia, myeloid neoplasm and lymphoid neoplasm. Chem Biol Interact. 2015;241:50–8.
108. Sakamoto KM, Grant S, Saleiro D, et al. Targeting novel signaling pathways for resistant acute myeloid leukemia. Mol Genet Metab. 2015;114(3):397–402.
109. Miranda-Filho AL, Piñeros M, Ferlay J, et al. Epidemiological patterns of leukaemia in 184 countries: a population-based study. Lancet Haematol. 2018;5(1):e14–24.
110. Shallis RM, Weiss JJ, Deziel NC, et al. Challenging the concept of de novo acute myeloid leukemia: Environmental and occupational leukemogens hiding in our midst. Blood Rev. 2021;47:100760.
111. Foucault A, Vallet N, Ravalet N, et al. Occupational pesticide exposure increases risk of acute myeloid leukemia: a meta-analysis of case–control studies including 3,955 cases and 9,948 controls. Sci Rep. 2021;11(1):1–13.
112. Brown LM, Blair A, Gibson R, et al. Pesticide exposures and other agricultural risk factors for leukemia among men in Iowa and Minnesota. Cancer Res. 1990;50(20):6585–91.
113. Nguyen A, Crespi CM, Vergara X, et al. Residential proximity to plant nurseries and risk of childhood leukemia. Environ Res. 2021;200:111388.
114. Patel DM, Gyldenkærne S, Jones RR, et al. Residential proximity to agriculture and risk of childhood leukemia and central nervous system tumors in the Danish national birth cohort. Environ Int. 2020;143:105955.
115. Park AS, Ritz B, Yu F, et al. Prenatal pesticide exposure and childhood leukemia–a California statewide case-control study. Int J Hyg Environ Health. 2020;226:113486.
116. Karalexi MA, Tagkas CF, Markozannes G, et al. Exposure to pesticides and childhood leukemia risk: a systematic review and meta-analysis. Environ Pollut. 2021;285:117376.
117. Bray F, Ren JS, Masuyer E, et al. Global estimates of cancer prevalence for 27 sites in the adult population in 2008. Int J Cancer. 2013;132(5):1133–45.
118. Farmanfarma KK, Mohammadian M, Shahabinia Z, et al. Brain cancer in the world: an epidemiological review. World Cancer Res J. 2019;6(5)
119. Johnson KJ, Cullen J, Barnholtz-Sloan JS, et al. Childhood brain tumor epidemiology: a brain tumor epidemiology consortium review. Cancer Epidemiol Biomark Prev. 2014;23(12):2716–36.
120. Bondy ML, Scheurer ME, Malmer B, et al. Brain tumor epidemiology: consensus from the Brain Tumor Epidemiology Consortium. Cancer. 2008;113(S7):1953–68.
121. Miranda-Filho AL, Koifman RJ, Koifman S, et al. Brain cancer mortality in an agricultural and a metropolitan region of Rio de Janeiro, Brazil: a population-based, age-period-cohort study, 1996–2010. BMC Cancer. 2014;14(1):1–9.

122. Carreón T, Butler MA, Ruder AM, et al. Gliomas and farm pesticide exposure in women: the Upper Midwest Health Study. Environ Health Perspect. 2005;113(5):546–51.
123. Lee WJ, Colt JS, Heineman EF, et al. Agricultural pesticide use and risk of glioma in Nebraska, United States. Occup Environ Med. 2005;62(11):786–92.
124. Provost D, Cantagrel A, Lebailly P, et al. Brain tumours and exposure to pesticides: a case–control study in southwestern France. Occup Environ Med. 2007;64(8):509–14.
125. Bhat AR, Wani MA, Kirmani AR, et al. Pesticides and brain cancer linked in orchard farmers of Kashmir. Indian J Med Paediatr Oncol. 2010;31(4):110–20.
126. Shim YK, Mlynarek SP, van Wijngaarden E. Parental exposure to pesticides and childhood brain cancer: US Atlantic coast childhood brain cancer study. Environ Health Perspect. 2009;117(6):1002–6.
127. Zhang L, Tai YT, Ho M, et al. Regulatory B cell-myeloma cell interaction confers immunosuppression and promotes their survival in the bone marrow milieu. Blood Cancer J. 2017;7(3):e547.
128. Huang J, Chan SC, Lok V, et al. The epidemiological landscape of multiple myeloma: a global cancer registry estimate of disease burden, risk factors, and temporal trends. Lancet Haematol. 2022;9(9):670–7.
129. Viel JF, Richardson ST. Lymphoma, multiple myeloma and leukaemia among French farmers in relation to pesticide exposure. Soc Sci Med. 1993;37(6):771–7.
130. Brown LM, Burmeister LF, Everett GD, et al. Pesticide exposures and multiple myeloma in Iowa men. Cancer Causes Control. 1993;4:153–6.
131. Orsi L, Delabre L, Monnereau A, et al. Occupational exposure to pesticides and lymphoid neoplasms among men: results of a French case-control study. Occup Environ Med. 2009;66(5):291–8.
132. Kachuri L, Demers PA, Blair A, et al. Multiple pesticide exposures and the risk of multiple myeloma in Canadian men. Int J Cancer. 2013;133(8):1846–58.
133. Pahwa P, Karunanayake CP, Dosman JA, et al. Multiple myeloma and exposure to pesticides: a Canadian case-control study. J Agromedicine. 2012;17(1):40–50.
134. Presutti R, Harris SA, Kachuri L, et al. Pesticide exposures and the risk of multiple myeloma in men: an analysis of the North American Pooled Project. Int J Cancer. 2016;139(8):1703–14.
135. Figgs LW, Holland NT, Rothman N, et al. Increased lymphocyte replicative index following 2,4-dichlorophenoxyacetic acid herbicide exposure. Cancer Causes Control. 2000;11:373–80.
136. Lavrik IN, Golks A, Krammer PH, et al. Caspases: pharmacological manipulation of cell death. J Clin Invest. 2005;115(10):2665–72.
137. Raab MS, Podar K, Breitkreutz I, et al. Multiple myeloma. Lancet. 2009;374(9686):324–39.
138. Turusov V, Rakitsky V, Tomatis L. Dichlorodiphenyltrichloroethane (DDT): ubiquity, persistence, and risks. Environ Health Perspect. 2002;110(2):125–8.
139. Randhawa N, Gulland F, Ylitalo GM, et al. Sentinel California sea lions provide insight into legacy organochlorine exposure trends and their association with cancer and infectious disease. One Health. 2015;1:37–43.
140. Li L, Zhang Y, Wang J, et al. History traces of HCHs and DDTs by groundwater dating and their behaviours and ecological risk in northeast China. Chemosphere. 2020;257:127212.
141. Mansouri A, Cregut M, Abbes C, et al. The environmental issues of DDT pollution and bioremediation: a multidisciplinary review. Appl Biochem Biotechnol. 2017;181:309–39.
142. Thakur M, Pathania D. Environmental fate of organic pollutants and effect on human health. In: Singh P, Kumar A, Borthakur A, editors. Abatement of environmental pollutants. Elsevier; 2020. p. 245–62.
143. Jaga K, Dharmani C. Global surveillance of DDT and DDE levels in human tissues. Int J Occup Med Environ Health. 2003;16(1):7–20.
144. John TJ, Dandona L, Sharma VP, et al. Continuing challenge of infectious diseases in India. Lancet. 2011;377(9761):252–69.
145. Van den Berg H. Global status of DDT and its alternatives for use in vector control to prevent disease. Environ Health Perspect. 2009;117(11):1656–63.

146. Garabrant DH, Held J, Langholz B, et al. DDT and related compounds and risk of pancreatic cancer. J Natl Cancer Inst. 1992;84(10):764–71.
147. Fryzek JP, Garabrant DH, Harlow SD, et al. A case-control study of self-reported exposures to pesticides and pancreas cancer in southeastern Michigan. Int J Cancer. 1997;72(1):62–7.
148. Cohn BA, Wolff MS, Cirillo PM, et al. DDT and breast cancer in young women: new data on the significance of age at exposure. Environ Health Perspect. 2007;115(10):1406–14.
149. McGlynn KA, Quraishi SM, Graubard BI, et al. Persistent organochlorine pesticides and risk of testicular germ cell tumors. J Natl Cancer Inst. 2008;100(9):663–71.
150. Persson EC, Graubard BI, Evans AA, et al. Dichlorodiphenyltrichloroethane and risk of hepatocellular carcinoma. Int J Cancer. 2012;131(9):2078–84.
151. Alavanja MC, Hofmann JN, Lynch CF, et al. Non-Hodgkin lymphoma risk and insecticide, fungicide and fumigant use in the agricultural health study. PLoS One. 2014;9(10):e109332.
152. Harada T, Takeda M, Kojima S, et al. Toxicity and carcinogenicity of dichlorodiphenyltrichloroethane (DDT). Toxicol Res. 2016;32(1):21–33.
153. Loomis D, Guyton K, Grosse Y, et al. Carcinogenicity of lindane, DDT, and 2, 4-dichlorophenoxyacetic acid. Lancet Oncol. 2015;16(8):891–2.
154. Kelce WR, Stone CR, Laws SC, et al. Persistent DDT metabolite p,p'-DDE is a potent androgen receptor antagonist. Nature. 1995;375:581–5.
155. Stejskal V, Vendl T, Aulicky R, et al. Synthetic and natural insecticides: Gas, liquid, gel and solid formulations for stored-product and food-industry pest control. Insects. 2021;12(7):590.
156. Saadati N, Abdullah MP, Zakaria Z, et al. Distribution and fate of HCH isomers and DDT metabolites in a tropical environment–case study Cameron Highlands–Malaysia. Chem Cent J. 2012;6(130):1–15.
157. Humphreys EH, Janssen S, Heil A, et al. Outcomes of the California ban on pharmaceutical lindane: clinical and ecologic impacts. Environ Health Perspect. 2008;116(3):297–302.
158. Ward MH, Colt JS, Metayer C, et al. Residential exposure to polychlorinated biphenyls and organochlorine pesticides and risk of childhood leukemia. Environ Health Perspect. 2009;117(6):1007–13.
159. Pattnaik M, Pany BK, Dena J, et al. Effect of organochlorine pesticides on living organisms and environment. Chem Sci Rev Lett. 2020;9:682–6.
160. Kasozi GN, Kiremire BT, Bugenyi FWB, et al. Organochlorine residues in fish and water samples from Lake Victoria, Uganda. J Environ Qual. 2006;35(2):584–9.
161. Parada H, Wolff MS, Engel LS, et al. Organochlorine insecticides DDT and chlordane in relation to survival following breast cancer. Int J Cancer. 2016;138(3):565–75.
162. Kachuri L, Beane Freeman LE, Spinelli JJ, et al. Insecticide use and risk of non-Hodgkin lymphoma subtypes: a subset meta-analysis of the North American Pooled Project. Int J Cancer. 2020;147(12):3370–83.
163. Lerro CC, Freeman LEB, DellaValle CT, et al. Pesticide exposure and incident thyroid cancer among male pesticide applicators in Agricultural Health Study. Environ Int. 2021;146:106187.
164. Mortazavi N, Asadikaram G, Ebadzadeh MR, et al. Organochlorine and organophosphorus pesticides and bladder cancer: a case-control study. J Cell Biochem. 2019;120(9):14847–59.
165. Khansari N, Shakiba Y, Mahmoudi M. Chronic inflammation and oxidative stress as a major cause of age-related diseases and cancer. Recent Patents Inflamm Allergy Drug Discov. 2009;3(1):73–80.
166. Breckenridge CB, Campbell JL, Clewell HJ, et al. PBPK-based probabilistic risk assessment for total chlorotriazines in drinking water. Toxicol Sci. 2016;150(2):269–82.
167. He H, Liu Y, You S, et al. A review on recent treatment technology for herbicide atrazine in contaminated environment. Int J Environ Res Public Health. 2019;16(24):5129.
168. Jones RM, Stayner LT, Demirtas H. Multiple imputation for assessment of exposures to drinking water contaminants: Evaluation with the Atrazine Monitoring Program. Environ Res. 2014;134:466–73.
169. Hopenhayn-Rich C, Stump ML, Browning SR. Regional assessment of atrazine exposure and incidence of breast and ovarian cancers in Kentucky. Arch Environ Contam Toxicol. 2002;42:127–36.

170. Inoue-Choi M, Weyer PJ, Jones RR, et al. Atrazine in public water supplies and risk of ovarian cancer among postmenopausal women in the Iowa Women's Health Study. Occup Environ Med. 2016;73(9):582–7.
171. Rull RP, Gunier R, Von Behren J, et al. Residential proximity to agricultural pesticide applications and childhood acute lymphoblastic leukemia. Environ Res. 2009;109(7):891–9.
172. Malagoli C, Costanzini S, Heck JE, et al. Passive exposure to agricultural pesticides and risk of childhood leukemia in an Italian community. Int J Hyg Environ Health. 2016;219(8):742–8.
173. McElroy JA, Gangnon RE, Newcomb PA, et al. Risk of breast cancer for women living in rural areas from adult exposure to atrazine from well water in Wisconsin. J Expo Sci Environ Epidemiol. 2007;17(2):207–14.
174. Tinfo NS, Hotchkiss MG, Buckalew AR, et al. Understanding the effects of atrazine on steroidogenesis in rat granulosa and H295R adrenal cortical carcinoma cells. Reprod Toxicol. 2011;31(2):184–93.
175. Urseler N, Bachetti R, Biolé F, et al. Atrazine pollution in groundwater and raw bovine milk: water quality, bioaccumulation and human risk assessment. Sci Total Environ. 2022;852:158498.
176. Sidhu GK, Singh S, Kumar V, et al. Toxicity, monitoring and biodegradation of organophosphate pesticides: a review. Crit Rev Environ Sci Technol. 2019;49(13):1135–87.
177. Mink PJ, Mandel JS, Sceurman BK, et al. Epidemiologic studies of glyphosate and cancer: a review. Regul Toxicol Pharmacol. 2012;63(3):440–52.
178. Benbrook CM. Trends in glyphosate herbicide use in the United States and globally. Environ Sci Eur. 2016;28(1):1–15.
179. Davoren MJ, Schiestl RH. Glyphosate-based herbicides and cancer risk: a post-IARC decision review of potential mechanisms, policy and avenues of research. Carcinogenesis. 2018;39(10):1207–15.
180. Samsel A, Seneff S. Glyphosate, pathways to modern diseases IV: cancer and related pathologies. J Biol Phys Chem. 2015;15(3):121–59.
181. Swanson NL, Leu A, Abrahamson J, et al. Genetically engineered crops, glyphosate and the deterioration of health in the United States of America. J Org Chem. 2014;9(2):6–37.
182. Alavanja MC, Sandler DP, McMaster SB, et al. The agricultural health study. Environ Health Perspect. 1996;104(4):362–9.
183. Engel LS, Hill DA, Hoppin JA, et al. Pesticide use and breast cancer risk among farmers' wives in the Agricultural Health Study. Am J Epidemiol. 2005;161(2):121–35.
184. Lee WJ, Sandler DP, Blair A, et al. Pesticide use and colorectal cancer risk in the Agricultural Health Study. Int J Cancer. 2007;121(2):339–46.
185. Andreotti G, Beane Freeman LE, Hou L, et al. Agricultural pesticide use and pancreatic cancer risk in the Agricultural Health Study Cohort. Int J Cancer. 2009;124(10):2495–500.
186. Dennis LK, Lynch CF, Sandler DP, et al. Pesticide use and cutaneous melanoma in pesticide applicators in the agricultural heath study. Environ Health Perspect. 2010;118(6):812–7.
187. Andreotti G, Koutros S, Hofmann JN, et al. Glyphosate use and cancer incidence in the agricultural health study. J Natl Cancer Inst. 2018;110(5):509–16.
188. De Roos A, Zahm SH, Cantor KP, et al. Integrative assessment of multiple pesticides as risk factors for non-Hodgkin's lymphoma among men. Occup Environ Med. 2003;60(9):e11.
189. Pahwa M, Beane Freeman LE, Spinelli JJ, et al. Glyphosate use and associations with non-Hodgkin lymphoma major histological sub-types. Scand J Work Environ Health. 2019;45(6):600–9.
190. Eriksson M, Hardell L, Carlberg M, et al. Pesticide exposure as risk factor for non-Hodgkin lymphoma including histopathological subgroup analysis. Int J Cancer. 2008;123(7):1657–63.
191. Miller K. Estrogen and DNA damage: the silent source of breast cancer? J Natl Cancer Inst. 2003;95(2):100–2.
192. Balkwill F, Charles KA, Mantovani A. Smoldering and polarized inflammation in the initiation and promotion of malignant disease. Cancer Cell. 2005;7(3):211–7.
193. Solomon KR. Glyphosate in the general population and in applicators: a critical review of studies on exposures. Crit Rev Toxicol. 2016;46(sup1):21–7.

194. Bonner MR, Coble J, Blair A, et al. Malathion exposure and the incidence of cancer in the agricultural health study. Am J Epidemiol. 2007;166(9):1023–34.
195. Eddleston M, Karalliedde L, Buckley N, et al. Pesticide poisoning in the developing world—a minimum pesticides list. The Lancet. 2002;360(9340):1163–7.
196. Gilliom RJ, Barbash JE, Crawford CG et al. 2007. Pesticides in the nation's streams and ground water 1992–2001. National Water Quality Assessment Program, Circular 1291. US Geological Survey. https://pubs.usgs.gov/circ/2005/1291/pdf/circ1291.pdf. Accessed 11 Feb 2023.
197. Koutros S, Harris SA, Spinelli JJ, et al. Non-Hodgkin lymphoma risk and organophosphate and carbamate insecticide use in the North American Pooled Project. Environ Int. 2019;127:199–205.
198. Latifovic L, Beane Freeman LE, Spinelli JJ, et al. Pesticide Use and Risk of Hodgkin Lymphoma: Results from the North American Pooled Project (NAPP). Cancer Causes Control. 2020;31:583–99.
199. Engel LS, Werder E, Satagopan J, et al. Insecticide use and breast cancer risk among farmers' wives in the Agricultural Health Study. Environ Health Perspect. 2017;125(9):097002.
200. Echiburu-Chau C, Calaf GM. Rat lung cancer induced by malathion and estrogen. Int J Oncol. 2008;33(3):603–11.
201. Calaf GM, Bleak TC, Roy D. Signs of carcinogenicity induced by parathion, malathion, and estrogen in human breast epithelial cells. Oncol Rep. 2021;45(4):24.
202. Bradberry SM, Proudfoot AT, Vale JA. Poisoning due to chlorophenoxy herbicides. Toxicol Rev. 2004;23:65–73.
203. Peterson MA, McMaster SA, Riechers DE, et al. 2, 4-D past, present, and future: a review. Weed Technol. 2016;30(2):303–45.
204. Ozkul M, Ozel CA, Yuzbasioglu D, et al. Does 2, 4-dichlorophenoxyacetic acid (2, 4-D) induce genotoxic effects in tissue cultured Allium roots? Cytotechnology. 2016;68:2395–405.
205. Miligi L, Costantini AS, Veraldi A, et al. Cancer and pesticides: an overview and some results of the Italian multicenter case–control study on hematolymphopoietic malignancies. Ann N Y Acad Sci. 2006;1076(1):366–77.
206. Burns C, Bodner K, Swaen G, et al. Cancer incidence of 2, 4-D production workers. Int J Environ Res Public Health. 2011;8(9):3579–90.
207. Smith AM, Smith MT, La Merrill MA, et al. 2,4-Dichlorophenoxyacetic acid (2,4-D) and risk of non-Hodgkin lymphoma: a meta-analysis accounting for exposure levels. Ann Epidemiol. 2017;27(4):281–9.
208. Bukowska B. Effects of 2, 4-D and its metabolite 2, 4-dichlorophenol on antioxidant enzymes and level of glutathione in human erythrocytes. Comp Biochem Physiol C Toxicol Pharmacol. 2003;135(4):435–41.
209. Troudi A, Ben Amara I, Samet AM, et al. Oxidative stress induced by 2, 4-phenoxyacetic acid in liver of female rats and their progeny: Biochemical and histopathological studies. Environ Toxicol. 2012;27(3):137–45.
210. Palmeira CM, Moreno AJ, Madeira VM. Thiols metabolism is altered by the herbicides paraquat, dinoseb and 2, 4-D: a study in isolated hepatocytes. Toxicol Lett. 1995;81(2–3):115–23.
211. Bukowska B. Toxicity of 2, 4-dichlorophenoxyacetic acid--molecular mechanisms. Pol J Environ Stud. 2006;15(3):365–74.
212. Lerro CC, Hofmann JN, Andreotti G, et al. Dicamba use and cancer incidence in the Agricultural Health Study: an updated analysis. Int J Epidemiol. 2020;49(4):1326–37.
213. Alavanja MC, Dosemeci M, Samanic C, et al. Pesticides and lung cancer risk in the agricultural health study cohort. Am J Epidemiol. 2004;160(9):876–85.
214. Samanic C, Rusiecki J, Dosemeci M, et al. Cancer incidence among pesticide applicators exposed to dicamba in the agricultural health study. Environ Health Perspect. 2006;114(10):1521–6.
215. Leon ME, Schinasi LH, Lebailly P, et al. Pesticide use and risk of non-Hodgkin lymphoid malignancies in agricultural cohorts from France, Norway and the USA: a pooled analysis from the AGRICOH consortium. Int J Epidemiol. 2019;48(5):1519–35.

216. Mesnage R, Brandsma I, Moelijker N, et al. Genotoxicity evaluation of 2, 4-D, dicamba and glyphosate alone or in combination with cell reporter assays for DNA damage, oxidative stress and unfolded protein response. Food Chem Toxicol. 2021;157:112601.
217. Espandiari P, Glauert HP, Lee EY, et al. Promoting activity of the herbicide dicamba (2-methoxy-3, 6-dichlorobenzoic acid) in two stage hepatocarcinogenesis. Int J Oncol. 1999;14(1):79–163.
218. Lanasa MC, Weinberg JB. Immunologic aspects of monoclonal B-cell lymphocytosis. Immunol Res. 2011;49:269–80.
219. Thatheyus AJ, Selvam AG. Synthetic pyrethroids: toxicity and biodegradation. Appl Ecol Environ Sci. 2013;1(3):33–6.
220. Boulware DR, Beisang AA. Passive prophylaxis with permethrin-treated tents reduces mosquito bites among North American summer campers. Wilderness Environ Med. 2005;16(1):9–15.
221. Gupta R, Gahlot P, Purohit P. Pyrethroids: exposure, toxicity, and their influence on human health. J Environ Sci Health C Toxicol Carcinog. 2013;31(4):363–88.
222. Rusiecki JA, Patel R, Koutros S, et al. Cancer incidence among pesticide applicators exposed to permethrin in the Agricultural Health Study. Environ Health Perspect. 2009;117(4):581–6.
223. Ferreira JD, Couto AC, Pombo-de-Oliveira MS, et al. In utero pesticide exposure and leukemia in Brazilian children< 2 years of age. Environ Health Perspect. 2013;121(2):269–75.
224. Boffetta P, Desai V. Exposure to permethrin and cancer risk: a systematic review. Crit Rev Toxicol. 2018;48(6):433–42.
225. Shrestha S, Parks CG, Umbach DM, et al. Use of permethrin and other pyrethroids and mortality in the Agricultural Health Study. Occup Environ Med. 2022;79(10):664–72.
226. Borkhardt A, Wilda M, Fuchs U, et al. Congenital leukemia after heavy abuse of permethrin during pregnancy. Arch Dis Child Fetal Neonatal Ed. 2003;88(5):F436–7.
227. LaFiura KM, Bielawski DM, Posecion NC Jr, et al. Association between prenatal pesticide exposures and the generation of leukemia-associated T (8; 21). Pediatr Blood Cancer. 2007;49(5):624–8.
228. Ye M, Beach J, Martin JW, et al. Occupational pesticide exposures and respiratory health. Int J Environ Res Public Health. 2013;10(12):6442–71.
229. Paiga P, Morais S, Correia M, et al. Determination of carbamate and urea pesticide residues in fresh vegetables using microwave-assisted extraction and liquid chromatography. Int J Environ Anal Chem. 2009;89(3):199–210.
230. Wei G, Li Y, Wang X. Comparison of efficiencies between single-drop microextraction and continuous-flow microextraction for the determination of methomyl in natural waters. Int J Environ Anal Chem. 2008;88(6):397–408.
231. Çelebi MS, Oturan N, Zazou H, et al. Electrochemical oxidation of carbaryl on platinum and boron-doped diamond anodes using electro-Fenton technology. Sep Purif Technol. 2015;3:996–1002.
232. World Health Organization. 2006. Pesticides and their application: for the control of vectors and pests of public health importance. https://apps.who.int/iris/handle/10665/69223.
233. Pesatori AC, Sontag JM, Lubin JH, et al. Cohort mortality and nested case-control study of lung cancer among structural pest control workers in Florida (United States). Cancer Causes Control. 1994;5(4):310–8.
234. Mahajan R, Blair A, Coble J, et al. Carbaryl exposure and incident cancer in the Agricultural Health Study. Int J Cancer. 2007;121(8):1799–805.
235. Landgren O, Kyle RA, Hoppin JA, et al. Pesticide exposure and risk of monoclonal gammopathy of undetermined significance in the Agricultural Health Study. Blood. 2009;113(25):6386–91.
236. Boccolini PD, Boccolini CS, Chrisman JD, et al. Non-Hodgkin lymphoma among Brazilian agricultural workers: a death certificate case-control study. Arch Environ Occup Health. 2017;72(3):139–44.

237. Acquavella JF, Alexander BH, Mandel JS, et al. Glyphosate biomonitoring for farmers and their families: results from the Farm Family Exposure Study. Environ Health Perspect. 2004;112(3):321–6.
238. Xia Y, Cheng S, Bian Q, et al. Genotoxic effects on spermatozoa of carbaryl-exposed workers. Toxicol Sci. 2005;85(1):615–23.
239. Peyre L, Zucchini-Pascal N, De Sousa G, et al. Potential involvement of chemicals in liver cancer progression: an alternative toxicological approach combining biomarkers and innovative technologies. Toxicol In Vitro. 2014;28(8):1507–20.
240. Alavanja MC. Pesticides use and exposure extensive worldwide. Environ Health Rev. 2009;24(4):303–9.
241. Damalas CA, Koutroubas SD. Farmers' exposure to pesticides: toxicity types and ways of prevention. Toxics. 2016;4(1):1.
242. Arcury TA, Grzywacz JG, Chen H, et al. Variation across the agricultural season in organophosphorus pesticide urinary metabolite levels for Latino farmworkers in eastern North Carolina: project design and descriptive results. Am J Ind Med. 2009;52(7):539–50.
243. Temkin AM, Uche UI, Evans S, et al. Racial and social disparities in Ventura County, California related to agricultural pesticide applications and toxicity. Sci Total Environ. 2022;853:158399.
244. Gee GC, Payne-Sturges DC, Martinez M. Environmental health disparities: a framework integrating psychosocial and environmental concepts. Environ Health Perspect. 2007;115(5):1645–55.
245. Zavala VA, Bracci PM, Carethers JM, et al. Cancer health disparities in racial/ethnic minorities in the United States. Br J Cancer. 2021;124(2):315–32.
246. Quandt SA, Hernández-Valero MA, Grzywacz JG, et al. Workplace, household, and personal predictors of pesticide exposure for farmworkers. Environ Health Perspect. 2006;114(6):943–52.
247. Zhang L, Rana I, Shaffer RM, et al. Exposure to glyphosate-based herbicides and risk for non-Hodgkin lymphoma: a meta-analysis and supporting evidence. Mutat Res Rev Mutat Res. 2019;781:186–206.

Environmental Justice, Equity and Cancer

9

Leticia Nogueira and Kristi E. White

Introduction

As has been discussed earlier in this book, cancer is one of the leading causes of death worldwide. The most frequently discussed approaches for addressing cancer have focused on individual behavioral changes (e.g. smoking, diet, physical activity) and access to care (e.g. screening, treatment) [1, 2]. However, environmental exposures are often overlooked as risk factors that worsen cancer risk and outcomes through increased exposure to carcinogens and disrupting access to cancer care [3]. Additionally, the risks, exposures, and disruptions are distributed inequitably across communities. In this chapter, we will discuss how discriminatory and oppressive practices place certain communities at greater risk of cancer disparities, the environmental justice implications of this relationship, and offer solutions that prioritize the most disproportionately impacted communities with an emphasis on intervening at root cause levels. Throughout the chapter, our intention is to use content and language that embodies the principles of antiracism, equity and justice. Thus, we follow the antiracism recommendations of members from communities that have been subjected to marginalization, which include the use of intentional language and centering the voices and scholarship of those who are directly impacted by discriminatory policies and practices. As recommended, we use terminology and language that centers the *constraints* imposed on these communities as the root cause of health inequities [4, 5]. Additionally, when referring to racialized groups, we intentionally chose to keep "white" lower case to de-center whiteness in favor of centering the voices of those who have been subjected to marginalization and also to avoid

L. Nogueira (✉)
American Cancer Society, Kennesaw, GA, USA
e-mail: leticia.nogueira@cancer.org

K. E. White
Department of Medicine, University of Minnesota, Minneapolis, MN, USA

E. H. Bernicker (ed.), *Environmental Oncology*,
https://doi.org/10.1007/978-3-031-33750-5_9

appropriating the significant labor of Black writers and scholars who worked to shift writing practices in favor of capitalizing Black when referring to racial identity.

Definitions

Environmental injustices are a result of systemic racism and contribute to a multitude of health disparities. We highlight a few important definitions before discussing the connection between environmental justice and cancer.

Health Disparities

Health disparities are a specific type of deleterious health difference that affects groups of people who have been subjected to systemic discriminatory or exclusionary social and/or economic obstacles to health [6, 7]. Race is a social construct without biological meaning [8], and the main characteristic historically linked to discrimination and exclusion in the US.

Race is a poor proxy for genetic background [9]; therefore, genetic differences should not be presented as a potential cause of racial health disparities. For example, Black women are more likely to be diagnosed with Estrogen Receptor (ER)-negative breast cancer than white women [10]. However, increases in incidence of this especially fatal type of breast cancer are not in line with the pace of change in phenotypic standard deviation per generation required for genetic background to be a plausible explanation for this increased risk [11]. Instead, such rapid phenotypic change can only be driven by factors exogenous to populations' genomes.

Because race is a social construct, exposure to different forms of racism is the only mechanism through which racial categorizations have biological consequences [12–14].

Racism

We have previously noted the different forms of racism and their impacts on health within the context of climate-related health inequities [7], which we will briefly highlight here. There are several different forms and levels of racism, all of which are rooted in discriminatory structures, policies, and practices, and all have health consequences [15–17].

Internalized racism consists of one's own beliefs about race, which are learned and internalized racism is associated with several negative health outcomes [18–20].

Interpersonal racism refers to race-driven discriminatory interactions between individuals, and also has negative consequences, including substandard cancer care and worse outcomes among individuals racialized as Black, Latinx, etc. [21–39]. For example, Black individuals are less likely to be included in clinical trials [40, 41]; and Black children are less likely to receive potentially superior radiation treatment even when enrolled in clinical trials where treatment assignment is highly standardized [23]. Additionally, the lack of diversity in the medical workforce [42–44] contributes to the inability of the healthcare system to demonstrate trustworthiness, which can further contribute to the detrimental consequences of interpersonal racism [45, 46].

Institutional racism includes prejudicial and biased policies and practices of institutions that result in unfair outcomes for communities subjected to race-based marginalization. For example, the Federal Emergency Management Agency (FEMA) is less likely to provide assistance to disaster survivors racialized as Black, and FEMA provides greater post-disaster financial assistance to those racialized as white, even when the amount of property damage is the same [47, 48].

Systemic racism refers to ways in which a variety of factors (institutional policies, individual practices, cultural representations) combine to systematically disadvantage communities that are subjected to race-based marginalization. For example, instead of Universal Healthcare, the Social Security Act of 1935 created a system of employment-based health insurance coverage in the US that deliberately excluded occupations predominantly held by Black individuals in the South [16]. This system, combined with current racial disparities in education and employment opportunities and discriminatory hiring practices, result in restricted access to health insurance coverage for individuals who are subjected to race-based marginalization, further leading to health inequities [6, 49, 50]. Health insurance coverage is one of the main determinants of access to quality cancer care in the US; therefore, disparities in health insurance coverage are one of the main determinants of racial disparities in cancer care and outcomes [51–53].

Structural racism is the aggregation of societal reinforcement of racial discrimination through inequitable systems. There are many ways in which structural racism impacts health [54]. For example, in 1933, the federal government established the Home Owner's Loan Corporation (HOLC) to expand homeownership as part of recovery from the great depression. HOLC staff used racial composition as part of its assessment of areas worthy of receiving loans and drew red lines (hence the term "redlining") around communities with large Black populations to denote these neighborhoods as hazardous investment areas [55]. HOLC maps captured how hierarchical racial categorizations of "worthiness" have long-term consequences. Institutional, interpersonal, and systemic racist practices within and outside of HOLC restricted access to home ownership (an asset that is central to intergenerational wealth transfer), undervalued real state in predominantly Black neighborhoods, and limited housing choices and residential mobility to individuals from communities targeted for marginalization [55, 56]. Physical separation of racialized groups facilitated intended and unintended unequal treatment, including exposure to health hazards and access to health resources [57–61]. Racial residential segregation created a platform for systemic disinvestment in neighborhoods targeted for marginalization [55, 62, 63], including substandard infrastructure (e.g. greenspaces, housing stock, roads) [64–70], services (e.g., clean and safe water, transportation, schools) [71–73], and a concentration of psychosocial stressors and barriers to health resources (e.g. availability of alcohol and tobacco outlets, unavailability of healthy foods and environments that facilitate physical activity) [74–81]. These conditions led to increased prevalence of chronic health conditions among individuals from communities targeted for marginalization [82–85]; one of the main determinants of disparities in cancer mortality [86–88]. Simply reporting racial health disparities, such as "higher prevalence of smoking and comorbidities among Black individuals" [89], without naming racism and centering the conditions imposed on individuals from communities targeted for marginalization renders the systems of oppression invisible and promotes racial stereotypes [90], further contributing to structural racism [12, 15, 16, 91–93].

Recognizing the social and environmental conditions imposed on racialized groups as the fundamental causes of health outcomes can help identify modifiable systemic factors contributing to health disparities and help dismantle structural racism [94–99].

Environmental Justice

Environmental justice is defined as the fair treatment and meaningful involvement of all people, regardless of race, color, national origin, or income, in the development, implementation, and enforcement of laws, regulations, and policies that affect the environment and public health. In the US, environmental injustices have primarily targeted groups racialized as Black, Latinx, and Native American [100–103].

Brief Overview of Environmental Justice History

In 1982, after identifying soil tainted with polychlorinated biphenyls (PCBs), which are known human carcinogens [104], the state of North Carolina evaluated nearly 100 potential sites for disposal. After selecting a site, the state of North Carolina had to apply for approval by the Environmental Protection Agency (EPA) sets the criteria for identifying feasible disposal sites. As the name of the Agency suggests, these criteria are in place for protecting the integrity of the environment (and subsequently of residents) at these sites.

When the state of North Carolina applied for approval of the disposal site in Warren County, the state requested that the EPA waive three of its criteria: (1) the requirement that groundwater be at least 50 feet below the landfill bottom (the groundwater was 7 feet below the landfill bottom at the chosen site); (2) the need for an artificial liner; and (3) the need for an underliner leachate collection system [56].

Not surprisingly, residents of Warren County began protesting the construction of the disposal site. In addition to the risks posed by waiving the three EPA's criteria as requested in the application, an independent consultant found that the soil at the Warren County site did not meet two additional criteria namely, the EPA's compaction criteria (i.e. would not form a protective layer when compressed) and the chemical-exchange capacity criteria. Moreover, Warren County had previously passed an ordinance banning the disposal of PCBs, noting the area was particularly unsuitable due to the high water table and highly permeable soils [56].

Despite residents' valid concerns and activism, the EPA waived its own requirements and approved construction of the landfill in Warren County. Further, lawsuits filed by residents after the EPA approval were struck down by the court, which argued that local jurisdictions could not take action that would stymie the "national goal of proper disposal of hazardous chemical substances" [56]. The repeated failure of authorities responsible for protecting the health and safety of the population to do so in Warren County demonstrates how current systems of power designate some communities as worthy of sacrifice. In fact, these geographic areas of higher exposure to toxins and pollutants that are concentrated in communities subjected to marginalization are sometimes referred to (including by individuals from these communities) as "sacrifice zones."

If the geomorphology of Warren County was inadequate for the siting of a PCB landfill, what motivated the state of North Carolina to select the site, the EPA to waive its own criteria to approve the application, and the court to strike down lawsuits, and together ultimately enable the construction of the landfill despite residents' protests? At the time the site was chosen, Black individuals comprised 22% of the population in North Carolina, but 60% of the population in Warren County. This type of institutional racism goes against environmental justice principles of fair treatment and meaningful involvement of all people, regardless of race, in the implementation and enforcement of laws, regulations, and policies that affect the environment and public health.

The Warren County incident is known as the birth of environmental justice in the US and led to the commissioning of two reports examining the relationship between race, class, and the siting of hazardous facilities [105]. The first study, conducted by the US General Accounting Office (GAO) focused on southern states [106]. The second study, conducted by the United Church of Christ Commission for Racial Justice (UCCRJ), was a national study [107]. Both studies found that hazardous facilities and toxic waste sites were more likely to be located in proximity to where individuals racialized as Black and Latinx reside. In fact, race was the strongest predictor of where hazardous facilities were located, providing yet another example of the designation of certain communities as "sacrifice zones."

Environmental Justice and Cancer

Cancer risk and cancer mortality is higher for individuals from communities targeted for marginalization than for white individuals [108–110]. Prostate, lung, colon, kidney, liver, and pancreatic cancer incidence rates are worse for Black men than for white men [10]. Breast cancer diagnosis occurs at a younger age and has worse survival outcomes for Black women than for white women [111–113]. Death rates for myeloma, and cancers of the stomach, prostate, and uterine corpus are twice as high for Black people compared to white people in the US [10]. Incidence and mortality rates for cancers of the liver, stomach, and kidney are two times higher for American Indian and Alaska Native populations, compared to individuals racialized as white [114]. These differences in cancer risk and mortality are examples of how systemic racism leads to health disparities.

In the previous sections, we highlighted how race is a social construct with no biological meaning and how discriminatory policies and practices determine access to health resources, including healthy environments (which leads to disparities in the prevalence of comorbidities, a main determinant of cancer survival), and quality healthcare. In the following section, we discuss how discriminatory policies and practices determine exposure to environmental hazards, which is a core environmental justice issue.

Sources of Environmental Hazards

Environmental hazards are any substance, state, or event that threatens or adversely impacts health, such as pollution and disasters [115]. In this section, we will focus on exposure to carcinogens, although exposure to other environmental hazards, such as extreme weather events, also influence cancer outcomes [116].

Individuals from communities targeted for marginalization are disproportionately exposed to carcinogens from many different sources, including hazardous waste treatment facilities, highway traffic, and industrial toxins [101, 117–139]; all of which are associated with increased cancer risk [140–144]. We will highlight two sources of environmental hazards that exemplify the shared root causes of environmental injustices and cancer health disparities; industrialized food systems and fossil fuel infrastructure [145].

Industrialized Food Systems

Although the Warren County incident became known as the birth of the Environmental Justice movement, environmental injustices have occurred since the beginning of American settler colonialism [146]. Settler seizure and interruption of Indigenous food systems was a form of environmental injustice deployed as a tool for subjugation, as it led to food shortages, forcing populations into dependency on colonizers [147]. Settler colonialism targeting of Indigenous food systems (a foundation of culture, the connection between people and place, and a main determinant of health) [148], exemplifies the hierarchical mindset guiding the establishment of structures of domination, in which Indigenous Peoples, Knowledge, and Systems were regarded "less than" the colonizers', and where disconnection with place and other life forms are at the root of so many current environmental justice issues [149, 150].

There are many ways in which settler colonialism and its associated philosophies have perpetuated environmental injustices that continue today. For example, land theft and the imposition of the reservation system led to a dramatic shift in food patterns as Native Peoples were confined to the artificial boundaries created by the treaties [151]. With dramatically reduced territories and contamination of food systems [148, 152, 153], (e.g. Mohawk food sources contaminated with PCBs and dioxins [154], mercury and PCBs contamination in Aleutians food sources) [152, 155, 156], Indigenous peoples were forced to rely on government rations (eventually replaced by the commodity food program), which consisted of distributing foods high in calories and low in nutritional content [157], fueling the disproportionate prevalence of chronic health conditions (e.g. obesity, diabetes) among Native populations [158]. These chronic health conditions are associated with increased cancer risk and worse cancer outcomes [159, 160]. Instead of improving Indigenous peoples' health, these food programs increased food insecurity, building on the settler colonialism structures of domination, as food was distributed under government terms [161].

The legacy of settler colonialism mentality influences most power and knowledge structures in the present day, including environmental and public health efforts. In addition to the commodity food program, the green revolution is another illustrative example of how well-intended efforts implemented with colonialist concepts of hegemony and domination can have detrimental effects [162, 163]. By focusing solely on increasing "yield" (reductionism is a characteristic of settler colonialism), the green revolution promoted monoculture of hybrid seeds that relied heavily on the use of pesticides, fertilizers, and irrigation [164]. This reductionist view of the food system ignored the importance of diversity (of seeds and food sources) as well as the role of food as a pillar of culture and connection between people and the environment, which had devastating impacts on existing farming systems [162]. The green revolution also resulted in forced dependency on the government, food insecurity, and worse health outcomes among Native communities, similar to the commodity food program.

Further, there is growing understanding of how the promotion of industrialized agriculture has been harmful to the environment and to public health beyond Native communities. For example, increased industrialization of food production has increased exposure to pesticides in the entire population through food consumption [165–167], residential contamination among communities residing near agricultural fields [168–171], and especially among farmworker populations [172, 173]. Several pesticides are known carcinogens [174–176], and exposure is associated with detrimental health consequences beyond increased cancer risk [177]. Additionally, the most highly subsidized crops are the most abundantly produced and consumed [178], and consumption of food derived from subsidized commodities is associated with increased cardiometabolic risk [179]. Cardiometabolic conditions are associated with worse cancer survival [180]; therefore, the current industrialized food system impacts cancer control efforts through the cancer control continuum, from prevention to survival, for the entire population.

Settler colonialism also required the enslavement of people stolen from their homeland and transported to labor the land stolen form Indigenous People under precarious conditions [181]. The same hierarchical mentality of colonial domination and disregard for life shapes the conditions currently imposed on farmworkers [182].

Farmwork labor still exploits displaced populations with limited choices in the current discriminatory systems of power [183]. For example, approximately 75% of farmworkers are immigrants from Latin America [184]; territories that were severely impacted by environmental degradations and uneven distribution of resources resulting from the green revolution and trade agreements [162]. Farmworkers and their families face numerous barriers to health and safety, including exposure to chemical hazards (pesticides, air pollutants, etc.) [185–187], physical hazards (injuries, heat, etc.) [188, 189], and inadequate working conditions [186, 187, 190–198]. These health hazards combined with psychosocial stressors stemming from policy and cultural contexts (e.g. cultural tropes and law enforcement activities that undermine the ability of the healthcare system to demonstrate trustworthiness, as well as some immigration documentation types being excluded from eligibility to health insurance coverage and social aid in the US) [199–201] may interact to amplify health disparities [188, 202–205].

In the past decade, the industrialization of agriculture has also led to a sharp increase in the number of concentrated animal feeding operations (CAFOs) [206–208], with similar occupational and community health hazards [209–220]. Intensive livestock operations are more likely to be localized in proximity to communities targeted for marginalization, highlighting another example of the existence of "sacrifice zones." [208, 221]

Beyond upholding structures of power that result in disproportionate exposure of communities targeted for marginalization to the health hazards from our industrialized food system, the legacy of settler colonialism also hinders our ability to identify and implement solutions that would benefit the entire population. For example, Indigenous, Black, and Latinx individuals have developed sustainable land stewardship practices in unfamiliar territories after being forcibly displaced. However, this valuable expertise has been overlooked in favor of technocratic solutions developed with the same reductionist and hegemonistic mentality of settler colonialism (more on this in "Solutions") [222–226].

Fossil Fuel Infrastructure

The association between air pollution from vehicles emissions and lung cancer risk is the most commonly recognized link between reliance on fossil fuels and cancer. The same pollutants that cause the greenhouse effect also increase lung cancer risk [227]. However, the extraction, processing, transportation, and waste management of fossil fuels release vast amounts of carcinogens in the environment beyond what is released when these are consumed. Therefore, reliance on fossil fuels is a shared cause of climate change and increase exposure to carcinogens across communities.

Fossil fuel infrastructure is dispersed throughout the US, frequently in proximity to residential areas [228–233]. In addition to furthering climate change, which has health consequences around the globe [234], the large number of fossil fuel operations located in communities throughout the US introduces health hazards that worsen cancer risk and outcomes locally. Even low-level exposures to carcinogenic pollutants from fossil fuels are relevant because of the multiplicity of substances, the involuntary nature of exposures, and the significant contribution to cancer burden when large numbers of people are exposed [235].

For example, residents of coal-producing counties have increased exposure to mercury, arsenic, chromium, and nickel [236–240], all known carcinogens [241, 242]. Cancer incidence and mortality rates are higher in coal-producing counties in the US, even after adjusting for other risk factors such as prevalence of obesity, smoking, and social determinants of health [243, 244].

Similarly, individuals residing near oil and gas extraction, transportation, processing, and waste management sites are exposed to air contaminated with arsenic [245], benzene [246–257], polycyclic aromatic hydrocarbons (PAHs) [258, 259], and fine particulate matter [260–266], all known carcinogens [227, 267–271]. Communities living near drilling sites can also be exposed to carcinogens in drilling muds and waste [272], such as trichloroethylene (TCE), and tetrachloroethylene (PCE), arsenic, beryllium, and cadmium [273, 274], through contamination of drinking water [275–291].

Due to discriminatory policies and practices that limit choices and political power [292–294], individuals from communities targeted for marginalization are more likely to reside in proximity to fossil fuel infrastructure [295–303], and these communities are also targeted for development of new fossil fuel infrastructure, another example of "sacrifice zones." [149] Importantly, there is an increase in emissions and neighborhood contamination during extreme weather events [304–312], an escalating hazard scenario that is becoming more frequent with climate change [313]. Further, individuals from communities targeted for marginalization are more likely to experience disaster-related unmet needs and adverse event experiences [314–319], including exposure to environmental hazards [309, 320–324], and are less likely to receive post-disaster government assistance [47, 48, 325]. Finally, communities residing in proximity to fossil fuel infrastructure also have increased exposure to noise and pollution from operations and traffic [326–329], as well as psychosocial stress and community disruption from the influx of workers [330, 331], which can result in cumulative health impacts [228].

Restricted Choices

Understanding causal channels is crucial for designing policies addressing environmental injustices [332]. Land-use decision-making mirrors the power arrangements of the dominant society, perpetuating the practice of placing hazardous developments and toxic sites in disenfranchised communities while restricting the choices available to these individuals [333]. This type of procedural environmental injustice, which deems environmental degradation as "necessary" and restricts the choices available to individuals, is a legacy of the settler colonialism mindset of domination through subjugation [334].

We previously discussed the role of redlining in restricting choices available to individuals racialized as Black [55], which continues to influence housing and mobility options to individuals racialized as Black, Latinx, and Indigenous, regardless of income levels [54, 65]. Municipal ordinances, which were first created to partition cities into commercial and residential zones, are another example of how power arrangements rooted in a settler colonialism mindset lead to procedural environmental injustice. Municipal ordinances were used in combination with intimidation and coercion tactics, rent increases, and private agreements between landowners and realtors to limit the housing options available to individuals racialized as Black [335]. Strips of major roads were zoned "for commercial use only," restricting residential construction even if there was no commercial or industrial development in the area. These "buffer strips" acted as barriers to population movement and are still effective in promoting racial residential segregation. Highways, rail lines, dead-ended streets, and industrial siting were also used to create buffer strips between racially segregated neighborhoods [336–338].

The partition of cities into commercial and residential zones, work together with other exclusionary zoning practices, such as specification of large lots, low-density development, and bans on the development of multifamily housing units, to control the type of residential development, and therefore the types of residents that can inhabit a particular area [339–345]. Discriminatory practices of realtors (through practices such as racial steering) [346, 347] and financial institutions (through discriminatory mortgage denial and predatory lending practices) [348–351] also restrict housing and mobility options available to individuals from communities targeted for marginalization [49, 352–354]. These systemically racist policies and practices obstruct mobility and eliminate the option for individuals in these communities to relocate to areas that would confer decreased risk of exposure to known carcinogens [55, 100, 146].

Structural racism also makes it difficult for individuals from communities targeted for marginalization to advocate for the closure and cleanup of existing hazardous facilities and to oppose siting of new hazardous facilities. For example, zoning is the most widely applied mechanism to regulate urban land use in the US [355, 356]. Restricted access to information, choice of inconvenient time and location of meetings, and inadequate notification of events are some of the practices that lead to underrepresentation of residents from communities targeted for marginalization in land-use-related planning, regulatory, and decision-making processes [357].

Additionally, important determinants of an institution's decision regarding where to locate a new hazardous facility include (1) economic cost, which includes the land price, (2) potential lawsuit cost, which is calculated using local population density, average household income, and average house value, and (3) residents effective collective action, which is usually calculated using percent of housing units occupied by homeowners, percent of residents whose English is not the primary language (because zoning hearings and city-council meetings are conducted in English in the US), and local voting rates [126]. Therefore, the discriminatory policies and practices previously discussed, which result in limited paths to home ownership and household income options among individuals from communities targeted for marginalization, also serve to hinder the ability of residents to oppose siting of new hazardous facilities and advocate for the closure and cleanup of existing ones [55, 56, 100, 358–360]. Similarly, voter suppression tactics and underfunding of government agencies (such as the US Census' population counts, which determine voting districts and allocation of federal funds) work to weaken collective action of neighborhoods targeted for marginalization [361, 362].

It is important to note that environmental degradation and pollution resulting from the development and siting of hazardous facilities is not contained within targeted neighborhoods, but have detrimental health consequences for the entire population (through increased exposure to carcinogens, for example) [145, 363]. Therefore, efforts to dismantle the power structures and halt discriminatory policies and practices that perpetuate environmental injustices and enable the development of polluting infrastructure would improve the health of the entire population [364].

For example, the same exclusionary zoning practices implemented to foster racial residential segregation also perpetuate the need to commute (due to the separation of residential and commercial zones), and high building-related energy consumption (to heat and cool large detached homes), contributing to increased exposure to air pollution (even in privileged neighborhoods) [365] and exacerbating climate change [366, 367]. Because single-vehicle transportation is prioritized over other types of transportation in the US [368], individual-level behavioral changes, such as walking or biking to places of employment, retail, recreation, and basic services is frequently an unfeasible or unsafe approach for overcoming the need to commute by car [369]. In contrast to resources invested in approaches to promote individual behavior changes (such as uptake of active transportation), approaches aimed at the structures of power and the discriminatory policies and practices are necessary for addressing health disparities and environmental injustices, and these system-level approaches have health co-benefits for the entire population.

There is a wealth of knowledge and solutions coming from communities targeted for marginalization, who suffer first and worst from the consequences of the current power structures and the resulting environmental degradation in their communities [149, 325, 370–373]. As an example, we discussed highways as a tool to foster racial residential segregation that has detrimental health consequences for the entire population. Communities targeted for marginalization were exposed first and worst to the hazards of highway pollution [117, 374]. Individuals from these communities have led the way in identifying strategies for opposing highway expansion and removing existing highways in a way that prioritizes the needs of the people in surrounding neighborhoods [375, 376], and decreases exposure to air pollution for the entire population [363].

Further, vulnerability is a dynamic state [377], a concept that has long been recognized by First Peoples [149]. For example, due to the physical, psychological, and socioeconomic consequences of cancer diagnosis and treatment, individuals who have been diagnosed with cancer are vulnerable to the health hazards of climate change, regardless of racial or socioeconomic privilege [3, 116, 378, 379]. Also, the same power structures enabling environmental injustices against predominantly Black, Latinx, and Indigenous communities work to limit the choices and power of predominantly white communities to avoid exposure to carcinogens, as seen with coal mining operations in the Appalachian region [380, 381]. Therefore, identifying and opposing unjust power structures and discriminatory policies and practices that lead to environmental degradation is in the best interest of everyone [363].

Solutions

As we seek to understand the link between environmental justice and cancer in order to identify solutions, it is important to recognize the shared causes of environmental injustices and health issues of the entire population. The colonial mindset of hegemony and assumed superiority was used to create the current system that subsidizes environmental degradation (and our disconnections with the land and other life forms), institutionalizes exposure to environmental contaminants, and exploits the vulnerabilities of disenfranchised communities [100]. Solutions proposed with the same framework of settler colonialism domination will continue to propagate environmental injustice and health disparities [181], and will also increase exposure to health hazards and worsen health outcomes for the entire population [363].

The Role of Public Health Professionals

Public health professionals have an important role to play in dismantling the structures at the root of environmental injustices and health hazards for the entire population. When considering methods for disrupting the mindsets of those who hold and uphold unequal systems of power, it is important to consider the approach to addressing these mindsets lest they become reinforced or further entrenched.

For example, public health professionals portraying the unbroken connection of Indigenous Peoples with the environment as a "vulnerability" (i.e. leading to increased risk of mental health consequences from witnessing environmental degradation and/or climate change) dismisses existing resilience knowledge crucial for identifying adaptive strategies for the entire population, and can also be used to justify outside intervention and control (i.e. colonial dominance) [382]. Instead, public health professionals should foster the recognition that we are all dependent on the health of the land and other life forms for our own health, an important knowledge and a strength (not a vulnerability) that has been preserved among Indigenous Peoples.

In addition to decreasing the overall knowledge pool (in understanding the problem and identifying solutions) and contributing to white supremacy narratives, continuing to apply colonialism mindsets to environmental, cancer, and public health research practices also diminishes public support for interventions that would improve the health of the entire population [363]. For example, increasing awareness of racial health disparities in COVID-19 infection, morbidity, and mortality decreased white individuals' apprehension of COVID-19 as well as support for public health measures that would protect individuals of all racialized groups [383]. Thus, simple racial disparity awareness-raising strategies may have a counter-productive effect, particularly if they reinforce the zero-sum mindset that any resources or benefits available to those disproportionately impacted come at the cost of white communities.

Investing resources into communities targeted for marginalization, which have experienced targeted disinvestment for over a century, is an important aspect of environmental justice. In addition to reparative investments, solutions will also need to address behaviors, policies, and practices that are rooted in settler colonialism and zero-sum ideological frameworks to ensure equitable and just outcomes [149, 384]. For example, climate or environmental gentrification can occur when affluent communities with economic power move to disadvantaged areas for the desirable land topography (e.g., Miami-Dade county) [385] or when interventions focus on investment in "green" infrastructure development in neighborhoods without meaningful engagement with residents [386, 387], thereby displacing the communities that would otherwise benefit from climate adaptation and resilience efforts. Additionally, attempting to implement environmental solutions with a "knowledge superiority" and domination mindset that ignores the knowledge and expertise within communities that are directly impacted can be disruptive and add to the injustices experienced by these communities [386], or justifiably rejected for perpetuating the discriminatory or untrustworthy practices of those in power [388].

Thus, without addressing the mindsets that lead to hierarchical categorization of knowledge and life forms and allow the settler colonialism paradigm of domination to become a guiding principle in society, any policies, practices, strategies, or interventions developed by those with access to the power and resources for implementing them will continue to displace communities subjected to marginalization even in the face of "green solutions."

Similarly, the practice of reinforcing the zero-sum mindset through simply reporting health disparities has long-reaching consequences. Current policies that tie wealth to local political power—and therefore resource allocation—help perpetuate structural disadvantages and promote racial stereotypes that undercut support for policies with the potential to improve economic well-being, environmental conditions, and health outcomes for individuals with low SES from all racialized groups [361, 363, 389, 390]. For example, because the Social Security Act of 1935 was designed to exclude Black individuals from health insurance coverage, racial resentment among white individuals is the main predictor of support for policies that would expand access to coverage in the US [391]. However, the majority of individuals without health insurance coverage in the US are racialized as white [391]. Therefore, identifying and addressing the root causes of the mindsets that perpetuate systemic inequities will be an important component of implementing solutions that improve cancer outcomes for the entire population.

Dismantling the Settler Colonialism Mindset

The hierarchical categorization of knowledge that permeates cancer research, clinical practice, public health, and environmental efforts is rooted in a settler colonialism mindset. Dehumanization of minoritized groups and ideologies of cultural and intellectual inferiority were strategies used to justify land dispossession and subjugation during settler colonialism [147], and again during the green revolution [162]. However, there is no evidence the knowledge and practices imposed through colonial mindsets of domination were superior. For example, polycultures cultivated by First Peoples outperform the yields of the same crops grown as monocultures in industrialized settings [164], without constant requirement for the external inputs (e.g. hybrid seeds, fertilizers, pesticides), and with the genetic diversity and knowledge of varieties that thrive at a wider range of environmental conditions [392].

Evidence of skillful, innovative, and ingenious pre-colonial land stewardship strategies abound, ranging from domestication of tree species for forest management in the Amazon [393], controlled burn strategies for wildfire prevention in the Yosemite Valley [161], to raised field systems for wetland agriculture in the Puebla Basin [394]. However, attempting to implement any of these strategies globally with a settler colonialist mentality of "scale up" hegemony would not be feasible. Most importantly, imposing this colonialist mindset of domination and reductionism renders any form of knowledge generated from a kinship mindset as "globally unfeasible" and perpetuates the hierarchical categorization of knowledge. Instead, solutions identified through "place-based" knowledge and developed with a kinship mindset need to be prioritized.

Recently, the value of knowledge generated from a kinship mindset, which recognizes that interests are interdependent, instead of the zero-sum-game belief system of settler colonialism [149, 384], is increasingly being acknowledged. Despite extensive land theft, Indigenous territories nourish nearly 40% of the world's protected lands and hold local knowledge crucial for environmentally responsible [395], sustainable, resilient, and productive land management practices [396]. Similarly, many forcibly displaced Afro-descendant and Latin American communities cultivate a rich repertoire of agroecological knowledge developed over centuries of hardship and experimentation in host territories [397–400]. This type of knowledge developed through a kinship mindset with the host territory and its life forms incorporates decentralized and responsible land management practices that are productive while fostering biodiversity [399, 401, 402]. These portfolios of evolving knowledge exemplify how recognizing our connection with the land and other life forms (in contrast with settler colonialism mentality of domination) is a strength [403], not a "vulnerability", especially during times of hardship [404], and is crucial for cultivating the responsible capacity to respond to constant change in the world [181].

Therefore, the non-hierarchical incorporation of knowledge generated from communities targeted for marginalization needs to go beyond considering the "scale up" feasibility of isolated strategies for environmental management or food production and promote a shift in mindset that fosters the recognition of interdependence with the land (i.e. place-based knowledge) and other life forms [149, 405, 406]. The only thing that should be done "to scale" is a shift in mindset; a shift towards a "kinship mindset".

Conclusion

The settler colonialism mindset that renders some places and life forms as worthy of sacrifice is at the root cause of environmental justice issues. The legacy of settler colonialism's paradigm of hierarchization of worth is present in current power structures and discriminatory policies and practices that perpetuate the disproportionate exposure to environmental hazards by limiting the choices and resources available to individuals from communities targeted for marginalization. Environmental injustices are a result of systemic racism rooted in settler colonialism and contribute to a multitude of health disparities. However, environmental hazards do not stay contained within communities targeted for marginalization and have detrimental health consequences for the entire population. For example, our reliance on fossil fuels and an industrialized food system leads to increased exposure to air pollution, pesticides, and the health hazards from climate change. These environmental hazards impact cancer control efforts throughout the care continuum, from cancer prevention, disruption is access to care, and survivorship. Further, the same power structures that allow for the development of hazardous site in communities targeted for marginalization also limits the choices available to individuals racialized as white, as seen with coal mining in the Appalachian region. Therefore, recognizing the value of shifting from a settler colonialism mindset of domination and hegemony towards a kinship mindset is fundamental for identifying solutions that do not perpetuate injustices and improve the health of the entire population.

References

1. Bandi P, et al. Updated review of major cancer risk factors and screening test use in the United States in 2018 and 2019, with a focus on smoking cessation. Cancer Epidemiol Biomark Prev. 2021;30(7):1287–99.
2. Miller KD, et al. Cancer treatment and survivorship statistics. CA Cancer J Clin. 2022.
3. Nogueira LM, Yabroff KR, Bernstein A. Climate change and cancer. CA Cancer J Clin. 2020;70(4):239–44.
4. Buchanan NT, et al. Upending racism in psychological science: strategies to change how science is conducted, reported, reviewed, and disseminated. Am Psychol. 2021;76(7):1097–112.
5. Cooper J. A call for a language shift: from covert oppression to overt empowerment; 2016. http://education.uconn.edu/2016/12/07/a-call-for-a-language-shift-from-covert-oppression-to-overt-empowerment/. Accessed 1 July 2020.
6. Alcaraz KI, et al. Understanding and addressing social determinants to advance cancer health equity in the United States: a blueprint for practice, research, and policy. CA Cancer J Clin. 2020;70(1):31–46.
7. Nogueira L, et al. The role of behavioral medicine in addressing climate change-related health inequities. Transl Behav Med. 2022;12(4):526–34.
8. DuBois WE. The health and physique of the negro American. 1906. Am J Public Health. 2003;93(2):272–6.
9. Cerdeña JP, Grubbs V, Non AL. Genomic supremacy: the harm of conflating genetic ancestry and race. Hum Genomics. 2022;16(1):18.
10. Giaquinto AN, et al. Cancer statistics for African American/black people 2022. CA Cancer J Clin. 2022;72(3):202–29.
11. Krieger N, et al. Breast cancer estrogen receptor status according to biological generation: US black and White women born 1915-1979. Am J Epidemiol. 2018;187(5):960–70.
12. Boyd RW, et al. On racism: a new standard for publishing on racial health inequities. In Health Affairs Blog. 2020.
13. Krieger N, Jahn JL, Waterman PD. Jim crow and estrogen-receptor-negative breast cancer: US-born black and white non-Hispanic women, 1992-2012. Cancer Causes Control. 2017;28(1):49–59.

14. Williams DR, Lawrence JA, Davis BA. Racism and health: evidence and needed research. Annu Rev Public Health. 2019;40:105–25.
15. Bailey ZD, Feldman JM, Bassett MT. How structural racism works - racist policies as a root cause of U.S. racial health inequities. N Engl J Med. 2020.
16. Bailey ZD, et al. Structural racism and health inequities in the USA: evidence and interventions. Lancet. 2017;389(10077):1453–63.
17. Jones CP. Levels of racism: a theoretic framework and a gardener's tale. Am J Public Health. 2000;90(8):1212–5.
18. Gale MM, et al. A meta-analysis of the relationship between internalized racial oppression and health-related outcomes. Couns Psychol. 2020;48(4):498–525.
19. James D. Health and health-related correlates of internalized racism among racial/ethnic minorities: a review of the literature. J Racial Ethn Health Disparities. 2020;7(4):785–806.
20. Paradies Y, et al. Racism as a determinant of health: a systematic review and meta-analysis. PLoS One. 2015;10(9):e0138511.
21. Shavers VL, et al. The state of research on racial/ethnic discrimination in the receipt of health care. Am J Public Health. 2012;102(5):953–66.
22. Flores G. Technical report--racial and ethnic disparities in the health and health care of children. Pediatrics. 2010;125(4):e979–e1020.
23. Bitterman DS, et al. Race disparities in proton radiotherapy use for cancer treatment in patients enrolled in children's oncology group trials. JAMA Oncol. 2020;6(9):1465–8.
24. Gangopadhyaya A. Black patients are more likely than White patients to be in hospitals with worse patient safety conditions; 2021. https://www.rwjf.org/en/library/research/2021/03/black-patients-are-more-likely-than-white-patients-to-be-in-hospitals-with-worse-patient-safety-conditions.html
25. Hoffman KM, et al. Racial bias in pain assessment and treatment recommendations, and false beliefs about biological differences between blacks and whites. Proc Natl Acad Sci U S A. 2016;113(16):4296–301.
26. Schulman KA, et al. The effect of race and sex on physicians' recommendations for cardiac catheterization. N Engl J Med. 1999;340(8):618–26.
27. Blair IV, et al. Assessment of biases against Latinos and African Americans among primary care providers and community members. Am J Public Health. 2013;103(1):92–8.
28. Puumala SE, et al. The role of bias by emergency department providers in care for American Indian Children. Med Care. 2016;54(6):562–9.
29. Cooper LA, et al. The associations of clinicians' implicit attitudes about race with medical visit communication and patient ratings of interpersonal care. Am J Public Health. 2012;102(5):979–87.
30. van Ryn M, et al. The impact of racism on clinician cognition, behavior, and clinical decision making. Du Bois Rev. 2011;8(1):199–218.
31. Popescu I, et al. Racial/ethnic and socioeconomic differences in colorectal and breast cancer treatment quality: the role of physician-level variations in care. Med Care. 2016;54(8):780–8.
32. Shen MJ, et al. The effects of race and racial concordance on patient-physician communication: a systematic review of the literature. J Racial Ethn Health Disparities. 2018;5(1):117–40.
33. Diaz A, et al. Association of historic housing policy, modern day neighborhood deprivation and outcomes after inpatient hospitalization. Ann Surg. 2021.
34. Gonzalez, D., et al. Perceptions of discrimination and unfair judgment while seeking health care. 2021
35. Nong P, et al. Patient-reported experiences of discrimination in the US health care system. JAMA Netw Open. 2020;3(12):e2029650.
36. Sinha S, et al. Disparities in electronic health record patient portal enrollment among oncology patients. JAMA Oncologia. 2021.
37. Warnecke RB, et al. Multilevel examination of health disparity: the role of policy implementation in neighborhood context, in patient resources, and in healthcare facilities on later stage of breast cancer diagnosis. Cancer Epidemiol Biomark Prev. 2019;28(1):59–66.

38. Krieger N. Embodying inequality: a review of concepts, measures, and methods for studying health consequences of discrimination. Int J Health Serv. 1999;29(2):295–352.
39. Tramontano AC, et al. Racial/ethnic disparities in colorectal cancer treatment utilization and phase-specific costs, 2000-2014. PLoS One. 2020;15(4):e0231599.
40. Al Hadidi S, et al. Participation of African American Persons in Clinical Trials Supporting U.S. Food and Drug Administration Approval of Cancer Drugs. Ann Intern Med. 2020.
41. Niranjan SJ, et al. Bias and stereotyping among research and clinical professionals: perspectives on minority recruitment for oncology clinical trials. Cancer. 2020;126(9):1958–68.
42. Alsan M, Garrick O, Graziani GC. Does diversity matter for health? Experimental evidence from Oakland. 2018. https://pubs.aeaweb.org/doi/pdfplus/10.1257/aer.20181446
43. Saha S, et al. Patient-physician racial concordance and the perceived quality and use of health care. Arch Intern Med. 1999;159(9):997–1004.
44. Winkfield KM, et al. American Society of Clinical Oncology strategic plan for increasing racial and ethnic diversity in the oncology workforce. J Clin Oncol. 2017;35(22):2576–9.
45. Arnett MJ, et al. Race, medical mistrust, and segregation in primary care as usual source of care: findings from the exploring health disparities in integrated communities study. J Urban Health. 2016;93(3):456–67.
46. Hill, A., Jones, D. and Woodworth, L., A doctor like me: physician-patient race-match and patient outcomes. 2018.
47. Billings SB, Gallagher E, Ricketts L. Let the rich be flooded: the unequal impact of hurricane Harvey on household debt. J Financ Econ. 2019;
48. Domingue SJ, Emrich CT. Social vulnerability and procedural equity: exploring the distribution of disaster aid across counties in the United States. Am Rev Public Adm. 2019;49(8):897–913.
49. Pager D, Shepherd H. The sociology of discrimination: racial discrimination in employment, housing, credit, and consumer markets. Annu Rev Sociol. 2008;34:181.
50. Katz RJ. Addressing the health care needs of American Indians and Alaska natives. Am J Public Health. 2004;94(1):13–4.
51. Ko NY, et al. Association of insurance status and racial disparities with the detection of early-stage breast cancer. JAMA Oncol. 2020;6(3):385–92.
52. Jemal A, et al. Factors that contributed to black-White disparities in survival among non-elderly women with breast cancer between 2004 and 2013. J Clin Oncol. 2018;36(1):14–24.
53. Aleshire ME, et al. Access to care as a barrier to mammography for black women. Policy Polit Nurs Pract. 2021;22(1):28–40.
54. Lynch EE, et al. The legacy of structural racism: associations between historic redlining, current mortgage lending, and health. SSM Popul Health. 2021;14:100793.
55. Rothstein R. The color of law: a forgotten history of how our government segregated America. New York, NY: Liveright Publishing; 2017.
56. Taylor DE. Toxic communities: environmental racism, industrial pollution, and residential mobility. NYU Press; 2014.
57. Reskin B. The race discrimination system. Annu Rev Sociol. 2012;38(1):17–35.
58. Schulz AJ, et al. Independent and joint contributions of economic, social and physical environmental characteristics to mortality in the Detroit metropolitan area: a study of cumulative effects and pathways. Health Place. 2020;65:102391.
59. Morello-Frosch R, Jesdale BM. Separate and unequal: residential segregation and estimated cancer risks associated with ambient air toxics in U.S. metropolitan areas. Environ Health Perspect. 2006;114(3):386–93.
60. Woo B, et al. Residential segregation and racial/ethnic disparities in ambient air pollution. Race Soc Probl. 2019;11(1):60–7.
61. Bravo MA, et al. Racial isolation and exposure to airborne particulate matter and ozone in understudied US populations: environmental justice applications of downscaled numerical model output. Environ Int. 2016;92–93:247–55.
62. Taylor D. Toxic communities, in toxic communities. New York University Press; 2014.

63. Demissie F, University D. Book review: residential apartheid: the American legacy. Los Angeles: Sage; 1995.
64. Greenberg MR, Renne J. Where does walkability matter the most? An environmental justice interpretation of New Jersey data. J Urban Health. 2005;82(1):90–100.
65. Coffey E, et al. Poisonous homes: the fight for environmental justice in federally assisted housing; 2020.
66. Feagin JR. A house is not a home: White racism and US housing practices. Residential Apartheid: The American Legacy. 1994;17:34–7.
67. Namin S, et al. The legacy of the home owners' loan corporation and the political ecology of urban trees and air pollution in the United States. Soc Sci Med. 2020;246:112758.
68. Hoffman JS, Shandas V, Pendleton N. The effects of historical housing policies on resident exposure to intra-urban heat: a study of 108 US urban areas. Climate. 2020;8(1)
69. Locke DH, et al. Residential housing segregation and urban tree canopy in 37 US cities. npj Urban Sustain. 2021;1(1):15.
70. Schell CJ, et al. The ecological and evolutionary consequences of systemic racism in urban environments. Science. 2020;369(6510):1446+.
71. Sharkey P. Stuck in place: urban neighborhoods and the end of progress toward racial equality. University of Chicago Press; 2013.
72. Benz TA. Toxic cities: neoliberalism and environmental racism in Flint and Detroit Michigan. Crit Sociol. 2019;45(1):49–62.
73. Fedinick KP, et al. Watered down justice. Natural Resource Defense Council Report; 2019.
74. Trangenstein PJ, et al. Alcohol outlet clusters and population disparities. J Urban Health. 2020;97(1):123–36.
75. Hilmers A, Hilmers DC, Dave J. Neighborhood disparities in access to healthy foods and their effects on environmental justice. Am J Public Health. 2012;102(9):1644–54.
76. Lee JG, et al. A systematic review of neighborhood disparities in point-of-sale tobacco marketing. Am J Public Health. 2015;105(9):e8–18.
77. John R, Cheney MK, Azad MR. Point-of-sale marketing of tobacco products: taking advantage of the socially disadvantaged? J Health Care Poor Underserved. 2009;20(2):489–506.
78. Hawes AM, et al. Disentangling race, poverty, and place in disparities in physical activity. Int J Environ Res Public Health. 2019;16(7)
79. Bower KM, et al. The intersection of neighborhood racial segregation, poverty, and urbanicity and its impact on food store availability in the United States. Prev Med. 2014;58:33–9.
80. Cooksey Stowers K, et al. Racial differences in perceived food swamp and food desert exposure and disparities in self-reported dietary habits. Int J Environ Res Public Health. 2020;17(19)
81. Mills SD, et al. Disparities in retail marketing for menthol cigarettes in the United States, 2015. Health Place. 2018;53:62–70.
82. Churchwell K, et al. Call to action: structural racism as a fundamental driver of health disparities: a presidential advisory from the American Heart Association. Circulation. 2020;142(24):e454–68.
83. Williams DR, Collins C. Racial residential segregation: a fundamental cause of racial disparities in health. Public Health Rep. 2001;116(5):404–16.
84. Acevedo-Garcia D, et al. Future directions in residential segregation and health research: a multilevel approach. Am J Public Health. 2003;93(2):215–21.
85. White K, Haas JS, Williams DR. Elucidating the role of place in health care disparities: the example of racial/ethnic residential segregation. Health Serv Res. 2012;47(3 Pt 2):1278–99.
86. Fong AJ, et al. Association of living in urban food deserts with mortality from breast and colorectal cancer. Ann Surg Oncol. 2021;28(3):1311–9.
87. Tammemagi CM, et al. Comorbidity and survival disparities among black and white patients with breast cancer. JAMA. 2005;294(14):1765–72.
88. Lam C, et al. Differences in cancer survival among white and black cancer patients by presence of diabetes mellitus: estimations based on SEER-Medicare-linked data resource. Cancer Med. 2018;7(7):3434–44.

89. Healton C, Nelson K. Reversal of misfortune: viewing tobacco as a social justice issue. Am J Public Health. 2004;94(2):186–91.
90. Egede LE, Walker RJ. Structural racism, social risk factors, and Covid-19—a dangerous convergence for black Americans. N Engl J Med. 2020;383(12):e77.
91. Rogers LO, Heard-Garris N. Documenting racial disparities or disrupting racism?: a call to center systems of power, privilege, and oppression in psychological and pediatric research. JAMA Pediatrics. 2022.
92. Bassett MT. #BlackLivesMatter--a challenge to the medical and public health communities. N Engl J Med. 2015;372(12):1085–7.
93. Joynt Maddox KE, James CV. How the Biden administration can improve health equity for racial and ethnic minority populations. JAMA. 2021;325(14):1387–8.
94. Krieger N. Structural racism, health inequities, and the two-edged sword of data: structural problems require structural solutions. Front Public Health. 2021;9:655447.
95. Corburn J. Confronting the challenges in reconnecting urban planning and public health. Am J Public Health. 2004;94(4):541–6.
96. Juarez T, Damian AJ. How federally qualified health centers can, and should, promote environmental justice. Health Affairs Forefront; 2022.
97. Scally BJ, Krieger N, Chen JT. Racialized economic segregation and stage at diagnosis of colorectal cancer in the United States. Cancer Causes Control. 2018;29(6):527–37.
98. Nelson B. How structural racism can kill cancer patients: black patients with breast cancer and other malignancies face historical inequities that are ingrained but not inevitable. In this article, the second of a 2-part series, we explore the consequences of and potential solutions to racism and inequality in cancer care. Cancer Cytopathol. 2020;128(2):83–4.
99. Pallok K, De Maio F, Ansell DA. Structural racism - a 60-year-old black woman with breast cancer. N Engl J Med. 2019;380(16):1489–93.
100. Bullard RD. Environmental justice in the 21st century: race still matters. Phylon (1960-). 2001;49(3/4):151–71.
101. Mohai P, Pellow D, Roberts JT. Environmental justice. Annu Rev Environ Resour. 2009;34(1):405–30.
102. Lazarus RJ. Environmental racism-that's what it is. U Ill L Rev. 2000:255.
103. Vig AC. Using title VI to salvage civil rights from waste: Chester residents concerned for quality living v. Seif, 132 f. 3d 925 (3d cir. 1997). U. Cin. L. Rev. 1998;67:907.
104. International Agency for Research on Cancer. Polychlorinated Biphenyls and Polybrominated Biphenyls; 2016. https://publications.iarc.fr/Book-And-Report-Series/Iarc-Monographs-On-The-Identification-Of-Carcinogenic-Hazards-To-Humans/Benzene-2018
105. Bullard RD. Dumping in Dixie: race, class, and environmental quality. Routledge; 1990.
106. Office USGA. Siting of hazardous waste landfills and their correlation with racial and economic status of surrounding communities: report. The Office; 1983.
107. Commission for Racial Justice and United Church of Christ. Toxic wastes and race in the United States: A national report on the racial and socio-economic characteristics of communities with hazardous waste sites; 1987.
108. Singh GK, Jemal A. Socioeconomic and racial/ethnic disparities in cancer mortality, incidence, and survival in the United States, 1950-2014: over six decades of changing patterns and widening inequalities. J Environ Public Health. 2017;2017:2819372.
109. Zavala VA, et al. Cancer health disparities in racial/ethnic minorities in the United States. Br J Cancer. 2021;124(2):315–32.
110. Ellis L, et al. Racial and ethnic disparities in cancer survival: the contribution of tumor, sociodemographic, institutional, and neighborhood characteristics. J Clin Oncol. 2018;36(1):25–33.
111. Hendrick RE, et al. Age distributions of breast cancer diagnosis and mortality by race and ethnicity in US women. Cancer. 2021;127(23):4384–92.
112. Robbins HA, et al. Age at cancer diagnosis for blacks compared with whites in the United States. J Natl Cancer Inst. 2015;107(3)

113. Cho B, et al. Evaluation of racial/ethnic differences in treatment and mortality among women with triple-negative breast cancer. JAMA Oncol. 2021;7(7):1016–23.
114. Kratzer TB, et al. Cancer statistics for American Indian and Alaska native individuals, 2022: including increasing disparities in early onset colorectal cancer. CA Cancer J Clin. 2022.
115. International Agency for Research on Cancer, I., IARC monographs on the evaluation of carcinogenic risks to humans: occupational exposures in insecticide application, and some pesticides. Vol. 53. 1991: IARC Lyon.
116. Nogueira LM, et al. Association between declared hurricane disasters and survival of patients with lung cancer undergoing radiation treatment. JAMA. 2019;322(3):269–71.
117. Tessum CW, et al. PM(2.5) polluters disproportionately and systemically affect people of color in the United States. Sci Adv. 2021;7(18)
118. Ash M, Fetter TR. Who lives on the wrong side of the environmental tracks? Evidence from the EPA's risk-screening environmental indicators model. Soc Sci Q. 2004;85(2):441–62.
119. Boer JT, et al. Is there environmental racism? The demographics of hazardous waste in Los Angeles County. Soc Sci Q. 1997;78(4):793–810.
120. Bolin B, et al. The ecology of technological risk in a sunbelt city. Environ Plann Econ Space. 2002;34(2):317–39.
121. Chakraborty J. Cancer risk from exposure to hazardous air pollutants: spatial and social inequities in Tampa Bay, Florida. Int J Environ Health Res. 2012;22(2):165–83.
122. Dolinoy DC, Miranda ML. GIS modeling of air toxics releases from TRI-reporting and non-TRI-reporting facilities: impacts for environmental justice. Environ Health Perspect. 2004;112(17):1717–24.
123. Faber DR, Krieg EJ. Unequal exposure to ecological hazards: environmental injustices in the commonwealth of Massachusetts. Environ Health Perspect. 2002;110:277–88.
124. Gilbert A, Chakraborty J. Using geographically weighted regression for environmental justice analysis: cumulative cancer risks from air toxics in Florida. Soc Sci Res. 2011;40(1):273–86.
125. Hipp JR, Lakon CM. Social disparities in health: disproportionate toxicity proximity in minority communities over a decade. Health Place. 2010;16(4):674–83.
126. Campbell HE, Peck LR, Tschudi MK. Justice for all? A cross-time analysis of toxics release inventory facility location. Rev Policy Res. 2010;27(1):1–25.
127. Pastor M Jr, Morello-Frosch R, Sadd JL. The air is always cleaner on the other side: race, space, and ambient air toxics exposures in California. J Urban Aff. 2005;27(2):127–48.
128. Ard K. Trends in exposure to industrial air toxins for different racial and socioeconomic groups: a spatial and temporal examination of environmental inequality in the U.S. from 1995 to 2004. Soc Sci Res. 2015(53):375–90.
129. Downey L, Hawkins B. Race, income, and environmental inequality in the UNITED STATES. Sociol Perspect. 2008;51(4):759–81.
130. Liévanos RS. Race, deprivation, and immigrant isolation: the spatial demography of air-toxic clusters in the continental United States. Soc Sci Res. 2015;54:50–67.
131. Mikati I, et al. Disparities in distribution of particulate matter emission sources by race and poverty status. Am J Public Health. 2018;108(4):480–5.
132. Bullard RD, et al. Toxic wastes and race at twenty 1987–2007: Grassroots struggles to dismantle environmental racism in the United States; 2007.
133. Cushing L, et al. Racial/ethnic disparities in cumulative environmental health impacts in California: evidence from a statewide environmental justice screening tool (CalEnviroScreen 1.1). Am J Public Health. 2015;105(11):2341–8.
134. Grineski SE, Collins T. Lifetime cancer risks from hazardous air pollutants in US public school districts. J Epidemiol Community Health. 2019;73(9):854–60.
135. Johnson G. Living on the frontline of environmental assault: lessons from the United States most vulnerable communities. J Educ Sci Math. 2013;3:33–61.
136. Lejano RP, Iseki H. Environmental justice: spatial distribution of hazardous waste treatment, storage and disposal facilities in Los Angeles. J Urban Plann Dev-Asce. 2001;127(2):51–62.

137. Sapolu C. Dumping on the Wai'ānae coast: achieving environmental justice through the Hawai'i state constitution; 2009. http://blog.hawaii.edu/aplpj/files/2011/11/APLPJ_11.1_sapolu.pdf
138. Younes L, Shaw A, Kofman A. How we created the most detailed map ever of cancer-causing industrial air pollution; 2021. https://www.propublica.org/article/how-we-created-the-most-detailed-map-ever-of-cancer-causing-industrial-air-pollution
139. Cheeseman MJ, et al. Disparities in air pollutants across racial, ethnic, and poverty groups at US public schools. Geohealth. 2022;6(12):e2022GH000672.
140. Wilson S, et al. Assessment of sociodemographic and geographic disparities in cancer risk from air toxics in South Carolina. Environ Res. 2015;140:562–8.
141. García-Pérez J, et al. Association between residential proximity to environmental pollution sources and childhood renal tumors. Environ Res. 2016;147:405–14.
142. Bulka C, et al. Relations between residential proximity to EPA-designated toxic release sites and diffuse large B-cell lymphoma incidence. South Med J. 2016;109(10):606–14.
143. Jephcote C, et al. A systematic review and meta-analysis of haematological malignancies in residents living near petrochemical facilities. Environ Health. 2020;19(1):53.
144. García-Pérez J, et al. Residential proximity to industrial pollution sources and colorectal cancer risk: a multicase-control study (MCC-Spain). Environ Int. 2020;144:106055.
145. Rudolph L, et al. Climate change, health, and equity: a guide for local health departments. Public Health Institute; 2018. https://www.apha.org/-/media/files/pdf/topics/climate/climate_health_equity.ashx
146. Maldonado JK, et al. The impact of climate change on tribal communities in the US: displacement, relocation, and human rights. Clim Chang. 2013;120(3):601–14.
147. Dunbar-Ortiz R. An indigenous peoples' history of the United States, vol. 3. Beacon Press; 2014.
148. Whyte K. Indigenous food systems, environmental justice, and settler-industrial states; 2016. https://kylewhyte.marcom.cal.msu.edu/wp-content/uploads/sites/12/2018/07/IP_Food_Systems__EJ_and_Settler_States1-1-16.pdf
149. Whyte K. Settler colonialism, ecology, and environmental injustice. Environ Soc. 2018;9:125–44.
150. Ferdinand M. Decolonial ecology: thinking from the Caribbean world. Wiley; 2021.
151. Whyte K, Meissner SN. Without land, decolonizing American philosophy is impossible. In: McCall C, McReynolds P, editors. Decolonizing American Philosophy; 2021. p. 37–61.
152. The Columbia River Inter-Tribal Fish Commission. A fish consumption survey of the Umatilla, Nez Perce, Yakama, and Warm Springs Tribes of the Columbia River Basin; 1994. https://critfc.org/reports/a-fish-consumption-survey-of-the-umatilla-nez-perce-yakama-and-warm-springs-tribes-of-the-columbia-river-basin/
153. Diné Policy Institute. Diné Food Sovereinty; 2018. https://www.dinecollege.edu/wp-content/uploads/2018/04/dpi-food-sovereignty-report.pdf
154. Hoover E. Cultural and health implications of fish advisories in a native American community. Ecol Process. 2013;2(1):4.
155. Burger J, et al. Mercury levels and potential risk from subsistence foods from the Aleutians. Sci Total Environ. 2007;384(1-3):93–105.
156. Harris SG, Harper BL. A native American exposure scenario. Risk Anal. 1997;17(6):789–95.
157. Chino M, Haff DR, Dodge Francis C. Patterns of commodity food use among American Indians. Pimatisiwin: a journal of aboriginal and indigenous. Community Health. 2009;7(2):279.
158. Warne D, Wescott S. Social determinants of American Indian nutritional health. Curr Dev Nutr. 2019;3(Suppl 2):12–8.
159. Lauby-Secretan B, et al. Body fatness and cancer--viewpoint of the IARC working group. N Engl J Med. 2016;375(8):794–8.
160. Giovannucci E, et al. Diabetes and cancer: a consensus report. Diabetes Care. 2010;33(7):1674–85.

161. Gilio-Whitaker D. As long as grass grows: the indigenous fight for environmental justice, from colonization to standing rock. Beacon Press; 2019.
162. Carlisle L. Healing grounds: climate, justice, and the deep roots of regenerative farming. Taylor & Francis; 2022.
163. International Panel of Experts on Sustainable Food Systems, I. Unravelling the Food–Health Nexus: addressing practices, political economy, and power relations to build healthier food systems. 2017. https://www.ipes-food.org/_img/upload/files/Health_FullReport(1).pdf
164. Gliessman S. Agroecology: growing the roots of resistance. Agroecol Sustain Food Syst. 2013;37(1):19–31.
165. National Research Council, Pesticides in the Diets of Infants and Children, in Pesticides in the Diets of Infants and Children. 1993, National Academies Press (US). Copyright 1993 by the National Academy of Sciences. All rights reserved: Washington (DC).
166. Freisthler MS, et al. Association between increasing agricultural use of 2,4-D and population biomarkers of exposure: findings from the National Health and nutrition examination survey, 2001-2014. Environ Health. 2022;21(1):23.
167. Eskenazi B, Bradman A, Castorina R. Exposures of children to organophosphate pesticides and their potential adverse health effects. Environ Health Perspect. 1999;107(Suppl 3):409–19.
168. Ward MH, et al. Proximity to crops and residential exposure to agricultural herbicides in Iowa. Environ Health Perspect. 2006;114(6):893–7.
169. Madrigal JM, et al. Contributions of nearby agricultural insecticide applications to indoor residential exposures. Environ Int. 2022;171:107657.
170. Deziel NC, et al. Relative contributions of agricultural drift, Para-occupational, and residential use exposure pathways to house dust pesticide concentrations: meta-regression of published data. Environ Health Perspect. 2017;125(3):296–305.
171. Deziel NC, et al. A review of nonoccupational pathways for pesticide exposure in women living in agricultural areas. Environ Health Perspect. 2015;123(6):515–24.
172. McCauley LA, et al. Studying health outcomes in farmworker populations exposed to pesticides. Environ Health Perspect. 2006;114(6):953–60.
173. Mills PK, Shah P. Cancer incidence in California farm workers, 1988-2010. Am J Ind Med. 2014;57(7):737–47.
174. Bassil KL, et al. Cancer health effects of pesticides: systematic review. Can Fam Physician. 2007;53(10):1704–11.
175. Settimi L, et al. Prostate cancer and exposure to pesticides in agricultural settings. Int J Cancer. 2003;104(4):458–61.
176. Humans, I.W.G.o.t.E.o.C.R.t. Some organophosphate insecticides and herbicides; 2017.
177. Myers JP, et al. Concerns over use of glyphosate-based herbicides and risks associated with exposures: a consensus statement. Environ Health. 2016;15
178. Hayes T, Kersha K. PRIMER: agriculture subsidies and their influence on the composition of U.S. Food Supply and Consumption; 2021. https://www.americanactionforum.org/research/primer-agriculture-subsidies-and-their-influence-on-the-composition-of-u-s-food-supply-and-consumption/
179. Do WL, et al. Consumption of foods derived from subsidized crops remains associated with cardiometabolic risk: an update on the evidence using the National Health and nutrition examination survey 2009-2014. Nutrients. 2020;12(11)
180. Simon MS, et al. Cardiometabolic risk factors and survival after cancer in the Women's Health Initiative. Cancer. 2021;127(4):598–608.
181. Whyte K. Against crisis epistemology. In: Hokowhitu B, et al., editors. Routledge handbook of critical indigenous studies; 2021. p. 52–64.
182. Nicholson E. 'Essential' but undervalued: industry must do more to protect farmworkers. J Agromedicine. 2022;27(1):4–5.
183. Arcury TA, Quandt SA. Latinx farmworkers in the eastern United States: health, safety, and justice. Springer; 2020.

184. Gold A, et al. Findings from the national agricultural workers survey (NAWS) 2019–2020: a demographic and employment profile of United States Farmworkers; 2022. https://www.dol.gov/sites/dolgov/files/ETA/naws/pdfs/NAWS%20Research%20Report%2016.pdf
185. Calvert GM, et al. Acute pesticide poisoning among agricultural workers in the United States, 1998-2005. Am J Ind Med. 2008;51(12):883–98.
186. Das R, et al. Pesticide-related illness among migrant farm workers in the United States. Int J Occup Environ Health. 2001;7(4):303–12.
187. Curl CL, et al. Synthetic pesticides and health in vulnerable populations: agricultural workers. Curr Environ Health Rep. 2020;7(1):13–29.
188. Moyce SC, Schenker M. Migrant workers and their occupational health and safety. Annu Rev Public Health. 2018;39:351–65.
189. Quandt SA, et al. Occupational health outcomes for workers in the agriculture, forestry and fishing sector: implications for immigrant workers in the southeastern US. Am J Ind Med. 2013;56(8):940–59.
190. Castillo F, et al. Environmental health threats to Latino migrant farmworkers. Annu Rev Public Health. 2021;42:257–76.
191. Arcury TA, et al. The abysmal organization of work and work safety culture experienced by North Carolina Latinx women in farmworker families. Int J Environ Res Public Health. 2022;19(8)
192. Sexsmith K, et al. Latino/a farmworkers' concerns about safety and health in the Pennsylvania mushroom industry. J Agromedicine. 2022;27(2):169–82.
193. Curl CL, Meierotto L, Som Castellano RL. Understanding challenges to well-being among Latina Farm Workers in rural Idaho using in an interdisciplinary, mixed-methods approach. Int J Environ Res Public Health. 2020;18(1)
194. López-Gálvez N, et al. Systematic literature review of the take-home route of pesticide exposure via biomonitoring and environmental monitoring. Int J Environ Res Public Health. 2019;16(12)
195. Gochfeld M. Occupational medicine practice in the United States since the industrial revolution. J Occup Environ Med. 2005;47(2):115–31.
196. Coronado GD, et al. Organophosphate pesticide exposure and work in pome fruit: evidence for the take-home pesticide pathway. Environ Health Perspect. 2006;114(7):999–1006.
197. Bradman A, et al. Pesticides and their metabolites in the homes and urine of farmworker children living in the Salinas Valley. CA J Expo Sci Environ Epidemiol. 2007;17(4):331–49.
198. Frank AL, et al. Issues of agricultural safety and health. Annu Rev Public Health. 2004;25:225–45.
199. Berk ML, Schur CL. The effect of fear on access to care among undocumented Latino immigrants. J Immigr Health. 2001;3(3):151–6.
200. Hardy LJ, et al. A call for further research on the impact of state-level immigration policies on public health. Am J Public Health. 2012;102(7):1250–4.
201. Sabo S, Lee AE. The spillover of US immigration policy on citizens and permanent residents of Mexican descent: how internalizing "illegality" impacts public health in the borderlands. Front Public Health. 2015:3.
202. Weigel MM, et al. The household food insecurity and health outcomes of U.S.-Mexico border migrant and seasonal farmworkers. J Immigr Minor Health. 2007;9(3):157–69.
203. Kiehne E, Mendoza NS. Migrant and seasonal farmworker food insecurity: prevalence, impact, risk factors, and coping strategies. Soc Work Public Health. 2015;30(5):397–409.
204. Hill BG, et al. Prevalence and predictors of food insecurity in migrant farmworkers in Georgia. Am J Public Health. 2011;101(5):831–3.
205. Castañeda H, et al. Immigration as a social determinant of health. Annu Rev Public Health. 2015;36:375–92.
206. Food & Water Watch. Factory Farm Nation; 2020. https://www.foodandwaterwatch.org/wp-content/uploads/2021/03/ib_2004_updfacfarmmaps-web2.pdf
207. Nicole W. CAFOs and environmental justice: the case of North Carolina. Environ Health Perspect. 2013;121(6):A182–9.

208. Wing S, Cole D, Grant G. Environmental injustice in North Carolina's hog industry. Environ Health Perspect. 2000;108(3):225–31.
209. Casey JA, et al. Industrial food animal production and community health. Curr Environ Health Rep. 2015;2(3):259–71.
210. Cole D, Todd L, Wing S. Concentrated swine feeding operations and public health: a review of occupational and community health effects. Environ Health Perspect. 2000;108(8):685–99.
211. Rasmussen SG, et al. Proximity to industrial food animal production and asthma exacerbations in Pennsylvania, 2005-2012. Int J Environ Res Public Health. 2017;14(4)
212. Kravchenko J, et al. Mortality and health outcomes in North Carolina communities located in close proximity to hog concentrated animal feeding operations. N C Med J. 2018;79(5):278–88.
213. Schultz AA, et al. Residential proximity to concentrated animal feeding operations and allergic and respiratory disease. Environ Int. 2019;130:104911.
214. Schinasi L, et al. Air pollution, lung function, and physical symptoms in communities near concentrated swine feeding operations. Epidemiology. 2011;22(2):208–15.
215. Guidry VT, et al. Hydrogen sulfide concentrations at three middle schools near industrial livestock facilities. J Expo Sci Environ Epidemiol. 2017;27(2):167–74.
216. Loftus C, et al. Estimated time-varying exposures to air emissions from animal feeding operations and childhood asthma. Int J Hyg Environ Health. 2020;223(1):187–98.
217. Domingo NGG, et al. Air quality-related health damages of food. Proc Natl Acad Sci U S A. 2021;118(20)
218. Heaney CD, et al. Source tracking swine fecal waste in surface water proximal to swine concentrated animal feeding operations. Sci Total Environ. 2015;511:676–83.
219. Ramos AK, et al. Health and well-being of Hispanic/Latino meatpacking workers in Nebraska: an application of the health belief model. Workplace Health Saf. 2021;69(12):564–72.
220. Peterson EM, Green FB, Smith PN. Pesticides used on beef cattle feed yards are aerially transported into the environment via particulate matter. Environ Sci Technol. 2020;54(20):13008–15.
221. Wing S, Wolf S. Intensive livestock operations, health, and quality of life among eastern North Carolina residents. Environ Health Perspect. 2000;108(3):233–8.
222. Penniman L. Farming while black: soul fire farm's practical guide to liberation on the land. Chelsea Green Publishing; 2018.
223. Kahrl A. Black people's land was stolen; 2019. https://www.nytimes.com/2019/06/20/opinion/sunday/reparations-hearing.html
224. Glave DD. Rooted in the earth: reclaiming the African American environmental heritage. Chicago Review Press; 2010.
225. White M. Preface for black agrarianism. Justine M. Williams & Eric Holt-Giménez. Land justice re-imagining land, food, and the commons in the United States. Oakland: Food First; 2017.
226. White MM. Freedom farmers: agricultural resistance and the black freedom movement. UNC Press Books; 2018.
227. International Agency for Research on Cancer. Outdoor Air Pollution; 2015. https://publications.iarc.fr/Book-And-Report-Series/Iarc-Monographs-On-The-Identification-Of-Carcinogenic-Hazards-To-Humans/Outdoor-Air-Pollution-2015
228. Adgate JL, Goldstein BD, McKenzie LM. Potential public health hazards, exposures and health effects from unconventional natural gas development. Environ Sci Technol. 2014;48(15):8307–20.
229. Czolowski ED, et al. Toward consistent methodology to quantify populations in proximity to oil and gas development: a national spatial analysis and review. Environ Health Perspect. 2017;125(8):086004.
230. McKenzie LM, et al. Population size, growth, and environmental justice near oil and gas wells in Colorado. Environ Sci Technol. 2016;50(21):11471–80.

231. Johnston JE, Lim E, Roh H. Impact of upstream oil extraction and environmental public health: a review of the evidence. Sci Total Environ. 2019;657:187–99.
232. Shonkoff SB, Hays J, Finkel ML. Environmental public health dimensions of shale and tight gas development. Environ Health Perspect. 2014;122(8):787–95.
233. Remy LL, et al. Hospital, health, and community burden after oil refinery fires, Richmond, California 2007 and 2012. Environ Health. 2019;18(1):48.
234. Intergovernmental Panel on Climate Change I, et al. Climate change 2022: impacts, adaptation and vulnerability. Cambridge University Press; 2022.
235. Ward EM, et al. Priorities for development of research methods in occupational cancer. Environ Health Perspect. 2003;111(1):1–12.
236. Johnson N, et al. Concentrations of arsenic, chromium, and nickel in toenail samples from Appalachian Kentucky residents. J Environ Pathol Toxicol Oncol. 2011;30(3):213–23.
237. Epstein PR, et al. Full cost accounting for the life cycle of coal. Ann N Y Acad Sci. 2011;1219:73–98.
238. National Research Council. Managing coal combustion residues in mines; 2006. https://nap.nationalacademies.org/catalog/11592/managing-coal-combustion-residues-in-mines
239. Stant J, et al. The impacts on water quality from placement of coal combustion waste in Pennsylvania coal mines; 2007. http://www.catf.us/publications/reports/PAMinefill.pdf
240. Environmental Integrity Project. Coal's Poisonous Legacy: Groundwater Contaminated by Coal Ash Across the United States. 2019. https://earthjustice.org/sites/default/files/files/national-coal-ash-report-7.11.19.pdf
241. International Agency for Research on Cancer. Chromium, nickel and welding. 1990. https://publications.iarc.fr/Book-And-Report-Series/Iarc-Monographs-On-The-Identification-Of-Carcinogenic-Hazards-To-Humans/Chromium-Nickel-And-Welding-1990
242. International Agency for Research on Cancer. Arsenic, Metals, Fibres, and Dusts. 2012. https://publications.iarc.fr/Book-And-Report-Series/Iarc-Monographs-On-The-Identification-Of-Carcinogenic-Hazards-To-Humans/Arsenic-Metals-Fibres-And-Dusts-2012
243. Dwyer-Lindgren L, et al. US County-level trends in mortality rates for major causes of death, 1980-2014. JAMA. 2016;316(22):2385–401.
244. Hendryx M, O'Donnell K, Horn K. Lung cancer mortality is elevated in coal-mining areas of Appalachia. Lung Cancer. 2008;62(1):1–7.
245. Schreiber ME, Cozzarelli IM. Arsenic release to the environment from hydrocarbon production, storage, transportation, use and waste management. J Hazard Mater. 2021;411:125013.
246. Hecobian A, et al. Air toxics and other volatile organic compound emissions from unconventional oil and gas development. Environ Sci Technol Lett. 2019;6(12):720–6.
247. Hildenbrand ZL, et al. Point source attribution of ambient contamination events near unconventional oil and gas development. Sci Total Environ. 2016;573:382–8.
248. Khalaj F, Sattler M. Modeling of VOCs and criteria pollutants from multiple natural gas well pads in close proximity, for different terrain conditions: a Barnett shale case study. Atmos Pollut Res. 2019;10(4):1239–49.
249. Marrero JE, et al. Estimating emissions of toxic hydrocarbons from natural gas production sites in the Barnett shale region of northern Texas. Environ Sci Technol. 2016;50(19):10756–64.
250. Macey GP, et al. Air concentrations of volatile compounds near oil and gas production: a community-based exploratory study. Environ Health. 2014;13(1):82.
251. Russo PN, Carpenter DO. Air emissions from natural gas facilities in New York State. Int J Environ Res Public Health. 2019;16(9):1591.
252. Brantley HL, Thoma ED, Eisele AP. Assessment of volatile organic compound and hazardous air pollutant emissions from oil and natural gas well pads using mobile remote and on-site direct measurements. J Air Waste Manage Assoc. 2015;65(9):1072–82.
253. Roest GS, Schade GW. Air quality measurements in the western Eagle Ford shale. Elementa: Science of the Anthropocene. 2020;8.

254. Chen H, Carter KE. Hazardous substances as the dominant non-methane volatile organic compounds with potential emissions from liquid storage tanks during well fracturing: a modeling approach. J Environ Manag. 2020;268:110715.
255. D'Andrea MA, Reddy GK. Detrimental health effects of benzene exposure in adults after a flaring disaster at the BP refinery Plant in Texas City. Disaster Med Public Health Prep. 2016;10(2):233–9.
256. D'Andrea MA, Reddy GK. Adverse health complaints of adults exposed to benzene after a flaring disaster at the BP refinery Facility in Texas City. Texas Disaster Med Public Health Prep. 2018;12(2):232–40.
257. Shaykevich A. Environmental justice and refinery pollution. 2021. https://environmentalintegrity.org/wp-content/uploads/2021/04/Benzene-report-4.28.21.pdf
258. Paulik LB, et al. Emissions of polycyclic aromatic hydrocarbons from natural gas extraction into air. Environ Sci Technol. 2016;50(14):7921–9.
259. Paulik LB, et al. Environmental and individual PAH exposures near rural natural gas extraction. Environ Pollut. 2018;241:397–405.
260. Banan Z, Gernand JM. Evaluation of gas well setback policy in the Marcellus shale region of Pennsylvania in relation to emissions of fine particulate matter. J Air Waste Manag Assoc. 2018;68(9):988–1000.
261. Fann N, et al. Assessing human health PM(2.5) and ozone impacts from U.S. oil and natural gas sector emissions in 2025. Environ Sci Technol. 2018;52(15):8095–103.
262. Long CM, Briggs NL, Bamgbose IA. Synthesis and health-based evaluation of ambient air monitoring data for the Marcellus shale region. J Air Waste Manag Assoc. 2019;69(5):527–47.
263. Roohani YH, et al. Impact of natural gas development in the Marcellus and Utica shales on regional ozone and fine particulate matter levels. Atmos Environ. 2017;155:11–20.
264. Prenni AJ, et al. Oil and gas impacts on air quality in federal lands in the Bakken region: an overview of the Bakken air quality study and first results. Atmos Chem Phys. 2016;16(3):1401–16.
265. Bean JK, et al. Formation of particulate matter from the oxidation of evaporated hydraulic fracturing wastewater. Environ Sci Technol. 2018;52(8):4960–8.
266. Bozlaker A, Peccia J, Chellam S. Indoor/outdoor relationships and anthropogenic elemental signatures in airborne PM2.5 at a high school: impacts of petroleum refining emissions on lanthanoid enrichment. Environ Sci Technol. 2017;51(9):4851–9.
267. International Agency for Research on Cancer. Benzene; 2018. https://publications.iarc.fr/Book-And-Report-Series/Iarc-Monographs-On-The-Identification-Of-Carcinogenic-Hazards-To-Humans/Benzene-2018
268. International Agency for Research on Cancer. Some non-heterocyclic polycyclic aromatic hydrocarbons and some related exposures; 2010. https://publications.iarc.fr/Book-And-Report-Series/Iarc-Monographs-On-The-Identification-Of-Carcinogenic-Hazards-To-Humans/Some-Non-heterocyclic-Polycyclic-Aromatic-Hydrocarbons-And-Some-Related-Exposures-2010
269. International Agency for Research on Cancer. Air Pollution and Cancer 2013. https://publications.iarc.fr/Book-And-Report-Series/Iarc-Scientific-Publications/Air-Pollution-And-Cancer-2013
270. Garcia-Gonzales DA, et al. Hazardous air pollutants associated with upstream oil and natural gas development: a critical synthesis of current peer-reviewed literature. Annu Rev Public Health. 2019;40:283–304.
271. Elliott EG, et al. Unconventional oil and gas development and risk of childhood leukemia: assessing the evidence. Sci Total Environ. 2017;576:138–47.
272. Stringfellow WT, et al. Identifying chemicals of concern in hydraulic fracturing fluids used for oil production. Environ Pollut. 2017;220(Pt A):413–20.
273. International Agency for Research on Cancer. Beryllium, Cadmium, Mercury, and Exposures in the Glass Manufacturing Industry; 1997. https://publications.iarc.fr/76
274. International Agency for Research on Cancer. Trichloroethylene, Tetrachloroethylene, and Some Other Chlorinated Agents. 2014. https://publications.iarc.fr/Book-And-Report-Series/

Iarc-Monographs-On-The-Identification-Of-Carcinogenic-Hazards-To-Humans/Trichloroethylene-Tetrachloroethylene-And-Some-Other-Chlorinated-Agents-2014
275. Environmental Protection Agency. Hydraulic Fracturing for Oil and Gas: Impacts from the Hydraulic Fracturing Water Cycle on Drinking Water Resources in the United States. 2016. https://cfpub.epa.gov/ncea/hfstudy/recordisplay.cfm?deid=332990
276. Agency for Toxic Substances and Disease Registry. JKLM natural gas well coudersport; 2019. https://www.atsdr.cdc.gov/HAC/pha/JKLMNaturalGas/JKLM_Energy_Natural_Gas_Well_Coudersport-508.pdf
277. Agency for Toxic Substances and Disease Registry. Panola county road 329; 2007. https://www.atsdr.cdc.gov/HAC/pha/PanolaCountyRoad329/PanolaCountyRoadHC080707.pdf
278. Agency for Toxic Substances and Disease Registry. Highway 18 Ground Water; 2020. https://www.atsdr.cdc.gov/HAC/pha/Highway18GroundWater/Highway_18_Ground%20Water_HC-508.pdf
279. Agency for Toxic Substances and Disease Registry. Sandy Beach Road Groundwater Plume. 2007. https://www.atsdr.cdc.gov/HAC/pha/SandyBeachRoadGroundwaterPlume/SandyBeachRoadPHA011707.pdf
280. Agency for Toxic Substances and Disease Registry. East 67th Street Groundwater Plume. 2008. https://www.atsdr.cdc.gov/HAC/pha/East67thStreetHC/East67thStreetHC030608.pdf
281. Agency for Toxic Substances and Disease Registry. Midessa Groundwater Plume; 2009. https://www.atsdr.cdc.gov/HAC/pha/MidessaGroundwaterPlume/MidessaGroundwaterPlumeFinalPHA05-21-09.pdf
282. Agency for Toxic Substances and Disease Registry. Circle Court Groundwater Plume; 2017. https://www.atsdr.cdc.gov/HAC/pha/CircleCourtGroundwaterPlume/Circle_Court_Groundwater_Plume_PHA_06-26-2017_508.pdf
283. Agency for Toxic Substances and Disease Registry. Camp Timberlake Public Water System; 2008. https://www.atsdr.cdc.gov/HAC/pha/CampTimberlakePublicWaterSystem/CampTimberlakePublicWaterSystem%20HC%206-17-2008.pdf
284. Agency for Toxic Substances and Disease Registry. Main Street Groundwater Plume Site; 2016. https://www.atsdr.cdc.gov/HAC/pha/MainStreetGroundwaterPlumeSite/Main_St_Gw_Plume_HC_2-22-2016_508.pdf
285. Agency for Toxic Substances and Disease Registry. Pelican Bay Public Water System; 2005. https://www.atsdr.cdc.gov/HAC/pha/PelicanBay/PelicanBayHC.pdf
286. Hildenbrand ZL, et al. A comprehensive analysis of groundwater quality in the Barnett shale region. Environ Sci Technol. 2015;49(13):8254–62.
287. Llewellyn GT, et al. Evaluating a groundwater supply contamination incident attributed to Marcellus Shale gas development. Proc Natl Acad Sci U S A. 2015;112(20):6325–30.
288. Agency for Toxic Substances and Disease Registry. West County Road 112. 2022. https://www.atsdr.cdc.gov/HAC/pha/WestCountyRoad112/WestCountyRd-HC-508.pdf
289. Maloney KO, et al. Unconventional oil and gas spills: materials, volumes, and risks to surface waters in four states of the U.S. Sci Total Environ. 2017;581-582:369–77.
290. Elliott EG, et al. A community-based evaluation of proximity to unconventional oil and gas wells, drinking water contaminants, and health symptoms in Ohio. Environ Res. 2018;167:550–7.
291. Colten CE. An incomplete solution: oil and water in Louisiana. J Am Hist. 2012;99(1):91–9.
292. Roberts JD, et al. "I Can't breathe": examining the legacy of American racism on determinants of health and the ongoing pursuit of environmental justice. Curr Environ Health Rep. 2022;9(2):211–27.
293. Jones R. Climate change and indigenous health promotion. Glob Health Promot. 2019;26(3_suppl):73–81.
294. Sultana F. The unbearable heaviness of climate coloniality. Polit Geogr. 2022:102638.
295. Casey JA, et al. Climate justice and California's methane superemitters: environmental equity assessment of community proximity and exposure intensity. Environ Sci Technol. 2021;55(21):14746–57.

296. Emanuel RE, et al. Natural gas gathering and transmission pipelines and social vulnerability in the United States. Geohealth. 2021;5(6):e2021GH000442.
297. Jonasson ME, et al. Oil pipelines and food sovereignty: threat to health equity for indigenous communities. J Public Health Policy. 2019;40(4):504–17.
298. Meaders JS. Health impacts of petrochemical expansion in Louisiana and realistic options for affected communities. Tul Env't LJ. 2021;34:113.
299. WindHorse Strategic Initiatives and Fire Heart Institute. Faulty Infrastructure and the Impacts of the Dakota Access Pipeline; 2022. https://climatejustice.ndncollective.org/dapl-report/
300. Johnston JE, Werder E, Sebastian D. Wastewater disposal wells, fracking, and environmental injustice in southern Texas. Am J Public Health. 2016;106(3):550–6.
301. Morey B. Environmental justice for native Hawaiians and Pacific islanders in Los Angeles County. Environ Just. 2014;7(1):9–17.
302. Madrid J, Delgado A. US Latinos and air pollution; 2011.
303. Gonzalez DJX, et al. Historic redlining and the siting of oil and gas wells in the United States. J Expo Sci Environ Epidemiol. 2023;33(1):76–83.
304. Nicole W. Wristbands for research: using wearable sensors to collect exposure data after hurricane Harvey. Environ Health Perspect. 2018;126(4):042001.
305. Ratnapradipa D, et al. Implications of hurricane Harvey on environmental public health in Harris County, Texas. J Environ Health. 2018;81(2):24–33.
306. Horney JA, et al. Comparing residential contamination in a Houston environmental justice neighborhood before and after hurricane Harvey. PLoS One. 2018;13(2):e0192660.
307. Kiaghadi A, Rifai HS. Physical, chemical, and microbial quality of floodwaters in Houston following hurricane Harvey. Environ Sci Technol. 2019;53(9):4832–40.
308. Vengosh A, et al. Evidence for unmonitored coal ash spills in Sutton Lake, North Carolina: implications for contamination of lake ecosystems. Sci Total Environ. 2019;686:1090–103.
309. Santella N, Steinberg LJ, Sengul H. Petroleum and hazardous material releases from industrial facilities associated with hurricane Katrina. Risk Anal. 2010;30(4):635–49.
310. Ruckart PZ, et al. Hazardous substances releases associated with hurricanes Katrina and Rita in industrial settings, Louisiana and Texas. J Hazard Mater. 2008;159(1):53–7.
311. Adair K, Miller S, Gage Witvliet M. An exploratory investigation of government air monitoring data after hurricane Harvey. Int J Environ Res Public Health. 2022;19(9)
312. Anenberg SC, Kalman C. Extreme weather, chemical facilities, and vulnerable communities in the U.S. Gulf Coast: a disastrous combination. Geohealth. 2019;3(5):122–6.
313. Madrigano J, et al. Fugitive chemicals and environmental justice: a model for environmental monitoring following climate-related disasters. Environ Just. 2018;11(3):95–100.
314. Neal D, Famira VE, Miller-Travis V. Now is the time: environmental injustice in the US and recommendations for eliminating disparities. Lawyers' Committee for Civil Rights, 2010.
315. Flores AB, et al. Social vulnerability to hurricane Harvey: unmet needs and adverse event experiences in greater Houston, Texas. Int J Disaster Risk Red. 2020;46
316. Kishore N, et al. Mortality in Puerto Rico after hurricane Maria. N Engl J Med. 2018;379(2):162–70.
317. Rodriguez-Rabassa M, et al. Impact of a natural disaster on access to care and biopsychosocial outcomes among Hispanic/Latino cancer survivors. Sci Rep. 2020;10(1):10376.
318. Santos-Burgoa C, et al. Ascertainment of the estimated excess mortality from hurricane Maria in Puerto Rico; 2018.
319. Lloréns H. Toxic racism in Puerto Rico's sacrifice zone. NACLA Rep Am. 2021;53(3):275–80.
320. Garcia-Lopez GA. The multiple layers of environmental injustice in contexts of (un)natural disasters: the case of Puerto Rico post-hurricane Maria. Environ Just. 2018;11(3):101–8.
321. Lancet Countdown. Compounding crises of our time during hurricane Laura. 2020. https://www.lancetcountdownus.org/wp-content/uploads/2020/12/cs-compounding.pdf
322. Méndez-Lázaro PA, et al. Environmental stressors suffered by women with gynecological cancers in the aftermath of hurricanes Irma and María in Puerto Rico. Int J Environ Res Public Health. 2021;18(21)

323. Schlanger Z. Oil refineries in hurricane Harvey's path are polluting Latino and low-income neighborhoods. 2017. https://qz.com/1066097/hurricane-harvey-oil-refineries-are-polluting-latino-and-low-income-neighborhoods/
324. Sharpe JD, Wolkin AF. The epidemiology and geographic patterns of natural disaster and extreme weather mortality by race and ethnicity, United States, 1999-2018. Public Health Rep. 2021:333549211047235.
325. National Academies of Sciences E, et al. The National Academies Collection: Reports funded by National Institutes of Health, in Communities, Climate Change, and Health Equity: Proceedings of a Workshop—in Brief, A. Reich, A. Ulman, and C. Berkower, Editors. 2022, National Academies Press (US). Copyright 2022 by the National Academy of Sciences. All rights reserved.: Washington, DC.
326. Blair BD, et al. Residential noise from nearby oil and gas well construction and drilling. J Expo Sci Environ Epidemiol. 2018;28(6):538–47.
327. Allshouse WB, et al. Community noise and air pollution exposure during the development of a multi-well oil and gas pad. Environ Sci Technol. 2019;53(12):7126–35.
328. Richburg CM, Slagley J. Noise concerns of residents living in close proximity to hydraulic fracturing sites in Southwest Pennsylvania. Public Health Nurs. 2019;36(1):3–10.
329. McCawley M. Air contaminants associated with potential respiratory effects from unconventional resource development activities. Semin Respir Crit Care Med. 2015;36(3):379–87.
330. Finn K, et al. Responsible resource development and prevention of sex trafficking: safeguarding native women and children on the Fort Berthold reservation. Harv Women's LJ. 2017;40:1.
331. Martin K, et al. Violent victimization known to law enforcement in the Bakken oil-producing region of Montana and North Dakota, 2006-2012. 2019: National Crime Statistics Exchange.
332. Banzhaf S, Ma L, Timmins C. Environmental justice: the economics of race, place, and pollution. J Econ Perspect. 2019;33(1):185–208.
333. Pinderhughes R. Who decides what constitutes a pollution problem? Race Gender Class. 1997;5(1):130–52.
334. Ishiyama N. Environmental justice and American Indian Tribal Sovereignty: case study of a land–use conflict in Skull Valley, Utah. Antipode. 2003;35(1):119–39.
335. Taylor DE. The environment and the people in American cities, 1600s-1900s. In: The environment and the people in American cities, 1600s-1900s. Duke University Press; 2009.
336. Avila E, Rose MH. Race, culture, politics, and urban renewal: an introduction. J Urban Hist. 2009;35(3):335–47.
337. Bayor RH. Racism as public policy in America's cities in the twentieth century. Crossing Boundaries: The Exclusion and Inclusion of Minorities in Germany and the United States; 2001. p. 70–82.
338. Blumberg L. Segregated housing, marginal location, and the crisis of confidence. Phylon (1960-). 1964;25(4):321–30.
339. Arnold CA. Fair and healthy land use. American Planning Association; 2007.
340. Clingermayer JC. Heresthetics and happenstance: intentional and unintentional exclusionary impacts of the zoning decision-making process. Urban Stud. 2004;41(2):377–88.
341. Collin RW. Environmental equity: a law and planning approach to environmental racism. Va Envtl LJ. 1991;11:495.
342. Connerly CE. "The most segregated city in America": City planning and civil rights in Birmingham, 1920-1980. University of Virginia Press; 2005.
343. Maantay J. Zoning law, health, and environmental justice: what's the connection? J Law Med Ethics. 2002;30(4):572–93.
344. Park LS-H, Pellow D. The slums of Aspen: immigrants vs. the environment in America's Eden. Vol. 2. NYU Press; 2013.
345. Ritzdorf M. Locked out of paradise: contemporary exclusionary zoning, the Supreme Court, and African Americans, 1970 to the present. Urban planning and the African American community in the shadows; 1997.
346. Slater G. The inventors of America's most dangerous idea. 2021. https://www.theatlantic.com/ideas/archive/2021/11/freedom-residential-segregation/620709/

347. Krysan M, Farley R, Couper MP. In the eye of the beholder: racial beliefs and residential segregation. Du Bois Rev. 2008;5(1):5–26.
348. Steil JP, et al. The social structure of mortgage discrimination. Hous Stud. 2018;33(5):759–76.
349. Bartlett R, et al. Consumer-lending discrimination in the FinTech era. J Financ Econ. 2022;143(1):30–56.
350. Hanson A, et al. Discrimination in mortgage lending: evidence from a correspondence experiment. J Urban Econ. 2016;92:48–65.
351. Munnell AH, et al. Mortgage lending in Boston: interpreting HMDA data. Am Econ Rev. 1996:25–53.
352. Liu F, et al. An updated review of the new and revised data points in HMDA. Washington, DC: Consumer Financial Protection Bureau; 2020. https://files.consumerfinance.gov/f/documents/cfpb_data-points_updated-review-hmda_report.pdf
353. Perry A, Rothwell J, Harshbarger D. The devaluation of assets in black neighborhoods. Library Catalog: www.brookings.edu, 2018.
354. Shapiro T, Meschede T, Osoro S. The roots of the widening racial wealth gap: explaining the black-white economic divide. 2013.
355. Bullard R. Examining the evidence of environmental racism. In land use forum: a journal of law, policy, and practice.
356. Takvorian, D. Toxics and neighborhoods don't mix. In land use forum: a journal of law, policy and practice. 1993.
357. Anderson JL, Sass E. Is the wheel unbalanced? A study of bias on zoning boards. Urban Lawyer. 2004:447–74.
358. Wing S, et al. Integrating epidemiology, education, and organizing for environmental justice: community health effects of industrial hog operations. Am J Public Health. 2008;98(8):1390–7.
359. Gibbs L. Citizen activism for environmental health: the growth of a powerful new grassroots health movement. Ann Am Acad Pol Soc Sci. 2002;584:97–109.
360. Auyero J, Hernandez M, Stitt ME. Grassroots activism in the belly of the beast: a relational account of the campaign against urban fracking in Texas. Soc Probl. 2017;66(1):28–50.
361. Krieger N. Climate crisis, health equity, and democratic governance: the need to act together. J Public Health Policy. 2020;41(1):4–10.
362. Wood A, Howarth M. How federal and state regulatory systems perpetuate environmental injustice in the United States: industrial ethylene oxide emissions as a case study. Environ Just. 2022.
363. McGhee H. The sum of us: what racism costs everyone and how we can prosper together. One World; 2022.
364. Ash M, et al. Is environmental justice good for white folks? Industrial air toxics exposure in urban America. Soc Sci Q. 2013;94(3):616–36.
365. Liu Y, Paciorek CJ, Koutrakis P. Estimating regional spatial and temporal variability of PM(2.5) concentrations using satellite data, meteorology, and land use information. Environ Health Perspect. 2009;117(6):886–92.
366. Tomer A, et al. We can't beat the climate crisis without rethinking land use. Brookings; 2021.
367. Goldstein B, Gounaridis D, Newell JP. The carbon footprint of household energy use in the United States. Proc Natl Acad Sci. 2020;117(32):19122–30.
368. Environmental Protection Agency, E. Inventory of US greenhouse gas emissions and sinks. 2022. https://www.epa.gov/system/files/documents/2022-04/us-ghg-inventory-2022-main-text.pdf
369. Wei T, Wu J, Chen S. Keeping track of greenhouse gas emission reduction progress and targets in 167 cities worldwide. Front Sustain Cities. 2021;3
370. Park H, County F, County W. Local policies for environmental justice: a national scan. 2019.
371. Raheja G, et al. Community-based participatory research for low-cost air pollution monitoring in the wake of unconventional oil and gas development in the Ohio River valley: empowering impacted residents through community science. Environ Res Lett. 2022;17(6):065006.

372. Stenclik D, Richwine M. Puerto Rico distributed solar integration study: island in the sun, a resilient, renewable grid through distributed solar and storage. IEEE Electrification Mag. 2022;10(3):45–54.
373. The Greenlining Institute. Climate Equity; 2022. https://greenlining.org/publications-resources/
374. Brugge D, et al. Improving health in communities near highways: design ideas from a charrette. 2014. https://sites.tufts.edu/cafeh/files/2011/10/CAFEH-Report-Final-2-26-15-hi-res1.pdf
375. Crowther B, Richards L. Freeways without futures. 2021. https://www.cnu.org/our-projects/highways-boulevards/freeways-without-futures
376. Vanderpool T. How to stop a highway. 2021. https://www.nrdc.org/stories/how-stop-highway?tkd=8003559&utm_campaign=TOM&utm_medium=link2&utm_source=body
377. Thomas K, et al. Explaining differential vulnerability to climate change: a social science review. Wiley Interdiscip Rev Clim Chang. 2019;10(2):e565.
378. Calo WA, et al. Disruptions in oncology care confronted by patients with gynecologic cancer following hurricanes Irma and Maria in Puerto Rico. Cancer Control. 2022;29:10732748221114691.
379. Man RX, et al. The effect of natural disasters on cancer care: a systematic review. Lancet Oncol. 2018;19(9):e482–99.
380. Bell SE. Fighting King coal the challenges to micromobilization in Central Appalachia. MIT; 2016.
381. Dodson W. A cloud of coal mine dust over a West Virginia community points to regulatory shortcomings. 2022. https://appvoices.org/2022/03/22/a-cloud-of-coal-mine-dust-over-a-west-virginia-community-points-to-regulatory-blindspots/
382. Ford JD, et al. The resilience of indigenous peoples to environmental change. One Earth. 2020;2(6):532–43.
383. Skinner-Dorkenoo AL, et al. Highlighting COVID-19 racial disparities can reduce support for safety precautions among White U.S. residents. Soc Sci Med. 2022;301:114951.
384. Rozycka-Tran J, Boski P, Wojciszke B. Belief in a zero-sum game as a social axiom: a 37-nation study. J Cross-Cult Psychol. 2015;46(4):525–48.
385. Keenan JM, Hill T, Gumber A. Climate gentrification: from theory to empiricism in Miami-Dade County, Florida. Environ Res Lett. 2018;13(5)
386. Triguero-Mas M, et al. Natural outdoor environments' health effects in gentrifying neighborhoods: disruptive green landscapes for underprivileged neighborhood residents. Soc Sci Med. 2021;279:113964.
387. Loughran K. Parks for profit: the high line, growth machines, and the uneven development of urban public spaces. City Community. 2014;13(1):49–68.
388. Carmichael CE, McDonough MH. Community stories: explaining resistance to street tree-planting programs in Detroit, Michigan, USA. Soc Nat Resour. 2019;32(5):588–605.
389. Cheng E, et al. Neighborhood and individual socioeconomic disadvantage and survival among patients with nonmetastatic common cancers. JAMA Netw Open. 2021;4(12):e2139593.
390. Interlandi J. Why doesn't America have universal health care? One word: race. NY Times Mag. 2019;14
391. Yearby R, Clark B, Figueroa J. Structural racism in historical and modern US health care policy. Health Aff. 2022;41(2):187–94.
392. Xolocotzi EH. Experiences leading to a greater emphasis on man in ethnobotanical studies. Econ Bot. 1987:6–11.
393. Levis C, et al. Persistent effects of pre-Columbian plant domestication on Amazonian forest composition. Science. 2017;355(6328):925–31.
394. Gliessman SR. Agroecology: the ecology of sustainable food systems. CRC; 2006.
395. Zimmerer KS, et al. The biodiversity of food and agriculture (agrobiodiversity) in the anthropocene: research advances and conceptual framework. Anthropocene. 2019;25
396. Garnett ST, et al. A spatial overview of the global importance of indigenous lands for conservation. Nat Sustain. 2018;1(7):369–74.

397. Carney JA. Subsistence in the Plantationocene: dooryard gardens, agrobiodiversity, and the subaltern economies of slavery. J Peasant Stud. 2021;48(5):1075–99.
398. Altieri MA, Toledo VM. The agroecological revolution in Latin America: rescuing nature, ensuring food sovereignty and empowering peasants. J Peasant Stud. 2011;38(3):587–612.
399. Steward A, Lima D. "We also preserve": Quilombola defense of traditional plant management practices against preservationist bias in Mumbuca, Minas Gerais, Brazil. J Ethnobiol. 2017;37
400. Carney J, Rosomoff RN. In the shadow of slavery: Africa's botanical legacy in the Atlantic world. University of California Press; 2011.
401. Barthel S, Crumley C, Svedin U. Bio-cultural refugia-safeguarding diversity of practices for food security and biodiversity. Glob Environ Change-Hum Policy Dimensions. 2013;23(5):1142–52.
402. Chappell MJ. Beginning to end hunger: food and the environment in Belo Horizonte, Brazil, and beyond. Univ of California Press; 2018.
403. Mughal, R., How the natural environment can support health and wellbeing through social prescribing. 2022.
404. White MM. Environmental reviews & case studies: D-town farm: African American resistance to food insecurity and the transformation of Detroit. Environ Pract. 2011;13(4):406–17.
405. Haraway D. Anthropocene, Capitalocene, Plantationocene, Chthulucene: making kin. Environ Hum. 2015;6(1):159–65.
406. Williams JM, Holt-Giménez E. Land justice: re-imagining land, food, and the commons. Food First Books; 2017.

Asbestos, Mining, Mesothelioma, and Lung Cancer

10

Oriana Salamo, Rosa M. Estrada-Y-Martin, and Sujith V. Cherian

Introduction

Asbestos is a naturally occurring mineral composed of flexible fibers of hydrated silicates that have a generous tensile strength, are thermochemical resistant, and have a low electrical conductivity [1]. The word asbestos comes from the ancient Greek term *"sasbestos"*, which means "inextinguishable" or "unquenchable", referring to the mineral's extraordinary characteristics [2]. Asbestos is abundant in nature and have been worldwide mined and commercially exploited because of their remarkable properties and endless applications, since they can be spun and woven into textiles and can also be easily incorporated into different materials. Asbestos has been utilized by civilizations for thousands of years, with the most remote use dating back to approximately 2500 B.C. for pottery and chinking of log homes in The Nordic Region. There is also evidence of asbestos use among different cultures over time, including the early Greek and Roman era where it was utilized as fire retardant in clothing and building materials, while the Egyptians employed such mineral for embalming purposes. It was in the 1720s when the development of asbestos textiles factories became popular in the Ural Mountains of Western Russia, home of one of the largest asbestos mines in the world. However, the formal industrial production of asbestos took place in the mid-nineteenth to early-twentieth century when the concept of mass-mining was introduced. Asbestos was admired for its electrical and thermal insulation against extreme temperatures and certain chemicals, becoming the most suitable building material. The world-wide commercial exploitation of asbestos initially took place in Russia, Canada, and Italy between

O. Salamo · R. M. Estrada-Y-Martin · S. V. Cherian (✉)
Department of Internal Medicine, Divisions of Critical Care, Pulmonary and Sleep Medicine, McGovern Medical School at The University of Texas Health Science Center, Houston, TX, USA
e-mail: oriana.k.salamocolletti@uth.tmc.edu; rosa.m.estrada.y.martin@uth.tmc.edu; Sujith.v.cherian@uth.tmc.edu

E. H. Bernicker (ed.), *Environmental Oncology*,
https://doi.org/10.1007/978-3-031-33750-5_10

1866 and 1890, followed by Australia and South Africa between World War I and II [3]. The commercial use of asbestos became attractive to the public in the 1950s, when it became part of daily life products such as clothing items, domestic products, and cigarette filters. The peak of asbestos production was between 1970 and 1980s, when it reached over 100 million tons, followed by a steady decline in production considering associations between asbestos exposure and an increased risk for malignant and non-malignant pulmonary diseases. In 2010, the World Health Organization (WHO) declared asbestos as one of ten substances of major public health concern causing over 100,000 death per year [4]. However, despite a dramatic decrease of asbestos production, data from the International Ban Asbestos Secretariat (IBAS) reported an estimated production of 2,026,000 tons of asbestos in 2015 [3]. Interestingly, over 50 countries have banned asbestos mining, importation, and use, while countries including Kazakhstan, China, India, and Russia, continue to engage with asbestos mining, production, and use, perpetuating the global incidence of exposure [5].

The Environmental Protection Agency (EPA) legally recognizes six types of asbestos fibers that can be broadly classified into two different categories based on their shapes:

- *Serpentine* fibers are typically long and curvy, forming a layered structure with a wavy morphology with splayed ends. Chrysotile is the sole member of this group, also referred as "white asbestos", and accounts for 90–95% of all commercial-used asbestos around the world. Chrysotile is composed of brucite and silicate layers covered by magnesium hydroxide. Moreover, chrysotile is considered less toxic than the *amphibole* types, since the outer layer can be easily dissolved by low pH levels, leaving an unstable structure behind that can be easily cleared by the alveolar macrophages [6].
- *Amphibole* fibers are usually long and straight structures with a rod-like appearance. The surface is made of silica, conferring a strong and durable structure that is acid resistant [6]. Exposure to *amphibole* fibers can be highly toxic, inducing the development of malignant and non-malignant lung diseases. Crocidolite, also known as "blue asbestos", and amosite, recognized as "brown asbestos", are the most commercially treasured type of asbestos from this category, accounting for 10% of asbestos utilized over the last century, while tremolite, anthophyllite, and actinolite, are considered non-commercial amphiboles.

Asbestos Exposure

Asbestos fibers can enter the body through different mechanisms, including ingestion, direct skin contact, and inhalation through primary and secondary exposures [7]. Primary exposure, also known as occupational exposure to asbestos, was the most common type of exposure in men in the twentieth century, and included activities such as asbestos mining, grinding, and bagging of asbestos-containing materials (ACMs), as well as different types of processes involved in the manufacturing of textiles and asbestos-cement products [8]. On the other hand, secondary exposure to asbestos was previously more common among women and children before strict regulations were enacted in the early 1970s, typically when asbestos fibers were brought home on contaminated clothes or objects from the workplace [9].

Environmental exposure to asbestos can also occur and represents a public health concern. Naturally occurring asbestos (NOA) refers to the presence of asbestos as component of rocks or soils. However, if NOA are left undisturbed, there are no health associated consequences. NOA can be found in four different types of rocks, including banded ironstones (rich in crocidolite and amosite), alpine-type ultramafic rock (containing chrysotile), stratiform ultramafic inclusions (composed of chrysolite and tremolite), and serpentinized limestone (made of chrysotile only) [1].

Asbestos Toxicity

The "mechanical interference mechanism", initially proposed in the 1980s by Olofsson *et al*, described how three different asbestos compounds (chrysotile, crocidolite, and amosite), triggered a numerical and/or structural abnormality in human mesothelial cells after 48 hours of exposure [10]. However, further experimental research revealed that asbestos toxicity is considerably more complex; it is not only related to the chemical composition of asbestos fibers but is also correlated to the fiber's size, dimension, and presence of transition metals [11, 12].

Size and Dimension

Alveolar macrophages are part of the innate immune cells in the lungs and are the first line of defense against asbestos fibers. Alveolar macrophages recognize inhaled asbestos fibers as foreign material and attempt its phagocytosis and further clearance. Asbestos fibers that are smaller than 5 μm in length can be more efficiently phagocytized and subsequently cleared by the alveolar macrophages without further disruption of the homeostasis, as reported by Schinwald *et al* [13]. Additionally, the dimension of the asbestos fiber is also critical for its clearance; a ratio > 3 (length/diameter) represents a challenge for the alveolar macrophages [14]. A "frustrated" or "incomplete" phagocytosis by alveolar macrophages takes place when asbestos fibers are longer than 5 μm in length. Those long fibers will promote the activation of the NOD-like receptor pyrin domain containing 3 (NLRP3) inflammasome and the transcription and subsequent production of interleukin-1ß, key mediator of chronic inflammatory response [15]. Furthermore, high mobility group box-1 protein (HMGB1) is secreted by the damaged or necrotic alveolar or mesothelial cells, inducing an amplification of the inflammatory response and secretion of tumor necrosis factor-α [16]. Finally, through the release of reactive oxygen species (ROS) and reactive nitrogen species (RNS), significant DNA damage takes place, promoting carcinogenesis in patients exposed to asbestos fibers [12].

Transition Metals

Iron is an essential metal for multiple intracellular processes. However, elevated levels of iron could be associated with the development of malignancies [14]. Previous studies have shown that intraperitoneal administration of ferric saccharate can induce malignant mesothelioma, raising the concern that excess iron could be involved in the carcinogenesis of mesothelioma [17]. Certain asbestos fibers, such as crocidolite and amosite, contain approximately 30% iron. Moreover, certain asbestos fibers, especially chrysotile, can attract iron from the surrounding environment through the induction of hemolysis [18]. Finally, DNA damage and apoptosis resistance can be induced by an increased iron mediated ROS and the generation of iron-triggered by the reaction of asbestos fibers rich in iron [19].

Asbestos-Related Lung Diseases

Asbestos was formally recognized as a health hazard in 1924, when the association of asbestos and lung fibrosis was first reported by W. E. Cooke in the *British Medical Journal* [20]. Multiple other case reports were published afterwards emphasizing the relationship between asbestos and non-malignant but debilitating and potentially fatal lung diseases. It was in the 1950s when the association between asbestos and lung cancer became apparent and more convincing to the general public [21]. Furthermore, it was not until the 1960s when mesothelioma was recognized as a new cancer entity related to asbestos exposure [22]. Unfortunately, currently there are no official guidelines for the screening of patients that have been exposed to asbestos in the past, representing a challenge for early detection in high- risk population. The purpose of this chapter is to expand on malignant asbestos-related lung diseases, particularly mesothelioma and lung cancer.

Malignant Pleural Mesothelioma

Malignant pleural mesothelioma (MPM) is a rare and aggressive cancer that arises from the mesothelial cells that form the serosal lining of the pleural cavity. The global incidence of MPM is estimated at 0.3 cases per 100,000 population and has a poor survival rate of approximately 38% at 1 year and 7% at 3 years [23, 24]. MPM typically affects individuals' older patients than 70 years of age. Interestingly, an association with previous asbestos exposure can be found in 70–80% of the cases, with a latency period as prolonged as 40 years [25]. 95% of the patients have unilateral disease, and the right hemithorax is affected in approximately 60% of the patients [26].

Risks Factors and Pathogenesis

Asbestos exposure is the most common risk factor for the development of MPM and is considered an occupational malignancy in most of the cases. However, cases of

MPM associated with secondary exposure to asbestos have also been reported, especially in the household of asbestos workers. Furthermore, the length of time of asbestos exposure is also considered a risk factor for the development of MPM, since prolonged periods can potentially increase the asbestos burden in the terminal airways, triggering the inflammatory cascade (described above) and subsequent carcinogenesis. MPM occurs predominantly in men, with a male to female ratio of 5:1, and the risk typically increases with age, with a mean age of diagnosis of 72 years [27]. Additionally, non-asbestos related risks factors for the development of MPM include prior chest wall radiation and erionite, a mineral typically found in the rocks of Turkey and Mexico, but also in some areas of the United States [28].

The pathogenesis of MPM is generally multifactorial. Asbestos fibers are typically inhaled and unable to be cleared by alveolar macrophages. Those fibers will then transverse the terminal airways, reaching the pleural space. As described above, repetitive DNA damage by ROS and RNS will lead to carcinogenesis of the mesothelial cells. Furthermore, germline mutation of genes can also be found in patients with MPM.

BRCA1-associated protein (BAP) 1 gene have been described as potential genetic component involved in cases of familial mesothelioma, as described by Testa *et al,* and approximately 60% of the patients with sporadic MPM will have BAP1 mutations [29]. BAP1 is implicated in DNA repair, cell cycle, and DNA damage response, also promoting apoptosis in damaged cells [30]. Moreover, frequently mutated and subsequent inactivated tumor suppressor genes, such as neurofibromin 2 (NF2), CDKN2A, CDKN2B, SETD2 and TP53, can also be found in patients with MPM [31].

Diagnostic Approach

Clinical Manifestations

Most of the patients with MPM will present with progressive shortness of breath. Pleural effusions can be found in up to 70% of the patients with MPM in early stages of the disease, and is generally the underlying etiology of the shortness of breath [32]. Chest pain, on the other hand, is also a common symptom in patients with MPM, and is typically associated to the tumor involving the chest wall, including nerves and bone structures, but can also be appreciated in patients with underlying pleural effusions. Other non-specific symptoms, such as low-grade fever, weight loss, anorexia, night sweats, and fatigue, are less commonly seen. Since the airways are typically spared, cough and hemoptysis are rarely reported. Superior vena cava syndrome secondary to local tumor invasion has also been described [33]. Less frequently, patients can also present with dysphagia and laryngeal nerve paralysis.

Imaging

Chest radiographs are generally the first diagnostic modality performed, and usually reveals a unilateral pleural effusion with or without pleural thickening or a localized mass. Moreover, computed tomography (CT) of the chest, ideally with intravenous contrast, is considered the gold standard for initial evaluation of patients with MPM, and will demonstrate an unilateral pleural effusion with diffuse or nodular pleural thickening, potentially involving the mediastinal pleural, with or without chest wall, mediastinum, or diaphragmatic invasion [34]. Unfortunately, the sensitivity and specificity of chest CT for pleural malignancy is roughly 68% and 78%, respectively [35]. However, mediastinal pleural involvement and circumferential pleural thickening are features generally associated with MPM (Figs. 10.1 and 10.2) and can help differentiate MPM from metastatic pleural tumors [28]. Furthermore, fluorodeoxyglucose (FDG)-positron emission tomography (PET)-CT scan can provide

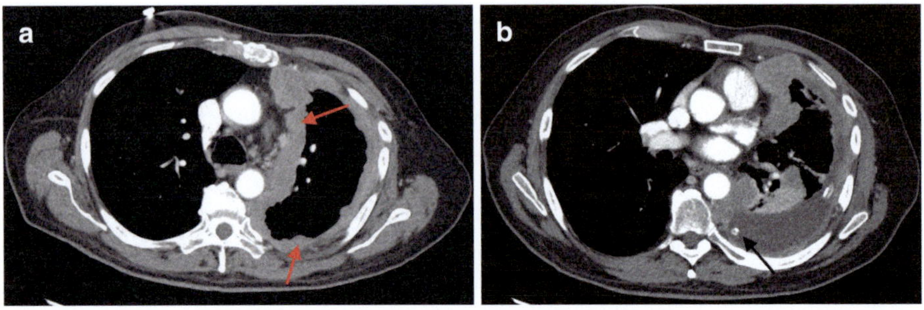

Fig. 10.1 CT images of a 68 year-old patient with malignant pleural mesothelioma. The patient reported working in construction with asbestos exposure for 30–40 years prior to presentation. Biopsy findings were consistent with epithelioid subtype. Notice the bumpy nodular deposits along the mediastinal and parietal pleura (**a** – arrows). The patient was started on chemotherapy with cisplatin and pemetrexed. An indwelling pleural catheter was placed for the pleural effusion, (**b** – arrow); which was subsequently removed 6 months after presentation. Of note, no track metastasis developed for this patient

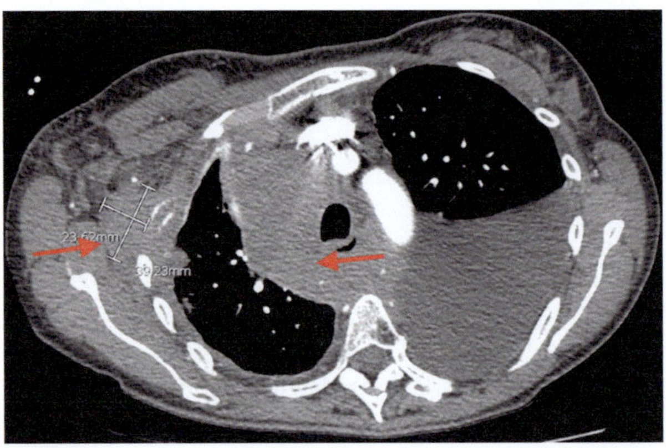

Fig. 10.2 CT images of 52-year-old lady with malignant pleural mesothelioma- which on CT guided biopsy was consistent with epithelioid subtype of malignant pleural mesothelioma. She denied any direct exposure to asbestos, however she was involved with washing clothes of people who had been working with asbestos. The patient had bilateral pleural involvement with mediastinal involvement (arrow) and right axillary lymph node involvement

important metabolic information and is largely used nowadays for staging purposes and treatment assessment response [36]. A standardized uptake value (SUV) of 2 or greater, in the right clinical context, favors the diagnosis of MPM over a benign pleural disease [37]. It is important to mention that false negative PET-CT can be seen in patients with small tumors or those with a low proliferative activity (commonly seeing in patients with MPM with epithelioid subtypes), while false positive results can be seen in inflammatory conditions with high metabolic activity, such as tuberculosis or rheumatoid arthritis involving the pleura, as well as prior pleurodesis [38]. Alternatively, magnetic resonance imaging (MRI) can be a valuable tool for soft tissue definition and for the delineation of the pleural tumor when invasion to the subclavian vessels, diaphragm, or chest wall is suspected [34].

Approach to Diagnosis

Patients with suspected MPM should undergo further investigation with tissue sampling. Pleural fluid cytology is generally the first step if a pleural effusion is present. However, the cytological fluid examination in patients with MPM is consistent with the disease in only one third of the cases [28]. Furthermore, a pleural biopsy is often required in most of the patients to make a final diagnosis and further characterize the histological subtype of MPM. A pleural biopsy can be obtained with either percutaneous fine needle aspiration, with or without radiological guidance, under direct visualization via pleuroscopy/medical thoracoscopy or video-assisted thoracic surgery (VATS), or with an open or surgical biopsy.

Blind pleural biopsies are generally not recommended as the first-line biopsy modality considering their low diagnostic yield. The sensitivity for malignant pleural diseases using the Abrams' needle was reported as 57% on the largest review available [39]. On the other hand, image-guided pleural biopsy, with ultrasound or CT assistance, demonstrated to be superior when compared to blind pleural biopsies with the Abrams' needle [40–42]. It is also worth mentioning that image-guided pleural biopsies can limit the number of unnecessary passes of the needle when compared to the blind approach, hence reducing the risk of albeit rare complications such as tumor seeding along the biopsy track [34]. In 2016, a randomized controlled trial assessed the diagnostic yield and procedural safety of CT-guided Abrams' needle pleural biopsy versus ultrasound-assisted cutting needle pleural biopsy for the diagnosis of patients with exudative pleural effusions [43]. This study recommended the use of CT-guided Abrams pleural biopsy in patients with underlying pleural effusion and associated pleural thickening, advocating the use of ultrasound-assisted cutting needle biopsy in those without pleural effusion but evidence of pleural thickening.

More invasive pleural interventions, such as medical thoracoscopy and VATS, can secure the pathological diagnosis due to direct pleural visualization, while allowing at the same time the evacuation and definitive management of associated pleural effusions with the placement of indwelling pleural catheter [44]. Medical thoracoscopy/pleuroscopy, is a minimally invasive procedure with low associated morbidity and mortality, generally performed by interventional pulmonologists, that provides a high diagnostic accuracy in patients with malignant pleural diseases. As expected, pleuroscopy has proven to be superior to blind pleural biopsies [45]. On the other hand, VATS is a more invasive procedure when compared to medical thoracoscopy but allows an additional benefit of performing more advanced interventions at the same time as the diagnostic procedure, such as tumor debulking or lung resection. It is important to mention that current trials comparing pleuroscopy versus VATS in regards with the sensitivity and specificity of pleural biopsies in patients with MPM are lacking, but current literature suggest that the sensitivity and specificity of the aforementioned pleural interventions are similar (95% and 100%, respectively) [34]. Finally, an open biopsy with thoracotomy may be considered in patients with locally advanced disease when VATS or medical thoracoscopy are not feasible options.

Tumor seeding at the site of biopsy has been reported to be high with incidences in the range of 10–15%, with open biopsies reported to have much higher incidences of tumor seeding, for which prophylactic radiation therapy has traditionally been recommended [43, 46, 47]. A large, randomized trial evaluating prophylactic irradiation after pleural biopsies in patients with MPM showed no significant difference in chest wall metastases between the groups receiving prophylactic radiation therapy vs not. Moreover, the incidence of tumor seeding at the site of biopsy was low (3.2% vs 5.3% in irradiated group vs group not receiving radiation therapy) [46], which is likely the result of newer chemotherapy regimens with carboplatin/cisplatin with pemetrexed [48]. It is pertinent to note that open thoracotomies were excluded in this study, and the risk is likely higher for tumor seeding given the wide incision margins [46].

Histopathological Diagnosis

In 2015, the WHO classified MPM into three different histologic subtypes: epithelioid, sarcomatoid, and mixed (also known as biphasic) [28]. The epithelioid subtype is the most common histologic type, accounting for 50 to 60% of the cases, and offers superior survival when compared to the sarcomatoid and mixed types. Those cells are generally oval, polygonal, or cuboidal in shape, often mimicking reactive mesothelial cells [49]. Necrosis and nuclear atypia are considered independent prognostic factors in patients with epithelioid MPM, allowing its further classification into low and high histological grades [50]. The sarcomatoid variant accounts for 10% of the cases, and is characterized of cells with a spindle shape that are distributed in a more disorganized fashion, typically showing some degree of cellular atypia [26]. Moreover, the sarcomatoid variant is considered to be highly invasive and also drug-resistant, representing a formidable challenge for its management [51]. Finally, the mixed or biphasic type accounts for the remaining 30–40% of the cases, and it is characterized by the mosaic mixture of at least 10% sarcomatoid and epithelioid cells each [26].

Unfortunately, the histopathological diagnosis of MPM is still considered a challenge for different reasons. As described above, the epithelioid subtype of MPM shares similar features with reactive mesothelial cells, including the presence of augmented cellularity and atypia. However, considerable amount of necrosis, certain specific vascular and growth patterns, and infiltration to adjacent structures can suggest MPM over other benign pleural diseases. Fluorescent in situ hybridization (FISH) and immunohistochemical (IHC) play a crucial role when trying to distinct the diagnosis of MPM from other malignancies, including pleural metastasis, with the detection of specific markers, such as loss of BAP1 expression, homozygous deletion of CDKN2A, and expression of methylthio-adenosine phosphorylase (MTAP) [26]. Unfortunately, there are no markers with a perfect sensitivity or specificity to be confidently used for the detection of MPM. Hence, the recommendation for the histopathological diagnosis of MPM includes a panel able to differentiate between MPM and adenocarcinoma, comprising at least two or more biomarkers for mesothelial cells (i.e. calretinin, cytokeratin 5/6, D2–40, or Wilms' tumor 1 antigen), as well as two or more biomarkers for adenocarcinoma cells (i.e. thyroid transcription factor 1, Ber-EP4, or carcinoembryonic antigen) detected by FISH and IHC [49, 51].

Staging

The first tumor, node, and metastasis (TNM) classification was introduced by the International Mesothelioma Interest Group (IMIG) in 1994 [52]. However, this first and subsequent staging system were found to have several limitations requiring further revisions based on the most updated clinical research available, leading to the most recent eighth TNM classification published in 2016. The overall purpose of the TNM classification is to provide prognosis, to guide therapy, and to allow the identification of patients that could potentially benefit from clinical trials. Nevertheless, despite novel advances in the field, the staging system for patients with MPM remains challenging, since post-mortem data often reveals a more widespread disease, raising the concern for under staging.

Tumor

The tumor (T) component of the TNM classification remains a challenge and is difficult to assess in patients with MPM. Unlike other solid tumors, patients with MPM typically present with diffuse nodularity with significant pleural thickening and underlying pleural effusion instead of a solid, well rounded, or circumferential mass. However, the most recent TNM classification simplified the T component of the TNM classification by collapsing the prior T1a (tumor only involving the parietal pleura) and T1b (tumor involving the parietal pleura with focal compromise of the visceral lining) into one combined T1 category, as appreciated in Table 10.1 [53]. The reasoning behind this change when compared to the seventh TNM classification relies on the difficulty to differentiate the involvement of the visceral pleural and the lack of statistical difference in regards with the survival of patients with T1a when compared to those with T1b.

Nodal Involvement

The nodal (N) component of the TNM classification was not modified since it was first introduced in 1995. However, the eighth TNM edition simplified the N component by grouping intra and extrapleural nodes into the N1 category, shifting the previous N3 into the N2 classification and finally eliminating N3, as appreciated in Table 10.2 [54]. Furthermore, in regards with the N descriptors, it was noted that the

Table 10.1 Eighth TNM classification showing T category in MPM

T category	Description
Tx	Not able to assess primary tumor
T0	No evidence of primary tumor
T1	Tumor limited to the ipsilateral parietal+/− visceral+/− mediastinal+/− diaphragmatic pleura
T2	Tumor involving each of the ipsilateral pleural surfaces (parietal, mediastinal, diaphragmatic, and visceral pleura) with at least one of the following features: – involvement of diaphragmatic muscle extension of tumor from visceral pleura into the underlying pulmonary parenchyma
T3	Describes locally advanced but potentially resectable tumor. Tumor involving all the ipsilateral pleural surfaces (parietal, mediastinal, diaphragmatic, and visceral pleura) with at least one of the following features: – Involvement of the endothoracic fascia – Extension into the mediastinal fat. – Solitary, completely resectable focus of tumor extending into the soft tissues of the chest wall – Non transmural involvement of the pericardium
T4	Describes locally advanced technically unresectable tumor. Tumor involving all of the ipsilateral pleural surfaces (parietal, mediastinal, diaphragmatic, and visceral pleura) with at least one of the following features: – Diffuse extension or multifocal masses of tumor in the chest wall, with or without associated rib destruction. – Direct transdiaphragmatic extension of tumor to the peritoneum – Direct extension of tumor to the contralateral pleura – Direct extension of tumor to mediastinal organs – Direct extension of tumor into the spine – Tumor extending through to the internal surface of the pericardium with or without a pericardial effusion – Tumor involving the myocardium

Table 10.2 Eighth TNM classification showing lymph node classification in MPM

N category	Description
Nx	Regional lymph nodes cannot be assessed
N0	No regional lymph node metastases
N1	Metastases in the ipsilateral bronchopulmonary, hilar, or mediastinal (including the internal mammary, peri diaphragmatic, pericardial fat pad, or intercostal lymph nodes) lymph nodes
N2	Metastases in the contralateral mediastinal, ipsilateral, or contralateral supraclavicular lymph nodes

survival of patients with MPM was not necessarily related to the specific anatomic location of the nodes involved, but it was associated with the number of nodes instead.

Distant Metastases

The distant metastases (M) component of the most recent TNM classification remained unchanged when compared to the prior edition, as appreciated in Table 10.3 [55].

Table 10.3 Eighth TNM classification of metastasis in MPM

M category	Description
M0	No distant metastasis
M1	Presence of distant metastasis

Staging

Patients diagnosed with MPM should initially undergo noninvasive staging, which includes an initial CT of the chest and abdomen typically followed by a PET-CT to assess distant metastatic disease. More invasive staging should only take place in patients that are likely suitable for surgical resection and/or chemotherapy based on the patient's functional status and preference, and generally consists of endobronchial ultrasound (EBUS) with fine needle aspiration (FNA) of mediastinal lymph nodes, mediastinoscopy, contralateral thoracoscopy, and laparoscopy, if applicable [34]. Moreover, EBUS or mediastinoscopy are generally recommended in patients that are eligible for surgical resection since mediastinal lymph node involvement was found to be a poor prognostic factor [56]. Table 10.4 describes the most recent staging system for patients with MPM based on the eighth TMN classification.

Prognostic Factors

In 1998, Herndon *et al* evaluated 337 patients with MPM and determined that, based on univariate and multivariate analysis, patients with an Eastern Cooperative Oncology Group (ECOG) functional status of 0, and Hemoglobin level equal or greater than 14.6, were found to have better survival [57]. Alternatively, the worst survival was appreciated in patients with functional status ECOG 1 and 2, and those with leukocytosis equal or greater than 15. On the other hand, a study from Taiwan assessed a total of 93 patients with MPM, and concluded that age, patient's performance status, clinical stage, and histological subtype (favoring the epithelioid class over the sarcomatoid and biphasic) were considered independent prognostic factors in patients with MPM [58]. Moreover, it was determined that surgery in potentially resectable cases of MPM and systemic chemotherapy conferred a survival benefit.

The modified Response Evaluation Criteria in Solid Tumors (mRECIST) for MPM was initially published in 2004 and it is a measurement of the tumor thickness from the chest wall to the pleural cortex in two positions at three separate levels, and it is currently used for initial assessment of the MPM mass as well as its response to therapy [59]. A recent study from 2021 by Guzman-Casta *et al*, described the characteristics of a total of 136 patients with MPM, concluding that the most determining prognostic factors for overall survival and progression-free survival included again the histological subtype, treatment chosen, and the mRECIST score [60].

As previously portrayed, chronic inflammation as the result of asbestos exposure has been associated with the carcinogenesis in patients with MPM. A study from 2015 analyzed several inflammation-based prognostic scores, including the lymphocyte to monocyte ratio (LMR) [61]. An elevated LMR equal or greater than 2.74 was found to be significantly associated with prolonged overall survival and was

Table 10.4 Stage grouping based on the 8th TNM classification

Stage	T	N	M
I			
IA	T1	N0	M0
IB	T2, T3	N0	M0
II	T1, T2	N1	M0
III			
IIIA	T3	N1	M0
IIIB	T1 – T3	N2	M0
IV	T4	N0–N2	M0
	Any T	Any N	M1

found to be superior to other inflammatory markers in regards with their prognostic ability.

Treatment

Management of patient with MPM remains a challenge due to the heterogeneity of the disease, late diagnosis, and its aggressive nature. Unfortunately, most of the patients will have an advanced disease at the time of diagnosis and are likely considered to have an unresectable disease in the majority of the cases. Early referral to specialized centers, however, is considered important in newly diagnosed patients with MPM since a multidisciplinary approach is always warranted. Management of recurrent malignant pleural effusion in patients with MPM is key for symptom control and can be achieved by talc pleurodesis or indwelling intrapleural catheter. The goal of this section is to briefly describe the overall management of patients with resectable and unresectable disease.

Resectable Disease

The approach for patients with resectable disease include a trimodality regimen with surgical resection of the tumor, radiation therapy, and perioperative platinum-based chemotherapy in combination with pemetrexed [62]. Macroscopic complete resection of the tumor is recommended in patients with resectable disease with a good functional status and acceptable cardiopulmonary reserve [63]. Interestingly, it remains unclear whether a cytoreductive surgery genuinely offers a prolonged survival in patients with MPM since there are no well-powered, randomized controlled clinical trials available to delineate a standard of care.

A macroscopic complete resection of the tumor is generally achieved by extrapleural pneumonectomy (EPP) or lung sparing pleurectomy with decortication (P/D). EPP is an aggressive *“en bloc”* technique that allows a complete removal of the gross tumor with the associated visceral and parietal pleura, affected portion of the lung, mediastinal lymph nodes, pericardium, and ipsilateral diaphragm [64]. On the other hand, P/D is considered a less invasive procedure that consists of the removal of the parietal pleura, pericardium, and diaphragm, as well as the visceral pleural in order to decorticate the ipsilateral lung [26]. In the last decade, a shift towards P/D was appreciated considering the high morbidity and mortality associated with the EPP technique [65]. Nevertheless, it is important to mention that the P/D procedure has been associated with a high rate of recurrence as reported by Ismail-Khan *et al* [66].

The use of radiation therapy in patients with MPM is generally targeted to patients with local symptoms for palliative purposes, as well as in the perioperative setting prior to the surgical intervention based on an expert consensus [67]. However, the data behind its use has not been supported in randomized, controlled trials. In 2015, a randomized, phase 2, multicenter trial compared neoadjuvant chemotherapy and EPP for patients with MPM with and without hemithoracic radiotherapy, known as the SAKK 17/04 trial [68]. Moreover, this trial failed to support the routine use of hemithoracic radiotherapy after neoadjuvant chemotherapy and surgical resection since a locoregional relapse free survival was not achieved. Furthermore, the use of prophylactic radiation to prevent chest wall invasion after pleural interventions was assessed in two randomized controlled trials. In 2016, the Surgical and Large-Bore Procedures in Malignant Pleural Mesothelioma and Radiotherapy Trial (SMART), a randomized, phase 3, multicenter trial, recommended against the use of prophylactic radiotherapy in all patients with mesothelioma after large-bore pleural interventions [69]. Lastly, in 2019, Bayman *et al* recommended against the use of prophylactic irradiation to the chest wall after pleural procedures in patients with MPM [46].

Neoadjuvant platinum-based chemotherapy plus pemetrexed became the standard of care and first line treatment for patients with MPM since 2003, when the EMPHACIS study showed a statistically increased overall survival with the combined regimen [48]. In 2020, the Food and Drug Administration (FDA) approved the use of the immune checkpoint inhibitors (ICH) nivolumab (programmed cell death 1) plus Ipilimumab (cytotoxic T-lymphocyte antigen 4) as first line treatment for unresectable disease based on a multicenter, randomized, open-label, phase 3 trial [70]. There are multiple ongoing clinical trials looking into further chemotherapy agents, immunotherapeutic options, as well as cellular therapy with genetically engineered T cells (chimeric antigen receptor T, known as CAR-T cells), and molecular stratified therapy, in order to provide a survival benefit for patients with MPM [51].

Unresectable Disease

The approach to patients with unresectable MPM relies on systemic and local therapy. As described above, front-line chemotherapy with a platinum-based agent, generally cisplatin, plus pemetrexed remains the standard of care. For elderly or unfit patients, carboplatin offers a similar benefit when used in combination with pemetrexed in regards with the overall survival, progression free survival, objective response rate, and disease control rate [63]. Additionally, the use of radiotherapy should be considered in patients with significant locoregional symptoms. However, nearly all patients with unresectable disease will progress while on first-line therapy, and consideration for ICH, targeted therapies, and enrollment in clinical trials should take place in specialized centers.

Lung Cancer

Lung cancer is the most common cause of cancer-related death worldwide, accounting for 23% of the cases [71].

Risks Factors and Pathogenesis

Lung cancer is the leading cause of cancer death in USA with mortality higher than colorectal, breast, pancreatic and prostate [72]. Lung cancer is strongly associated with smoking prevalence. Although smoking has decreased in past years in men, still >80% of lung cancer cases are related to smoking [72, 73]. Asbestos is the single most important cause of occupation related lung cancer mortality in the world and account for 30% of occupation related lung cancer deaths globally in 2015 [74]. Asbestos bans have contributed to the decline in asbestos consumption in most of countries except for Asia (Russia, China, Kazakhstan, and India) and Brazil. These countries accounted for 99% of the annual consumption in 2013 [75].

Asbestos exposure has been related to lung cancer since the 1930's, initially identified as anecdotal autopsy case reports in workers with asbestosis [76–78]. An incidence of 12–14% of lung cancer was shown in several autopsy studies published 20 years later, of asbestos workers all of which were also seen to have asbestosis [79]. The synergistic effect of cigarette smoking and asbestos exposures on causing lung cancer was recognized in the 1960's, where Selikoff et al. showed a 90-fold increased risk of death due to lung cancer among insulators as compared with never smokers and no exposure to asbestos [80]. Large scale cohort studies done after these findings, in North America showed a fourfold increase in lung cancer incidence among insulators compared to the US general population [81]. The Helsinki Criteria purported to develop guidelines for diagnosis of asbestos related lung conditions further highlighted this association [82, 83] (see below). A landmark cohort study of asbestos insulator workers by Markowitz et al., also confirmed these findings. Indeed, the risk of developing lung cancer was 3.6 times higher in the asbestos exposed non-smokers as compared to non-asbestos exposed workers. Combined asbestos exposure and cigarette smoking was associated with a 14 fold increased risk of lung cancer mortality, while the presence of asbestosis was associated with the 37 fold increase in lung cancer mortality in smoking asbestos insulator workers [84]. Thus, the asbestos-smoking interaction is greater than additive but not multiplicative [85, 86].

Risk of lung cancer is generally much greater after exposure to the commercial amphiboles than to commercial chrysolite [87]. Moreover, the risk to chrysolite is largely dependent on the contamination with varying fibers of tremolite [88]. The above-mentioned Helsinki criteria has been used by various societies to define exposure risk when alluding to asbestos exposure for Asbestos related Lung Cancer (ARLC); which include the following [82]:

A. One year of heavy exposure (e.g., manufacturing of asbestos products, asbestos spraying, insulation work with asbestos materials, demolition of old buildings) or 5–10 years of moderate exposure (e.g., construction, shipbuilding)
B. Estimated cumulative exposure to mixed (amphibole plus chrysolite) asbestos fibers of 25 fibers/milliliter/year (fiber-years)

C. A lung fiber burden within the range recorded for asbestosis within the same laboratory.
D. Retained fiber levels of two million amphibole fibers (> 5 micron) per gram of dry lung tissue, as determined by electron microscopic analysis.
E. Asbestos body concentrations determined by light microscopic analysis greater than 10,000 per gram of lung tissue.

ARLC is defined as the development of lung cancer within the backgrounds of asbestos exposure which is evaluated by using one or all the defined Helsinki criteria [89]. Extensive research has been performed to elucidate the mechanisms of asbestos exposure and development of lung cancer. The most widely accepted explanation is that asbestos participates in both the initiation and proliferation phases of tumor development. Carcinogenesis by asbestos fibers occurs at different levels including a) altered expression or function of genes arising from genetic or epigenetic alterations, b) altered cell proliferation, c) altered regulation of apoptosis and d) chronic, persistent inflammation. Asbestos fibers cause increase uptake of carcinogenic components within tobacco smoke- polycyclic aromatic hydrocarbons to lung epithelial cells. In addition, cigarette smoking increases the binding of asbestos fibers to lung epithelial cells [79].

Controversy exists about asbestosis being the major risk factor for the development of lung cancer, rather than asbestos exposure alone. It is generally agreed that the presence of asbestosis increases the risk of lung cancer; however, this is more likely reflective of the increased asbestos exposure. Of note, pleural plaques have not shown to infer an increased risk of lung cancer [83, 90–92].

ARLC can occur in any lobe of the lung with a comparable distribution of all histopathologic lung cancer types [93]. In a pooled case control study, an association between small cell lung cancer and asbestos exposure was seen among nonsmokers. However, among smokers, associations with all lung cancer types were found. Moreover, no significant differences exist in immunophenotypes or molecular-genetic profiles of ARLC compared to non-exposed patients [79, 94].

Screening

Trials using low dose CT scan (LDCT) for cancer screening such as The US National Lung Cancer Screening Trial (NLST) [95] and the Dutch- Belgian Randomized Lung Cancer Screening trial (NELSON) [96] improved mortality in smokers at risk of lung cancer. Adding asbestos exposure to lung cancer screening was cost-effective in France [97]. Adding occupational exposure, more so asbestos exposure, may be feasible and recommended even in individuals with less smoking years [98]. Several reports show that LDCT screening of asbestos exposed workers detect lung cancers with yields like those of high-yield smokers with a baseline prevalence of around 1% [99, 100]. Moreover, proportions with Stage 1 lung cancer (57%) [99] were similar to the rates observed in the NLST and the NELSON trials and thus potentially could reduce mortality among these patients which may be more significant in asbestos exposed smokers.

In the absence of any randomized trials or recommendations from societies, it seems plausible that LDCT should be offered to workers aged 50 years and older who have a history of 5 or more years of asbestos exposure (beginning 20 or more years previously), who have had a smoking history of 10 pack years and for former smokers—irrespective of when they quit. Workers with heavy exposure and less than 5 years of exposure are candidates who should be considered for screening. Similarly, workers with underlying lung conditions such as COPD, ILD, asbestosis, family history of cancer, previous history of cancer should also be offered LDCT screening. Future research may help guide lung cancer screening in workers without any of these risk factors and never smokers. Currently, it is not recommended in this patient population. As discussed above, pleural plaques do not imply an increased risk of malignancy, however- it may help to identify patient populations with asbestos exposure who may not be able to remember it [98, 101].

Clinical Manifestations and Prognosis

The clinical presentation of ARLCs is not different with relation to the presentation in non -asbestos related lung cancers. Moreover, the prognosis of ARLC does not differ from that of other lung cancers [85]. The clinical presentation, diagnosis, and management of the different types of lung cancers have been reviewed elsewhere [102–104]. Of mention, pseudo-mesotheliomatous adenocarcinoma has been a cancer subtype more recognized within this cohort, with Koss et al. reporting an association with asbestos exposure in 21% of patients with this malignancy. Pseudomesotheliomatous adenocarcinoma has been mostly reported in men in the sixth to seventh decade with clinical presentation as chest pain, shortness of breath and presence of pleural effusions. Prognosis invariably has been poor [105].

Conclusion

Asbestos is the single most important cause of occupation related lung and pleural malignancy. While the association with asbestos exposure and malignant pleural mesothelioma is without question, the association between asbestos exposure and lung cancers is not that clearly defined. Malignant pleural mesotheliomas unfortunately present in more advanced stages and consequently are associated with poor prognosis. Tobacco abuse and asbestos exposure has a supra-additive effect in the development of lung cancer. There are no differences in asbestos related lung cancers as compared to patients without known asbestos exposure. Considering the long latency between asbestos exposure and the development of thoracic malignancies with considerable morbidity and mortality, a total asbestos ban should be strongly supported and advocated worldwide. Future research and establishment of screening programs are paramount to investigate the prevalence of these asbestos related malignancies.

References

1. Sporn TA. Mineralogy of asbestos. Recent Results Cancer Res. 2011;189:1–11.
2. Reid G, Klebe S, van Zandwijk N, George AM. Correction to asbestos and zeolites: from A to Z via a common ion. Chem Res Toxicol. 2021;34(6):1694.
3. Pira E, Donato F, Maida L, Discalzi G. Exposure to asbestos: past, present and future. J Thorac Dis. 2018;10(Suppl 2):S237–S45.
4. Leong SL, Zainudin R, Kazan-Allen L, Robinson BW. Asbestos in Asia. Respirology. 2015;20(4):548–55.
5. Emmett EA. Asbestos in high-risk communities: public health implications. Int J Environ Res Public Health. 2021;18:4.
6. Bernstein DM. The health risk of chrysotile asbestos. Curr Opin Pulm Med. 2014;20(4):366–70.
7. Cugell DW, Kamp DW. Asbestos and the pleura: a review. Chest. 2004;125(3):1103–17.
8. Thives LP, Ghisi E, Thives Junior JJ, Vieira AS. Is asbestos still a problem in the world? A current review. J Environ Manag. 2022;319:115716.
9. Tlotleng N, Sidwell Wilson K, Naicker N, Koegelenberg CF, Rees D, Phillips JI. The significance of non-occupational asbestos exposure in women with mesothelioma. Respirol Case Rep. 2019;7(1):e00386.
10. Olofsson K, Mark J. Specificity of asbestos-induced chromosomal aberrations in short-term cultured human mesothelial cells. Cancer Genet Cytogenet. 1989;41(1):33–9.
11. Bayram M, Bakan ND. Environmental exposure to asbestos: from geology to mesothelioma. Curr Opin Pulm Med. 2014;20(3):301–7.
12. Kuroda A. Recent progress and perspectives on the mechanisms underlying Asbestos toxicity. Genes Environ. 2021;43(1):46.
13. Schinwald A, Murphy FA, Prina-Mello A, Poland CA, Byrne F, Movia D, et al. The threshold length for fiber-induced acute pleural inflammation: shedding light on the early events in asbestos-induced mesothelioma. Toxicol Sci. 2012;128(2):461–70.
14. Toyokuni S. Iron addiction with ferroptosis-resistance in asbestos-induced mesothelial carcinogenesis: toward the era of mesothelioma prevention. Free Radic Biol Med. 2019;133:206–15.
15. Dostert C, Petrilli V, Van Bruggen R, Steele C, Mossman BT, Tschopp J. Innate immune activation through Nalp3 inflammasome sensing of asbestos and silica. Science. 2008;320(5876):674–7.
16. Carbone M, Yang H. Molecular pathways: targeting mechanisms of asbestos and erionite carcinogenesis in mesothelioma. Clin Cancer Res. 2012;18(3):598–604.
17. Okada S, Hamazaki S, Toyokuni S, Midorikawa O. Induction of mesothelioma by intraperitoneal injections of ferric saccharate in male Wistar rats. Br J Cancer. 1989;60(5):708–11.
18. Nagai H, Ishihara T, Lee WH, Ohara H, Okazaki Y, Okawa K, et al. Asbestos surface provides a niche for oxidative modification. Cancer Sci. 2011;102(12):2118–25.
19. Pascolo L, Gianoncelli A, Schneider G, Salome M, Schneider M, Calligaro C, et al. The interaction of asbestos and iron in lung tissue revealed by synchrotron-based scanning X-ray microscopy. Sci Rep. 2013;3:1123.
20. Cooke WE. Fibrosis of the lungs due to the inhalation of Asbestos dust. Br Med J. 1924;2(3317):147–2.
21. Cureton RJ. Squamous cell carcinoma occurring in asbestosis of the lung. Br J Cancer. 1948;2(3):249–53.
22. Wagner JC, Sleggs CA, Marchand P. Diffuse pleural mesothelioma and asbestos exposure in the North Western Cape Province. Br J Ind Med. 1960;17:260–71.
23. Binazzi A, Di Marzio D, Verardo M, Migliore E, Benfatto L, Malacarne D, et al. Asbestos exposure and malignant mesothelioma in construction workers-epidemiological remarks by the Italian National Mesothelioma Registry (ReNaM). Int J Environ Res Public Health. 2021;19:1.

24. Asciak R, George V, Rahman NM. Update on biology and management of mesothelioma. Eur Respir Rev. 2021;30:159.
25. Attanoos RL, Churg A, Galateau-Salle F, Gibbs AR, Roggli VL. Malignant mesothelioma and its non-Asbestos causes. Arch Pathol Lab Med. 2018;142(6):753–60.
26. Hajj GNM, Cavarson CH, Pinto CAL, Venturi G, Navarro JR, Lima VCC. Malignant pleural mesothelioma: an update. J Bras Pneumol. 2021;47(6):e20210129.
27. Tsao AS, Wistuba I, Roth JA, Kindler HL. Malignant pleural mesothelioma. J Clin Oncol. 2009;27(12):2081–90.
28. Bibby AC, Tsim S, Kanellakis N, Ball H, Talbot DC, Blyth KG, et al. Malignant pleural mesothelioma: an update on investigation, diagnosis and treatment. Eur Respir Rev. 2016;25(142):472–86.
29. Testa JR, Cheung M, Pei J, Below JE, Tan Y, Sementino E, et al. Germline BAP1 mutations predispose to malignant mesothelioma. Nat Genet. 2011;43(10):1022–5.
30. Wadowski B, De Rienzo A, Bueno R. The molecular basis of malignant pleural mesothelioma. Thorac Surg Clin. 2020;30(4):383–93.
31. Carbone M, Ly BH, Dodson RF, Pagano I, Morris PT, Dogan UA, et al. Malignant mesothelioma: facts, myths, and hypotheses. J Cell Physiol. 2012;227(1):44–58.
32. Rudd RM. Malignant mesothelioma. Br Med Bull. 2010;93:105–23.
33. Ragalie GF, Varkey B, Choi H. Malignant pleural mesothelioma presenting as superior vena cava syndrome. Can Med Assoc J. 1983;128(6):689–91. 740
34. Scherpereel A, Opitz I, Berghmans T, Psallidas I, Glatzer M, Rigau D, et al. ERS/ESTS/EACTS/ESTRO guidelines for the management of malignant pleural mesothelioma. Eur Respir J. 2020;55:6.
35. Hierholzer J, Luo L, Bittner RC, Stroszczynski C, Schroder RJ, Schoenfeld N, et al. MRI and CT in the differential diagnosis of pleural disease. Chest. 2000;118(3):604–9.
36. Taralli S, Giancipoli RG, Caldarella C, Scolozzi V, Ricciardi S, Cardillo G, et al. The prognostic value of (18)F-FDG PET imaging at staging in patients with malignant pleural mesothelioma: a literature review. J Clin Med. 2021;11:1.
37. Benard F, Sterman D, Smith RJ, Kaiser LR, Albelda SM, Alavi A. Metabolic imaging of malignant pleural mesothelioma with fluorodeoxyglucose positron emission tomography. Chest. 1998;114(3):713–22.
38. Treglia G, Sadeghi R, Annunziata S, Lococo F, Cafarotti S, Bertagna F, et al. Diagnostic accuracy of 18F-FDG-PET and PET/CT in the differential diagnosis between malignant and benign pleural lesions: a systematic review and meta-analysis. Acad Radiol. 2014;21(1):11–20.
39. Pereyra MF, San-Jose E, Ferreiro L, Golpe A, Antunez J, Gonzalez-Barcala FJ, et al. Role of blind closed pleural biopsy in the management of pleural exudates. Can Respir J. 2013;20(5):362–6.
40. Heilo A, Stenwig AE, Solheim OP. Malignant pleural mesothelioma: US-guided histologic core-needle biopsy. Radiology. 1999;211(3):657–9.
41. Diacon AH, Schuurmans MM, Theron J, Schubert PT, Wright CA, Bolliger CT. Safety and yield of ultrasound-assisted transthoracic biopsy performed by pulmonologists. Respiration. 2004;71(5):519–22.
42. Maskell NA, Gleeson FV, Davies RJ. Standard pleural biopsy versus CT-guided cutting-needle biopsy for diagnosis of malignant disease in pleural effusions: a randomised controlled trial. Lancet. 2003;361(9366):1326–30.
43. Metintas M, Yildirim H, Kaya T, Ak G, Dundar E, Ozkan R, et al. CT scan-guided Abrams' needle pleural biopsy versus ultrasound-assisted cutting needle pleural biopsy for diagnosis in patients with pleural effusion: a randomized. Controlled Trial Respiration. 2016;91(2):156–63.
44. Greillier L, Cavailles A, Fraticelli A, Scherpereel A, Barlesi F, Tassi G, et al. Accuracy of pleural biopsy using thoracoscopy for the diagnosis of histologic subtype in patients with malignant pleural mesothelioma. Cancer. 2007;110(10):2248–52.

45. Maturu VN, Dhooria S, Bal A, Singh N, Aggarwal AN, Gupta D, et al. Role of medical thoracoscopy and closed-blind pleural biopsy in undiagnosed exudative pleural effusions: a single-center experience of 348 patients. J Bronchology Interv Pulmonol. 2015;22(2):121–9.
46. Bayman N, Appel W, Ashcroft L, Baldwin DR, Bates A, Darlison L, et al. Prophylactic irradiation of tracts in patients with malignant pleural mesothelioma: an open-label, multicenter, phase III randomized trial. J Clin Oncol. 2019;37(14):1200–8.
47. Agarwal PP, Seely JM, Matzinger FR, MacRae RM, Peterson RA, Maziak DE, et al. Pleural mesothelioma: sensitivity and incidence of needle track seeding after image-guided biopsy versus surgical biopsy. Radiology. 2006;241(2):589–94.
48. Vogelzang NJ, Rusthoven JJ, Symanowski J, Denham C, Kaukel E, Ruffie P, et al. Phase III study of pemetrexed in combination with cisplatin versus cisplatin alone in patients with malignant pleural mesothelioma. J Clin Oncol. 2003;21(14):2636–44.
49. Husain AN, Colby TV, Ordonez NG, Allen TC, Attanoos RL, Beasley MB, et al. Guidelines for pathologic diagnosis of malignant mesothelioma 2017 update of the consensus statement from the international mesothelioma interest group. Arch Pathol Lab Med. 2018;142(1):89–108.
50. Kadota K, Suzuki K, Colovos C, Sima CS, Rusch VW, Travis WD, et al. A nuclear grading system is a strong predictor of survival in epitheloid diffuse malignant pleural mesothelioma. Mod Pathol. 2012;25(2):260–71.
51. Janes SM, Alrifai D, Fennell DA. Perspectives on the treatment of malignant pleural mesothelioma. N Engl J Med. 2021;385(13):1207–18.
52. Rusch VW. A proposed new international TNM staging system for malignant pleural mesothelioma. From the International Mesothelioma Interest Group. Chest. 1995;108(4):1122–8.
53. Nowak AK, Chansky K, Rice DC, Pass HI, Kindler HL, Shemanski L, et al. The IASLC mesothelioma staging project: proposals for revisions of the T descriptors in the forthcoming eighth edition of the TNM classification for pleural mesothelioma. J Thorac Oncol. 2016;11(12):2089–99.
54. Rice D, Chansky K, Nowak A, Pass H, Kindler H, Shemanski L, et al. The IASLC mesothelioma staging project: proposals for revisions of the N descriptors in the forthcoming eighth edition of the TNM classification for pleural mesothelioma. J Thorac Oncol. 2016;11(12):2100–11.
55. Berzenji L, Van Schil PE, Carp L. The eighth TNM classification for malignant pleural mesothelioma. Transl Lung Cancer Res. 2018;7(5):543–9.
56. Carbone M, Adusumilli PS, Alexander HR Jr, Baas P, Bardelli F, Bononi A, et al. Mesothelioma: scientific clues for prevention, diagnosis, and therapy. CA Cancer J Clin. 2019;69(5):402–29.
57. Herndon JE, Green MR, Chahinian AP, Corson JM, Suzuki Y, Vogelzang NJ. Factors predictive of survival among 337 patients with mesothelioma treated between 1984 and 1994 by the Cancer and Leukemia Group B. Chest. 1998;113(3):723–31.
58. Wu TH, Lee LJ, Yuan CT, Chen TW, Yang JC. Prognostic factors and treatment outcomes of malignant pleural mesothelioma in eastern Asian patients—a Taiwanese study. J Formos Med Assoc. 2019;118(1 Pt 2):230–6.
59. Byrne MJ, Nowak AK. Modified RECIST criteria for assessment of response in malignant pleural mesothelioma. Ann Oncol. 2004;15(2):257–60.
60. Guzman-Casta J, Carrasco-CaraChards S, Guzman-Huesca J, Sanchez-Rios CP, Riera-Sala R, Martinez-Herrera JF, et al. Prognostic factors for progression-free and overall survival in malignant pleural mesothelioma. Thorac Cancer. 2021;12(7):1014–22.
61. Yamagishi T, Fujimoto N, Nishi H, Miyamoto Y, Hara N, Asano M, et al. Prognostic significance of the lymphocyte-to-monocyte ratio in patients with malignant pleural mesothelioma. Lung Cancer. 2015;90(1):111–7.
62. Lopez-Castro R, Recondo G, Gorria T, Mezquita L. A new pretreatment mesothelioma risk score: integrating clinical and molecular factors for predicting outcomes in malignant pleural mesothelioma. J Thorac Oncol. 2021;16(11):1782–4.

63. Viscardi G, Di Liello R, Morgillo F. How I treat malignant pleural mesothelioma. ESMO Open. 2020;4(Suppl 2):e000669.
64. Rusch VW, Piantadosi S, Holmes EC. The role of extrapleural pneumonectomy in malignant pleural mesothelioma. A Lung Cancer Study Group trial. J Thorac Cardiovasc Surg. 1991;102(1):1–9.
65. Flores RM, Pass HI, Seshan VE, Dycoco J, Zakowski M, Carbone M, et al. Extrapleural pneumonectomy versus pleurectomy/decortication in the surgical management of malignant pleural mesothelioma: results in 663 patients. J Thorac Cardiovasc Surg. 2008;135(3):620–6. 6 e1-3
66. Ismail-Khan R, Robinson LA, Williams CC Jr, Garrett CR, Bepler G, Simon GR. Malignant pleural mesothelioma: a comprehensive review. Cancer Control. 2006;13(4):255–63.
67. Gomez DR, Rimner A, Simone CB 2nd, Cho BCJ, de Perrot M, Adjei AA, et al. The use of radiation therapy for the treatment of malignant pleural mesothelioma: expert opinion from the National Cancer Institute Thoracic Malignancy Steering Committee, International Association for the Study of Lung Cancer, and Mesothelioma Applied Research Foundation. J Thorac Oncol. 2019;14(7):1172–83.
68. Stahel RA, Riesterer O, Xyrafas A, Opitz I, Beyeler M, Ochsenbein A, et al. Neoadjuvant chemotherapy and extrapleural pneumonectomy of malignant pleural mesothelioma with or without hemithoracic radiotherapy (SAKK 17/04): a randomised, international, multicentre phase 2 trial. Lancet Oncol. 2015;16(16):1651–8.
69. Clive AO, Taylor H, Dobson L, Wilson P, de Winton E, Panakis N, et al. Prophylactic radiotherapy for the prevention of procedure-tract metastases after surgical and large-bore pleural procedures in malignant pleural mesothelioma (SMART): a multicentre, open-label, phase 3, randomised controlled trial. Lancet Oncol. 2016;17(8):1094–104.
70. Baas P, Scherpereel A, Nowak AK, Fujimoto N, Peters S, Tsao AS, et al. First-line nivolumab plus ipilimumab in unresectable malignant pleural mesothelioma (CheckMate 743): a multicentre, randomised, open-label, phase 3 trial. Lancet. 2021;397(10272):375–86.
71. Siegel RL, Miller KD, Jemal A. Cancer statistics, 2020. CA Cancer J Clin. 2020;70(1):7–30.
72. Siegel RL, Miller KD, Fuchs HE, Jemal A. Cancer statistics, 2022. CA Cancer J Clin. 2022;72(1):7–33.
73. Devesa SS, Bray F, Vizcaino AP, Parkin DM. International lung cancer trends by histologic type: male:female differences diminishing and adenocarcinoma rates rising. Int J Cancer. 2005;117(2):294–9.
74. Global, regional, and national comparative risk assessment of 79 behavioural, environmental and occupational, and metabolic risks or clusters of risks, 1990-2015: a systematic analysis for the Global Burden of Disease Study 2015. Lancet. 2016;388(10053):1659–724.
75. Allen LP, Baez J, Stern MEC, Takahashi K, George F. Trends and the economic effect of Asbestos bans and decline in Asbestos consumption and production worldwide. Int J Environ Res Public Health. 2018;15:3.
76. Gloyne SR. Two cases of squamous carcinoma of the lung occurring in asbestosis. Tubercle. 1935;17:5–10.
77. Gloyne SR. A case of oat cell carcinoma of the lung occurring in asbestosis. Tubercle. 1936;18(3):100–IN1.
78. Egbert DS, Geiger AJ. Pulmonary asbestosis and carcinoma: report of a case with necropsy findings. Am Rev Tuberc. 1936;34(1):143–50.
79. Klebe S, Leigh J, Henderson DW, Nurminen M. Asbestos, smoking and lung cancer: an update. Int J Environ Res Public Health. 2020;17(1):258.
80. Selikoff IJ, Hammond EC, Churg J. Asbestos exposure, smoking, and neoplasia. JAMA. 1968;204(2):106–12.
81. Seidman H, Hammond E, Selikoff I. Short-term asbestos work exposure and long-term observation. Ann N Y Acad Sci. 1979;330:61–89.
82. Asbestos, asbestosis, and cancer: the Helsinki criteria for diagnosis and attribution. Scand J Work Environ Health. 1997;23(4):311–6.

83. Henderson DW, Rödelsperger K, Woitowitz HJ, Leigh J. After Helsinki: a multidisciplinary review of the relationship between asbestos exposure and lung cancer, with emphasis on studies published during 1997–2004. Pathology. 2004;36(6):517–50.
84. Markowitz SB, Levin SM, Miller A, Morabia A. Asbestos, asbestosis, smoking, and lung cancer. New findings from the North American insulator cohort. Am J Respir Crit Care Med. 2013;188(1):90–6.
85. Nielsen LS, Bælum J, Rasmussen J, Dahl S, Olsen KE, Albin M, et al. Occupational asbestos exposure and lung cancer—a systematic review of the literature. Arch Environ Occup Health. 2014;69(4):191–206.
86. Balmes JR. Asbestos and lung cancer: what we know. Am J Respir Crit Care Med. 2013;188(1):8–9.
87. Hessel PA, Gamble JF, McDonald JC. Asbestos, asbestosis, and lung cancer: a critical assessment of the epidemiological evidence. Thorax. 2005;60(5):433–6.
88. McDonald JC, McDonald AD. Chrysotile, tremolite and carcinogenicity. Ann Occup Hyg. 1997;41(6):699–705.
89. Uguen M, Dewitte JD, Marcorelles P, Loddé B, Pougnet R, Saliou P, et al. Asbestos-related lung cancers: a retrospective clinical and pathological study. Mol Clin Oncol. 2017;7(1):135–9.
90. Brims FJH, Kong K, Harris EJA, Sodhi-Berry N, Reid A, Murray CP, et al. Pleural plaques and the risk of lung cancer in Asbestos-exposed subjects. Am J Respir Crit Care Med. 2020;201(1):57–62.
91. Weiss W. Asbestos-related pleural plaques and lung cancer. Chest. 1993;103(6):1854–9.
92. Weiss W. Asbestosis: a marker for the increased risk of lung cancer among workers exposed to asbestos. Chest. 1999;115(2):536–49.
93. Kamp DW. Asbestos-induced lung diseases: an update. Transl Res. 2009;153(4):143–52.
94. Olsson AC, Vermeulen R, Schüz J, Kromhout H, Pesch B, Peters S, et al. Exposure-response analyses of Asbestos and lung cancer subtypes in a pooled analysis of case-control studies. Epidemiology. 2017;28(2):288–99.
95. Aberle DR, Adams AM, Berg CD, Black WC, Clapp JD, Fagerstrom RM, et al. Reduced lung-cancer mortality with low-dose computed tomographic screening. N Engl J Med. 2011;365(5):395–409.
96. de Koning HJ, van der Aalst CM, de Jong PA, Scholten ET, Nackaerts K, Heuvelmans MA, et al. Reduced lung-cancer mortality with volume CT screening in a randomized trial. N Engl J Med. 2020;382(6):503–13.
97. Gendarme S, Pairon JC, Andujar P, Laurent F, Brochard P, Delva F, et al. Cost-effectiveness of an organized lung cancer screening program for Asbestos-exposed subjects. Cancers (Basel). 2022;14:17.
98. Markowitz SB. Lung cancer screening in Asbestos-exposed populations. Int J Environ Res Public Health. 2022;19:5.
99. Ollier M, Chamoux A, Naughton G, Pereira B, Dutheil F. Chest CT scan screening for lung cancer in asbestos occupational exposure: a systematic review and meta-analysis. Chest. 2014;145(6):1339–46.
100. Brims F. Identifying occupationally exposed populations for lung cancer screening: it is about time. Occup Environ Med. 2019;76(3):135–6.
101. Harber P. Asbestos, pleural plaques, and lung cancer: untangling the relationships. Am J Respir Crit Care Med. 2020;201(1):4–6.
102. Rodriguez-Canales J, Parra-Cuentas E, Wistuba II. Diagnosis and molecular classification of lung cancer. Cancer Treat Res. 2016;170:25–46.
103. Nasim F, Sabath BF, Eapen GA. Lung cancer. Med Clin North Am. 2019;103(3):463–73.
104. Jones GS, Baldwin DR. Recent advances in the management of lung cancer. Clin Med (Lond). 2018;18(Suppl 2):s41–s6.
105. Koss M, Travis W, Moran C, Hochholzer L. Pseudomesotheliomatous adenocarcinoma: a reappraisal. Semin Diagn Pathol. 1992;9(2):117–23.

Legal Issues in Cancer

11

Lois Schiffer

This Chapter addresses the ways in which cancer and the law intersect. The laws that may have a bearing on cancer, including cancer research, and the ways in which cancer affects the law, for example, the statutes to deter pollution, are as complex and varied as cancer itself. To give readers a flavor of the breadth and scope of the law-cancer relationship, this chapter will focus on examples, and use those examples to explain the cancer and law connection.

"The law" is a body of customs or rules recognized as binding or enforced by a controlling authority.[1] The rule or custom may be recognized as binding because, for example, it is put in place by a formal process (Congress or a state legislature passes a law and the President or Governor signs it), or because it is accepted as custom through long use and observance, or in other ways. Implementation by a controlling authority may similarly be formal—a civil or criminal enforcement case brought by a designated enforcement authority—or (on the positive side) a funding grant made by a federal or local agency—or by an authority recognized by the parties such as a peer-review entity. For the United States, the law may be federal, state, or local, as well as court-developed (common law), or profession-based (medical ethics). The United States and those who work here act internationally as well, through a variety of formal or less formal organizations.

Cancer intersects with this complex network of law in myriad ways. The goal of the chapter is to outline some of those ways; some readers may want to pursue some of the topics further. This is not a guide for lawyers, and will focus on explanation and not technical legal points. Some of the topics will be more familiar to readers, such as malpractice cases or funding streams for research, and others will be less so. So keep your general knowledge close and intuition in hand as we move through these topics.

[1] Merriam Webster on-line dictionary: https://www.merriam-webster.com/dictionary/law.

L. Schiffer (✉)
Washington, DC, USA

E. H. Bernicker (ed.), *Environmental Oncology*,
https://doi.org/10.1007/978-3-031-33750-5_11

Important functions of the law are to encourage good conduct and to deter problematic behavior. The intersections of law and cancer this chapter covers are grouped into those categories, though they have some overlap.[2]

Laws that Seek to Encourage Good Conduct

The United States Congress Enacts Many Laws Related to Cancer, with a Primary Focus on Funding Research

The National Cancer Act of 1971, signed by President Nixon, is hugely significant for cancer research, treatment and prevention. It is summarized by the National Cancer Institute:[3]

> This [law] strengthened NCI [the National Cancer Institute] in carrying out the national effort against cancer by creating the National Cancer Program. It mandated the following:
>
> - The program be developed by the NCI director with the advice of the National Cancer Advisory Board (NCAB), a presidentially appointed committee of 18 members, including both distinguished scientists and laypersons from the general public. The NCAB also includes ex officio members from other government agencies.
> - A three-member panel, the President's Cancer Panel (PCP), review the program by holding periodic public hearings and submitting an annual progress report directly to the President.
> - The annual budget of NCI, called the bypass budget, be submitted directly to the President, bypassing traditional approval by the NIH or the Department of Health and Human Services required of other NIH institutes.
> - The NCI director and members of the NCAB and PCP be presidential appointees.
> - The director of NCI was given additional authorities, in consultation with NCAB, under the act that include:
>   - Create new cancer centers and manpower training programs
>   - Appoint advisory committees, allowing the director to explore new issues/opportunities
>   - Expand the physical location at NIH and other research facilities across the country
>   - Award contracts for research
>   - Collaborate with other federal, state, or local public agencies and private industry
>   - Conduct cancer control activities
>   - Establish an international cancer research data bank that collects, catalogues, stores, and disseminates results of cancer research
>   - Award research grants

[2] ***Note***: the material in this Chapter is written as of December 15, 2022. Because the law develops, is interpreted, and changes through new statutes, it is important to note the date as of which the material is current. This is especially important for the material in Section D.8 regarding *Dobbs v. Jackson Women's Health Organizatio*n, the 2022 Supreme Court case on abortion overruling *Roe v. Wade*. That law is changing literally daily.

[3] The law is P.L. 92-218. This summary and much other information about the Act is set forth at the National Cancer institute website: https://www.cancer.gov/about-nci/overview/history/national-cancer-act-1971#ui-id-2.

At the time of the fiftieth anniversary of this Act in December 2021 the National Institutes of Health issued a review of the Act.[4] By providing funding and a structure for research, the Act has supported treatment, prevention, and a better understanding of cancer through basic research. It clearly fits as a law to encourage good behavior. A book developed from a conference on the policy, politics, and law of cancer sponsored by Yale Law School and the Yale Cancer Center examines the successes and failures of the Act, including advances and setbacks in cancer research, care, access, and policy.[5] Both the NCI Report and the Yale conference book demonstrate the importance of this law and note that there is room for improvement.

The Cancer Moonshot has been another source of attention to cancer including through federal funding. In 2016, President Obama announced the original Cancer Moonshot.[6] In February 2022, President Biden issued a Proclamation underscoring the National Cancer Act's success in addressing cancer, and on the matter of room for improvement, also announced the rekindling of the 2016 Cancer Moonshot. Biden issued a statement of progress in implementation of this reinvigorated Cancer Moonshot in September, 2022.[7]

Here is the Association of Clinical Oncology's summary of federally funding cancer research: https://old-prod.asco.org/get-involved/advocacy/advocacy-agenda-initiatives/federally-funded-cancer-research. It is also useful to note that Congress funds research related to cancer through the work of a number of federal agencies, including for example the National Institutes of Health (NIH) and the Centers for Disease Control (CDC).[8]

The US Congress has also passed numerous other laws related to cancer prevention and treatment. These include as an example the Patient Protection and Affordable Care Act, 42 USC Sec. 18001 et seq., extending health insurance coverage broadly, and specifically providing among many other parts that most insurance plans (as well as Medicare and Medicaid) provide screening mammograms to woman over 40 and in some instances to younger women at no cost.[9] The Affordable

[4] https://nihrecord.nih.gov/2021/12/10/impact-1971s-national-cancer-act--marked#:~:text=President%20Richard%20Nixon%20signs%20the%20National%20Cancer%20Act,and%20improved%20the%20lives%20of%20the%20American%20people.

[5] *A New Deal for Cancer: Lessons from a 50 Year War*, Abbe Gluck and Carl Fuchs ed. : https://law.yale.edu/yls-today/news/war-cancer-50.

[6] See https://obamawhitehouse.archives.gov/cancermoonshot.

[7] The 2022 Moonshot announcement is here: https://www.whitehouse.gov/briefing-room/statements-releases/2022/02/02/fact-sheet-president-biden-reignites-cancer-moonshot-to-end-cancer-as-we-know-it/. September statement of steps is here: https://www.whitehouse.gov/briefing-room/statements-releases/2022/09/12/fact-sheet-president-biden-details-cancer-moonshot-progress-and-new-initiatives-on-60th-anniversary-of-president-kennedys-moonshot-address/. The National Cancer Institute Website also includes statements of action steps taken: https://www.cancer.gov/research/key-initiatives/moonshot-cancer-initiative.

[8] https://www.cdc.gov/cancer/index.htm.

[9] See https://www.medicare.gov/coverage/mammograms (Medicare); https://www.cancer.columbia.edu/news/medicaid-expansion-led-higher-rates-mammography-and-insurance-coverage: (study contrasting states with expanded Medicaid and those without it).

Care Act also provided an additional $1.8 billion of funding over 7 years for cancer research.[10] For other laws related to prevention and treatment see the National Cancer Institute website.[11]

At the State level, governments also have funded research into cancer. An example is Texas, which in 2007 established the Cancer Prevention and Research Institute of Texas with $3 billion state grant program for research, product development, and prevention.[12]

Governments at every level have the capacity to take further steps to address cancer prevention, identification, and treatment. A helpful list is provided in this abstract of an article in Lancet, a weekly peer-reviewed medical journal:

> Abstract: The world population is ageing and increasing in size. As a result, the numbers of people diagnosed with and dying of cancer are increasing. Cancer is also a growing problem in developing countries. Government, be it local, state, provincial, national, or even a union of nations, has clear roles in the control of cancer. It is widely appreciated that much of the research that has defined the causes and treatment of cancer was, and is, government funded. Less appreciated, the body of work about how to control cancer shows the importance of an environment that encourages individuals to adopt healthy behaviours, and government has a vitally important role. Through regulation, education, and support programmes, governments can create an environment in which tobacco use is reduced and citizens maintain good levels of physical activity, healthy bodyweight, and good nutrition. Cancer prevention and the creation of a culture of health is an essential mission of government, beyond that of the traditional health-focused departments such as health ministries; it is in the domain of governmental agencies involved in environmental protection, occupational safety, and transportation. Cancer prevention and health promotion are also in the realm of the zoning board, the board of education, and the board of health. " Copyright © 2017 Elsevier Ltd. All rights reserved.[13]

Note: other laws related to prevention including deterrence of problematic conduct such as smoking are described below.

It is also important to note growing concern for disparities in treatment and access to treatment particularly in disadvantaged communities and efforts to address that problem. For example, the National Library of Medicine lists a number of articles and studies. https://pubmed.ncbi.nlm.nih.gov/32901135/. The National Cancer Institute has a useful analysis as well. https://www.cancer.gov/about-cancer/understanding/disparities. President Biden in his announcement of the reinvigorated Moonshot in 2022 noted a report on this topic from a government panel: https://prescancerpanel.cancer.gov/report/cancerscreening/.

[10] Fact Sheet from White House announcing reinvigorated Moon Shot against Cancer (2022): https://www.whitehouse.gov/briefing-room/statements-releases/2022/02/02/fact-sheet-president-biden-reignites-cancer-moonshot-to-end-cancer-as-we-know-it/

[11] For recent laws, see generally National Cancer Institute website: https://www.cancer.gov/about-nci/legislative/recent-public-laws

[12] https://www.cprit.state.tx.us/

[13] Brawley OW. The role of government and regulation in cancer prevention. Lancet Oncol. 2017 Aug;188):e438-3493. Doi: 10.1016/S1470-2045(17)30374-1. Epub 2017 Jul 26. PMID: 28759387.

In sum, laws at every level have intersected with cancer to encourage "good conduct" by moving forward research, providing better access to care, and helping to develop treatments.

Food and Drug Laws for Approval of Drugs and Devices

Another set of laws that seek to assure "good conduct" are the United States' food and drug laws administered by the US Food and Drug Administration.[14] Medical professionals who work with diagnosis and treatment of cancer patients and the patients themselves may be familiar with regulation by the federal Food & Drug Administration ("FDA"). Under the federal Food, Drug, and Cosmetic Act, 21 USC Sec.301 et seq., FDA must approve a drug as safe and effective before it can be marketed. Its process for doing so, including evaluation of drug company clinical trials, is set forth in detailed regulations. Because cancer patients are treated with chemotherapies, and increasingly innovative forms of drug therapy, the process and what is approved is an important component of the cancer/law connection.[15] As a general matter, a drug company submits an application for approval, including its clinical studies (under established FDA requirements). The drug application is evaluated by FDA Advisory Committees and then by FDA staff. Evaluation includes an analysis of whether the benefits of the drug outweigh its risks. The process for approval is laid out in understandable detail on the FDA website. See, e.g. https://www.fda.gov/drugs/development-approval-process-drugs; https://www.drugs.com/fda-approval-process.html.

The process includes analysis of the target condition and available treatments; assessment of benefits and risks from clinical data; and strategies for managing risks.[16] There is a process for accelerated approval;[17] there is also a process for allowing "compassionate use" when a new, unapproved drug is made available to a seriously ill patient when no other treatments are available.[18] Also, once a drug is approved for a particular use, it may lawfully be prescribed for other uses (off-label use) if the prescriber determines that there is reasonable evidence the drug is useful for that other purpose.[19] FDA also requires post-market safety surveillance by drug companies.

Because of the complexities of cancer and cancer drugs, FDA has an Oncological Drugs Advisory Committee—https://www.fda.gov/advisory-committees/

[14] Website is https://www.fda.gov/.

[15] https://www.fda.gov/drugs/development-approval-process-drugs/laws-regulations-policies-and-procedures-drug-applications

[16] https://www.fda.gov/drugs/development-approval-process-drugs

[17] Ibid.

[18] https://www.cancer.org/treatment/treatments-and-side-effects/clinical-trials/compassionate-drug-use.html#:~:text=Compassionate%20drug%20use%20means%20making%20a%20new%2C%20unapproved,who%20are%20taking%20part%20in%20a%20clinical%20trial.

[19] www.medicinenet.com/indications_for_drugs__approved_vs_non-approved/views.htm
Was this helpful?

human-drug-advisory-committees/oncologic-drugs-advisory-committee. In addition, FDA has an oncology Center of Excellence that undertakes specific projects. See e.g. https://friendsofcancerresearch.org/news/pink-sheet-us-fda-project-protect-oncology-safety-program-in-broad-use-after-quiet-launch/.

While this paper addresses the role of FDA and its drug approval process as one encouraging good behavior, FDA also enforces the limitations the law sets on promotion and advertising of drugs, and other FDA rules and requirements, in the interest of assuring compliance with those provisions and protecting the public. Enforcement tools include warning letters, seeking court injunctions, and criminal prosecution. Other federal agencies have a role in other enforcement actions related to advertising and marketing.

Medical Ethics

Doctors who practice medicine must comply with medical ethics—a strong and important approach to "doing good." While the American Medical Association in its preamble to the Standards of Ethics notes that they are not laws but are standards of conduct", they meet the definition of law we have set out at the outset of this chapter—they are a set of customs or rules that are enforced. Indeed, they guide physician conduct, and therefore are included here.

In general, these rules derive from the four pillars of medical ethics:[20]

> Beneficence (doing good); Non-maleficence (to do no harm); autonomy (giving the patient the freedom to choose freely, where they are able); and Justice (ensuring fairness).

Implementing these principles, The American Medical Association states:

> "Since its adoption at the founding meeting of the American Medical Association in 1847, the AMA Code of Medical Ethics has articulated the values to which physicians commit themselves as members of the medical profession. Together, the Principles of Medical Ethics and the Opinions of the AMA's Council on Ethical and Judicial Affairs make up the Code."[21] The Code notes that the relationship between law and ethics is complex, and suggests that ethical obligations may exceed (be stronger than) legal obligations.

Doctors are licensed by State Medical Boards,[22] usually established by a state medical Act, and it is those Boards, through application of state standards, that "enforce" ethical standards through licensing and relicensing. See, for example, https://journalofethics.ama-assn.org/article/role-state-medical-boards/2005-04. Licensing is an important component of the law as it applies to doctors and therefore to doctors

---

[20] https://www.ncbi.nlm.nih.gov/pmc/articles/PMC3528420/.

[21] The AMA website provides links: Review *why the Code is important* and read the *Code Preface and Preamble*."

[22] https://journalofethics.ama-assn.org/article/role-state-medical-boards/2005-04#:~:text=These%20regulations%20are%20laid%20out%20in%20a%20state,the%20needs%20of%20the%20public%20begins%20with%20licensure.

who work to prevent, diagnose, and treat cancer. With some of the difficult decisions attendant on working with cancer, understanding the role of medical ethics standards is essential.

Particular complexities in meeting standards of both medical ethics and the law is presented by the recent US Supreme Court decision in *Dobbs v. Jackson Women's Health Organization* overruling the constitutional right to abortion. See below.

Law Related to Prevention of Cancer

While there is some overlap with encouraging research and cancer treatment on the one hand, and cancer prevention on the other, we here address four particular examples of US laws designed to take into account, and when possible reduce or prevent, cancer risk. These are laws related to and restricting sale tobacco, laws related to toxic chemicals, medical malpractice laws, and laws that prohibit discrimination and confer other rights on patients. The broad scope of these topics underscores the many ways that the law and cancer intersect.

Tobacco

The connection between tobacco/smoking and cancer has been well-established. In 1964, the U.S. Surgeon General with an Advisory Committee issued a report making clear the connection between smoking and cancer, as well as other diseases.[23] Details about the very clear connection is now on the Centers for Disease Control website: See for example the Centers for Disease Control (CDC) website at https://www.cdc.gov/cancer/tobacco/index.htm.

Before and for a time after the Report, states regulated sale of cigarettes, mostly by prohibiting sales to minors. After that Report, various steps were taken, including a requirement that cigarette manufacturers put a warning label on cigarette packages.[24] In 1996, the Food and Drug Administration issued regulations under authority of the food & drug laws. In the year 2000, the Supreme Court struck down the regulation on the basis that the Food and Drug Act did not provide authority to FDA to regulate tobacco. In 2009, Congress passed and President Obama signed the Family Smoking Prevention and Tobacco Control Act giving FDA the explicit authority to regulate tobacco.[25] The FDA now extensively regulates tobacco

[23] https://www.pbs.org/newshour/health/first-surgeon-general-report-on-smokings-health-effects-marks-50-year-anniversary.

[24] For a summary history of warning label requirements in the US see https://www.cdc.gov/tobacco/data_statistics/sgr/2000/highlights/labels/index.htm.

[25] See generally regarding this history https://en.wikipedia.org/wiki/Regulation_of_tobacco_by_the_U.S._Food_and_Drug_Administration.

products, and in 2019 issued a report about the success of tobacco regulation in the previous ten years.[26]

While the regulatory approach to addressing the health harms of tobacco, including cancer, was making its way through the agency and courts, states and the federal government were bringing lawsuits to stop the tobacco companies and protect public health through prevention of exposure to smoking.

In 1994, the State of Mississippi sued, and that began a series of lawsuits by state attorneys general across the country. This litigation sought in part to demonstrate that the tobacco companies for years had funded and/or had access to health studies linking tobacco to cancer and other health problems, and had deceived the public about the "safety" of smoking. Attorneys General from 46 states agreed to a settlement under which tobacco companies have paid $368 billion over 25 years.[27] Materials related to that settlement are here: https://www.stateag.org/initiatives/the-tobacco-settlement. Some of these cases are summarized at https://lawyerinc.com/biggest-tobacco-lawsuits/.

In 1999, following the litigation brought by the States, the US Department of Justice sued tobacco companies under the federal Racketeering law (RICO) to allege that they had for many years known about the dangers of smoking and continued deceptive practices. After motions and an extensive trial, in 2006 Judge Gladys Kessler of the US District Court for the District of Columbia issued a detailed 1683- page opinion in the case *United States v. Philip Morris.*[28] The decision stated that the tobacco companies "have marketed and sold their lethal products with zeal, with deception, with a single-minded focus on their financial success, and without regard for the human tragedy or social costs that success exacted." The Judge also found that the deception continued. After further proceedings, the companies were required to and did begin to provide specified corrective statements.

However, an agreement on corrective statements at point of sale was reached only in 2022.[29] As noted in the Justice Department's statement,[30] signs to be displayed in retail stores near the cigarettes include:

- "Smoking cigarettes causes numerous diseases and on average 1,200 American deaths every day;

[26] On August 20, 2019 FDA issued a report on the decade of tobacco regulation. See https://www.fda.gov/news-events/fda-voices/achievements-tobacco-regulation-over-past-decade-and-beyond.

[27] https://money.cnn.com/1997/06/20/companies/tobacco_settlement/#:~:text=June%2020%2C%201997%3A%204%3A50%20p.m.%20ET%20State%20negotiators,of%20tobacco%20litigation%20and%20initiate%20tough%20anti-smoking%20programs.

[28] Summaries related to the case may be found at https://www.justice.gov/civil/case-4 and https://www.tobaccofreekids.org/what-we-do/industry-watch/doj as well as at many other websites.

[29] https://www.tobaccofreekids.org/press-releases/2022_04_02_racketeering-lawsuit-pos-corrective.

[30] https://www.justice.gov/opa/pr/court-issues-order-requiring-cigarette-companies-post-corrective-statements-resolves-historic (December 6, 2022).

- The nicotine in cigarettes is highly addictive and that cigarettes have been designed to create and sustain addiction;
- So-called light, low-tar and natural cigarettes are just as harmful as regular cigarettes; and
- Secondhand smoke causes disease and death in people who do not smoke."

A significant move toward prevention of cancer, but a very slow process.

In addition to government action via the courts, private individuals have filed lawsuits over the years seeking money damages for their injuries from smoking. Some of those cases are summarized here (also linked above regarding some State cases): https://lawyerinc.com/biggest-tobacco-lawsuits/. Also, all of these cases rely on experts—doctors and scientists who have studied the connection between tobacco (smoking) and serious adverse health effects including cancer.

On the federal legislative front, on Dec. 20, 2019, President Biden signed legislation amending the Federal Food, Drug, and Cosmetic Act, and raising the *federal* minimum age for sale of tobacco products from 18 to 21 years.[31] This legislation (known as "Tobacco 21" or "T21") became effective immediately, and it is now illegal for a retailer to sell any tobacco product—including cigarettes, cigars, and e-cigarettes—to anyone under 21. The new federal minimum age of sale applies to all retail establishments and persons with no exceptions. https://www.fda.gov/tobacco-products/retail-sales-tobacco-products/tobacco-21.

The FDA summarizes its regulation of tobacco as a step toward protecting deaths from tobacco use: https://www.fda.gov/news-events/fda-voices/achievements-tobacco-regulation-over-past-decade-and-beyond. A recent assessment shows mixed results but movement forward: https://www.tandfonline.com/doi/full/10.1080/01947648.2020.1868940. A current issue—regulation of e-cigarettes ("vaping")—continues to show the intersection of the law with cancer and particularly cancer prevention. States as well as the federal government are taking steps related to vaping. California voters by referendum banned the sale of flavored tobacco products in California; a tobacco company sought from the US Supreme Court an injunction pending appeal to hold up its implementation while the courts considered whether a federal law regulating tobacco preempted that state law, and on December 12, 2022 the US Supreme Court did not grant that stay, so the referendum law goes into effect.[32]

[31] https://www.pbs.org/newshour/health/first-surgeon-general-report-on-smokings-health-effects-marks-50-year-anniversary.

[32] Supreme Court denial of stay, see https://www.scotusblog.com/case-files/cases/r-j-reynolds-tobacco-company-v-bonta/. The facts underlying the State referendum and related actions are set forth in the State of California opposition to the stay filed in the Supreme Court on December 7, 2022: https://www.supremecourt.gov/DocketPDF/22/22A474/249235/20221206165512415_RJ%20Reynolds%20v%20Bonta%20Case%20No.%2022A474%20Opposition.pdf.

Regulation of Toxic Chemicals

A number of federal and state laws require the reduction or limitation of risk to human health and to the environment caused by toxic chemicals. These are mostly regulatory statutes. Identifying and limiting cancer risk is a measure used in applying these laws, and the evaluation and application of risk assessments relies on research related to cancer. The primary goal is to reduce or prevent health risks including the risk of cancer; the implementation of these laws is squarely at the intersection of law and cancer.

This analysis focuses on several of these laws, and a discussion of toxic tort litigation as an alternative remedial and deterrent approach. The examples here are: the federal Clean Air Act, implemented in part through State Implementation Plans; the federal Superfund law (CERCLA); the federal Toxic Substances Control Act, designed to keep toxic substances off the market or allow them on the market only with appropriate controls; and toxic torts. Other laws, including such State laws as California's Proposition 65, also address such concerns but are not the focus here

The Clean Air Act

The federal Clean Air Act, passed approximately its present framework in 1970, and amended several times including extensively in 1990, is a complicated statute.[33] Its goal is for pollution prevention; its approach includes steps to protect public health and welfare (including the environment).

The Act requires EPA to establish standards for "criteria pollutants" –pollutants that, if emitted into the ambient air endanger public health and welfare and come from widespread multiple sources. Currently there are six: carbon monoxide (CO), lead (Pb), nitrogen dioxide (NO 2), ozone (O 3), particulate matter (PM), and sulfur dioxide (SO 2).[34] The standards must be established for pollutants that the standards for each air pollutant that may reasonably be anticipated to endanger public health or welfare, and results from numerous or diverse mobile or stationary source. CAA Sec. 108 (42 USC Se. 7408(a). The EPA Administrator establishes these National Ambient Air Quality Standards –as s/he determines, based on criteria established by EPA and allowing an adequate margin of safety are requisite to protect public health (primary standards) and public welfare (including the environment) (secondary standards). Clean Air Act Sec. 109, 42 USC 7409 (b). The standards are to be reviewed and if necessary, revised every 5 years.[35] Of these criteria pollutants, the one most directly related to cancer is that for particulate matter (particles of both 10.0 and 2.5 microns are now regulated). There is an ongoing reexamination of the particulate matter NAAQs standards that is considering among other factors what standard are necessary to protect appropriately against cancer risk.[36] The science

[33] Clean Air At, 42 U.S Code Sec. 7401 et seq.

[34] See generally: https://www.epa.gov/criteria-air-pollutants.

[35] This EPA website is helpful: https://www.epa.gov/naaqs.

[36] In December 2020 as part of the 5 year review the Trump administration determined not to change the prior NAAQS for particulate matter. On June 10, 2021, the new Administrator of EPA announced

that evaluates air exposure and cancer risk is an important part of the administrative records that the Administrator uses to set these standards and, if they are challenged in court, the courts use to review the Administrator's determinations. Court review is whether the Administrator, on the record before her or him, was not arbitrary and capricious in making the decision.

These standards are implemented through State Implementation Plans – the States develop plans for facilities incorporating the NAAQS. Federal facilities must also meet these standards.[37]

Under the Act, EPA is also required to regulate "air toxics", formally called "hazardous air pollutants". The 1970 Clean Air Act required EPA to identify and regulate hazardous air pollutants. Because industry considered such potential regulation expensive and burdensome, EPA regulated very few, despite their impact on public health. Then, in the 1990 Clean Air Act amendments, Congress listed 189 hazardous air pollutants, and required regulation on a particular schedule. Clean Air Act, Section 112 (42 USC 7412).[38] EPA may add toxic pollutants "which present, or may present, through inhalation or other routes of exposure, a threat of adverse human health effects (including but not limited to substances which are known to be, r may reasonably be anticipated to be, carcinogenic…." Or meet other specified health criteria. The statute defines carcinogenic effect as follows: "Unless revised, the term 'carcinogenic effect' shall have the meaning provided by the Administrator under guidelines for Carcinogenic Risk Assessment as of the date of enactment. Any revision in the existing Guidelines shall be subject to notice and opportunity for comment." Under current regulations, the standard is a complex one, summarized in the linked document:

https://www.epa.gov/fera/risk-assessment-carcinogenic-effects. The longer full risk assessment document is set forth here: https://www.epa.gov/sites/default/files/2013-09/documents/cancer_guidelines_final_3-25-05.pdf. Cleanly the intent of these detailed and specific requirements in the statute is to assure that EPA actually regulates air toxics, and of course part of the effect of that regulation is to reduce cancer risk from air emissions exposure. While EPA notes that it has made progress with this regulation, a 2021 report acknowledges challenges and establishes a strategy. https://www.epa.gov/sites/default/files/2021-04/documents/oaqps_air_toxics_strategy_public_facing_document_final.pdf.

The Clean Air Act has many other requirements, including regulation of "mobile sources" (cars, trucks, diesel engines, other off-road vehicles). The component

that EPa would begin the process of reconsidering that decision. https://www.epa.gov/newsreleases/epa-reexamine-health-standards-harmful-soot-previous-administration-left-unchanged That reconsideration is in process. A supplemental policy analysis is here: https://www.epa.gov/system/files/documents/2022-05/Final%20Policy%20Assessment%20for%20the%20Reconsideration%20of%20the%20PM%20NAAQS_May2022_0.pdf; and science analyses are here: https://www.epa.gov/air-quality-analysis/particulate-matter-naaqs-review-analyses-and-data-sets.

[37] See https://www.epa.gov/fedfacts/resource-conservation-and-recovery-act-rcra.

[38] See https://www.epa.gov/haps/initial-list-hazardous-air-pollutants-modifications.

important for this chapter is that cancer science and cancer prevention are important components of implementing the Act.

Hazardous Substances and Hazardous Waste Laws

A second example of laws designed to prevent or reduce the risk of cancer among other conditions, and thus operating at the intersection of cancer and pollution laws, are two federal laws designed to address hazardous waste and hazardous substances. These are the Comprehensive Environmental Response, Compensation, and Liability Act, commonly known as CERCLA or Superfund;[39] and the law directly regulating hazardous wastes, the Resource Conservation and Recovery Act, or RCRA.[40] The goal of these laws overlaps.[41] In general, RCRA regulates ongoing disposal of hazardous waste, provides a framework for states to handle non-hazardous waste, and is an antidote to the years of companies dumping tox wastes into landfills or soil where the chemicals leached into drinking water or caused injury to people and the environment through exposure. The statute and implementing regulations provide detailed requirements for how such wastes must be identified, treated, and disposed of to protect public health and the environment. The regulatory system is cradle-to-grave—from the time the waste is generated through disposal and monitoring of the disposal.

The Superfund Act provides authority for the federal government to clean up sites contaminated with hazardous substances, primarily by holding owners, operators, transporters, and generators of the hazardous substances responsible for the cleanup. Because the approach of the Superfund law is to have the polluter pay for the cleanup, one effect of has been deterrence—because of the expense of cleaning up past contamination, companies are deterred from creating new superfund sites, and handle hazardous substances more carefully, or reduce the amount of waste generated in a production process.

Under both statutes, waste or hazardous substances are characterized to determine hazard, using a set of standards and tests that includes risk assessment of the substance for causing cancer. A waste under RCRA can be listed, or is otherwise characterized as hazardous if it meets one of four tests: "toxicity"; corrosiveness, reactivity, and ignitability. At a Superfund site, establishing the standards for a cleanup is a site by site process that takes into account chemicals causing the contamination, future use of the site, and extensive health risk assessment with a particular focus on risk assessment for cancer. The Superfund cleanup program is a federal program. Under both laws, evaluation of risk including toxicity takes into account the risk for cancer. Several EPA documents reflect the role for cancer risk assessment research in applying these laws. These include: (1) a summary of the

[39] 42 USC Sec. 9601-9765. Superfund was enacted in late 1980, and amended extensively in 1986.

[40] 42 USC Sec. 6901 et seq. See for EPA description of the statute: https://www.epa.gov/rcra.

[41] A helpful and mostly accurate comparison of the application of the two laws is here: https://www.actenviro.com/rcra-vs-cercla/.

Environmental Protection Agency's risk assessment approach that has a specific section on cancer risk assessment: https://www.epa.gov/risk/superfund-risk-assessment; (2) EPA Guidelines for Carcinogen Risk Assessment (March, 2005): https://www.epa.gov/sites/default/files/2013-09/documents/cancer_guidelines_final_3-25-05.pdf; and (3) and EPA Region 4 and Techlaw document "Risk Assessments in RCRA : https://archive.epa.gov/epawaste/hazard/web/pdf/risk-as.pdf. (undated).[42]

As is evident from the face of these two statutes as well as their application, an important goal is reducing exposure to chemicals that pose a risk of cancer.

The express provisions of RCRA permit States to apply more stringent standards to addressing hazardous waste. In addition, States may impose more stringent requirements on hazardous substances (no requirement to prove they are "wastes." A well-known example of more stringent regulation is California's Proposition 65. As noted on the website of the California Office of Environmental Health Hazard Assessment: "Proposition 65, officially known as the Safe Drinking Water and Toxic Enforcement Act of 1986, was enacted as a ballot initiative in November 1986. The proposition protects the state's drinking water sources from being contaminated with chemicals known to cause cancer, birth defects or other reproductive harm, and requires businesses to inform Californians about exposures to such chemicals. Proposition 65 requires the state to maintain and update a list of chemicals known to the state to cause cancer or reproductive toxicity."[43] This proposition focuses among other components on causes of cancer. For Superfund, some states have their own statutes requirement cleanup of contaminated sites including those less contaminated than are the focus of the federal Superfund program. A helpful review of all 50 states is here: https://www.eli.org/research-report/analysis-state-superfund-programs-50-state-study-2001-update.

With all of these federal laws related to pollution prevention and control, , the federal government is now focused on environmental justice in their use and application. The Environmental Protection Agency defines environmental justice as: "the fair treatment and meaningful involvement of all people regardless of race, color, national origin, or income, with respect to the development, implementation, and enforcement of environmental laws, regulations, and policies. This goal will be achieved when everyone enjoys: The same degree of protection from environmental and health hazards, and Equal access to the decision-making process to have a healthy environment in which to live, learn, and work."[44]

---

[42] In general, under CERCLA, EPA seeks to limit risk to below 10 to the -6.

[43] https://oehha.ca.gov/proposition-65. More information about Prop 65 is included on that website. For example, restaurants and other businesses in California have relevant notices at the door.

[44] https://www.epa.gov/environmentaljustice/factsheet-epas-office-environmental-justice

Toxic Substances Control Act as Amended

The Toxic Substances Control Act of 1976 requires EPA to protect the public from "unreasonable risk of injury to health or the environment" by regulating the manufacture and sale of chemicals through a regulatory system of reporting, record-keeping, and testing requirements, and where necessary prohibiting their sale. The 1976 law, not vigorously implemented or enforced, was extensively amended and updated by in 2016 by the Frank R. Lautenberg Chemical Safety for the 21st Century Act.[45] Under the law, on July 20, 2017 EPA issue a risk assessment rule for determining whether a chemical substance presents and unreasonable risk of injury to health or the environment.[46] The Procedures throughout specify that EPA may evaluate both cancer and non-cancer risks. The goal is to carry out the purposes of the statute to regulate chemicals to protect against health and environmental risks including the risk of cancer.

Toxic Torts

Finally, hazardous wastes and hazardous substances may be addressed through toxic tort litigation. Essentially a toxic tort case is a personal injury lawsuit in which the person or group of persons claim injury or disease caused by exposure to a chemical or dangerous substance caused by the person or company being sued. Such ort claims seek damages or restitution after the injury, so are less preventive than the federal and state statutory approaches outlined above. However, substantial tort injury judgements may encourage companies or individuals to take more protective steps to avoid such payments in the future. Two of the most well-known toxic tort cases have been the subject of movies. They are, first, A Civil Action, is a 1998 movie based on a book about a suit seeking redress for industrial trichloroethylene contamination in Woburn Massachusetts that allegedly caused cancer in plaintiffs.[47] The second movie is Erin Brockovich, a 2000 movie reflecting an actual case challenging Pacific Gas and Electric for groundwater contamination that allegedly caused cancer.[48]

The connection between statutes to prevent and address pollution and prevention or reduction of cancer and cancer risk is a critical element the link between cancer and the law.

---

[45] TSCA is at 15 U.S.C. Sec. 6701 et seq.; see also https://www.epa.gov/laws-regulations/summary-toxic-substances-control-act.. The updated provisions from 2016 are described at https://www.epa.gov/assessing-and-managing-chemicals-under-tsca/frank-r-lautenberg-chemical-safety-21st-century-act.

[46] 82 FR 33726 (July 20, 2017).

[47] https://en.wikipedia.org/wiki/A_Civil_Action_(film)

[48] https://en.wikipedia.org/wiki/Erin_Brockovich_(film).

Medical Malpractice

Some medical personnel, when asked about "cancer and the law" may focus on medical malpractice—am I being sufficiently cautious and careful to avoid a malpractice claim against me. The internet (worldwide web) is filled with law firm materials outlining the elements of malpractice suits and encouraging those who have concerns about the way they have been treated to seek a legal consultation. Malpractice suits in all areas of medicine have raised malpractice insurance rates with attendant public attention and in some states, laws that put strict time or money limits on such claims. In addition, most medical personnel and medical employers carry some form of malpractice insurance, so patients and their legal counsel may find themselves dealing with the insurers in resolving these matters. While medical malpractice is not unique to cancer, medical practitioners who work with cancer patients are likely mindful of medical malpractice.

In general, medical malpractice law is a function of state law—developed either through state common law (precedents developed and applied by courts) or by state statutes.[49] While the details vary state by state, these laws typically reflect general principles. Under much common malpractice law, the plaintiff (the person complaining of problematic care or treatment) must prove "by a preponderance of the evidence" (a legal standard that roughly translates to more than 50%) four elements: that the person being sued (e.g. the doctor or the hospital or both) owed a duty of care to the plaintiff; that the person being sued deviated from the applicable standard of care; that the plaintiff suffered damages; and the damages were directly caused by the deviation of the person being sued. The professional (the person or entity being sued) has a duty to exercise the level of care that a reasonably prudent person in their line of work would exercise. Each of these elements has had extensive legal interpretation in court decisions.

For cancer care, in general, medical malpractice cases focus on failure to diagnosis, delayed diagnosis, misdiagnosis, and wrong treatment.[50] Some states have imposed caps on amounts a plaintiff may recover, and in all states there is a statute of limitations—a deadline by which a malpractice case must be filed. An extremely thorough article on medical malpractice in the United States is by Sonny Bal, an Introduction to Medical Malpractice in the United States (on the Library of Medicine at NIH website):

https://www.ncbi.nlm.nih.gov/pmc/articles/PMC2628513/#:~:text=In%20the%20United%20States%2C%20medical%20malpractice%20law%20has,that%20substandard%20medical%20care%20resulted%20in%20an%20injury.

Malpractice insurers provide guidance to medical practitioners about how to avoid malpractice suits. In general, their advice is to pay attention to the patient,

[49] For those treated in federal facilities a special variant of the Federal Tort Claims Act applies. See 38 USC 1151, federal law on disabililty compensation.

[50] The web is filled with summaries. An example is at: https://www.simonsonlegal.com/blog/2021/november/what-you-should-know-about-medical-malpractice-a/.

listen to and talk to the patient, and show care for the patient and his/her family. While described defensively, this advice is a matter of common sense, and could better be described to the medical practitioner as having and communicating care for and to the patient. See HCP National (insurance company)(1922), How to Avoid Being Sued for Medical Malpractice: https://hcpnational.com/how-to-avoid-being-sued-for-medical-malpractice/.

The goal of a malpractice suit is to compensate the patient for failure of proper timely diagnosis and/or treatment. The effect also is to deter such problematic conduct in the future by the medical professional who is sued, and by others who may not be careful about carrying out the duty they owe to their patients—patients with cancer or otherwise. Resolution of a case can punish improper treatment and provide remediation—it also can have the effect of deterring such conduct in the future.

Disability Rights and Other Laws that Prohibit Discriminatory Conduct and Confer Certain Rights to Protection

Patients with serious health conditions face a number of challenges to which disability rights laws may apply. While these disability rights laws focus more broadly and not just on person with cancer, they can be important tools for cancer patients and a key intersection of the law with cancer. These laws confer certain rights on all people meeting legal standards of "disability". For example, the federal Americans with Disabilities Act of 1990, 42 USC Sec. 12101 et seq., prohibits discrimination on the basis of disability in employment, State and local government, public accommodations, commercial facilities, transportation, and telecommunications. This law built on the provisions of Section 501 of the Rehabilitation Act of 1973, 29 USC Sec. 791, which extended those protections to those employed by the federal government.

Disability is defined broadly under these laws to include physical or mental impairment that substantially limits one or more major life activities, including the operation of major bodily functions, which in turn include but are not limited to "functions of h immune system, normal cell growth, digestive, bowel, bladder, neurological, brain, respiratory, circulatory, endocrine, and reproductive functions."[51] To be entitled to the protection of these laws a person need not be permanently disabled. Those who meet the definition of disability under these laws must be given "reasonable accommodation". Thus, for example, an employee who needs a schedule adjustment to enable cancer treatments may have rights to such adjustments as a reasonable accommodation. More generally, rights of patients under these laws may serve to enable a cancer patient to keep their job, and any employment benefits such as health insurance that come with that job.

While the complexities of determining what constitutes reasonable accommodations are beyond the scope of this chapter, many materials are available to assist. For these and other federal laws that confer rights on disabled people, the U.S Department

[51] Definitions at 42 USC Sec. 12102(1) and (2).

of Justice has issued a Guide to Disability Rights Law that outlines a number of statutes and the protections they provide.[52] Other laws, both federal and state, provide for access to healthcare and government benefits including sick leave and family leave that may be important for cancer patients.

It is important to note that State laws also provide rights that may benefit cancer patients. A helpful chart of relevant states laws is available at https://triagecancer.org/state-laws. For example, state laws providing medical leave may confer rights or benefits to cancer patients or their families or caretakers.

In addition to the Justice Department Guide above, several groups are available to help patients with these matters, including, for example, the Cancer Legal Resource Center, https://thedrlc.org/cancer. That Center also provides a patient legal handbook in both English and Spanish that furnish information about cancer-related legal issues from diagnosis through survivorship.[53]

The Unexpected: Laws that May Imperil Cancer Care Progress

US Supreme Court Overrule of Federal Constitutional Right to Abortion Established in Roe v. Wade in 1973

In June, 2022, the US Supreme Court overturned the almost 50 year old interpretation of the US Constitution conferring the right to an abortion in *Roe v. Wade*. The right was later modified to permit state laws limiting abortion so long as they did not impose an "undue burden" on the exercise of that constitutional right.[54] In *Dobbs v. Jackson Women's Health Organization*, 597 U.S. ___(June 24, 2022), the U.S. Supreme Court upset decades of expectations about the availability of reproductive care; as a result, in many states abortion became unavailable. The overruling of Roe and the federal constitutional right it established has significantly compromised the delivery of reproductive medical care, and medical care more broadly, including cancer care in those states that have now imposed serious bans on abortion care. As of November 23, 2022, approximately 13 US States ban abortion, and serious restrictions are imposed in other states, with yet others seeking to impose limits. See continually updating charts in the New York Times.[55] Because some of the state laws punish doctors who perform abortions, as well as anyone assisting women to access abortion, the impact has been to create chaos and uncertainty, making access to appropriate are more difficult for patients and their families as well as medical professionals that care for them.

Articles have specifically identified the many complications this case has caused for care of cancer patients in particular. For example, a statement by the National

---

[52] https://www.ada.gov/resources/disability-rights-guide/

[53] https://thedrlc.org/cancer/publications-webinars/patient-legal-handbook/.

[54] *Roe v. Wade,* 410 U.S. 113 (1973); reaffirmed with limits in *Planned Parenthood v. Casey*, 505 U.S. 833 1992).

[55] https://www.nytimes.com/interactive/2022/us/abortion-laws-roe-v-wade.html.

Coalition for Cancer Survivorship shortly after *Dobbs* was decided gives specific examples and outlines some of the problems.[56] A statement by the American Cancer Society shortly after the decision highlights additional difficulties the failure to protect reproductive rights and abortion poses for cancer patients.[57] An article from the September 2022 issue of the Oncologist, authored by a number of doctors, further outlines the difficulties.[58] Further, particular problems for fairly treating adolescents and young adults with a cancer diagnosis is described in an October 2022 article in The Lancet—it describes how fertility preservation, crucial to care in this age group, can be affected and hampered.[59]

Dobbs also raises serious ethical quandaries for doctors, as the American Medical Association and a number of other organizations stated in a brief they filed in the Supreme Court in the Dobbs case: "'The ban forces clinicians to make an impossible choice between upholding their ethical obligations and following the law'" [60]

Importantly, in recognition of these many difficulties and dilemmas, on November 16, 2022, the American Medical Association announced new adopted policies related to reproductive health care that address some of the difficult legal and ethical problems that medical professionals face because of Dobbs. In a press release with linked policies the AMA[61], the AMA sets out its new policies, that opposing criminalization for pregnancy loss resulting from medically necessary care; clarifying ethical guidance regarding abortion bans; expanding support for access to abortion care; and preserving access to abortion training for physicians in training.

The overruling in *Dobbs* of the federally recognized constitutional right to abortion is an unfortunate but real example of the law intersecting with cancer in an unexpected, deeply problematic way-- not well thought through, not based in medical science, and raising series ethical concerns for health care provider and grave unfairness and challenges for patients.

[56] https://canceradvocacy.org/nccs-reacts-dobbs-v-jackson-decision-impact-on-access-to-quality--cancer-care/#:~:text=The%20decision%20in%20Dobbs%20v.%20Jackson%20Women%E2%80%99s%20Health,cancer%20patients%20that%20supports%20patient-centered%20decision-making%20and%20care..

[57] https://www.fightcancer.org/releases/american-cancer-society-highlights-impact-dobbs-v-jackson-ruling-cancer-patients-and-their.

[58] Andrew Shuman et al., "Supporting patients with cancer after Dobbs v. Jackson Women's Health Organization, https://www.ncbi.nlm.nih.gov/pmc/articles/PMC9438903/.

[59] https://www.thelancet.com/journals/lanonc/article/PIIS1470-2045(22)00562-9/fulltext.

[60] Quote from brief filed by American Medical Association, Association of Obstetricians and Gynecologists, and other medical groups in Dobbs case, quoted in NPR, All Things Considered, June 24, 2022: "For doctors, abortion restrictions create an 'impossible choice' when providing care. '"

[61] https://www.ama-assn.org/press-center/press-releases/ama-announces-new-adopted-policies-related-reproductive-health-care.

Conclusion

The law, a body of customs or rules recognized as binding or enforced by a controlling authority, affects and is affected by cancer in many ways. This chapter by example identifies and explains how cancer research and treatment can benefit from the law, how the law can seek to prevent or reduce the risk of cancer, and how cancer risk assessments are relied on in developing the application of some of that law. Because the law is fluid, it can also affect cancer patients and medical practitioners in unexpected and sometimes unevaluated ways. Effective and sound communication among law practitioners, policymakers and legislators, medical practitioners, researchers, and patients may increase the effectiveness and soundness of the cancer-law intersection.

Lois Schiffer is an environmental lawyer with extensive experience in the United States federal government, non-profit organizations, and private practice. Her work includes Assistant Attorney General for the Environment and Natural Resources Division, US Department of Justice; an honorary degree from Vermont Law School; and General Counsel, National Oceanic and Atmospheric Administration. She is a member of the American Law Institute, and a Fellow of the American Bar Foundation. She is an active member of the District of Columbia Bar, and admitted to practice before the United States Supreme Court and several other United States courts. Her recognition and awards include Phi Beta Kappa; Edmund Randolph Award from US Department of Justice; and the Lifetime Achievement award from the Section of Energy, Environment and Resources of the American Bar Association. She received her B.A. from Radcliffe College and her J.D. from Harvard Law School.

Part IV

War and Cancer

Agent Orange and Cancer

12

Eric H. Bernicker

Introduction

It is a somewhat trite to note that often technological advances walk a line between enhancing human flourishing and threatening human well-being. It is a rare development that does not also bring an opportunity for weaponization. Fritz Haber's discovery of the Haber-Bosch process—which allowed large scale synthesizing of ammonia and thus enhanced production of both fertilizers and explosives—is forever linked to his development of chlorine gas and the horrors of World War One chemical warfare in the trenches of Europe. Likewise, the development of herbicides was initially touted as a significant advance to enhance farming and improve crop yields. The phenoxy compounds 2,4 -dichlorophenoxyacetic acid (2,4-D) and 2,4,5-tri-chlorophenoxyacetic acid (2,4,5-T) were discovered in the 1940s and were soon found to be among the most effective herbicides, primarily as they stopped plant growth rather than promote growth, and they preferentially killed common weeds rather than staple crops. These agents were commonly used in the US by farmers and foresters to control plant growth prior to their identification as having a role to play in warfare.

Agent orange was a 1:1 mix of 2,4-D and 2,4,5-T. The process by which Agent Orange was produced often caused the final product to be contaminated with 2,3,7,8-tetrachlorodienzo-p-dioxin (TCDD) [1]. In addition, Agent Orange was often mixed with gasoline and other aromatic hydrocarbons to facilitate its application.

Dioxins and especially TCDD is the most toxic of the dioxins and is both teratogenic as well as carcinogenic. TCDD has been determined by the International Agency for Research on Cancer (IARC) as a Group 1 known carcinogen in 1997.

E. H. Bernicker (✉)
The Neal Cancer Center, Houston Methodist Hospital, Houston, TX, USA
e-mail: EricBernicker@Centura.org

E. H. Bernicker (ed.), *Environmental Oncology*,
https://doi.org/10.1007/978-3-031-33750-5_12

Dioxins as a group of chemical compounds are classified as persistent organic pollutants and are often generated as byproducts of burning or other industrial processes—as they were in the chemical formation of Agent Orange and other herbicides used by the military in Vietnam. Dioxins are highly lipid soluble, stable, and enter the food chain through animal fats. TCDD is a chronic carcinogen that causes malignancies in laboratory animals in doses as low as 0.001 micrograms/kg/day [2]. The half life in humans is long—almost a decade. It can also be excreted in breast milk and semen. The long half life as well as the entry of these agents into the food chain lead some researchers to claim that dioxins are a multigenerational toxin.

The exact mechanism by which TCDD induces carcinogenesis is not known; it is postulated that is exerts its malignant effects through binding to the aryl hydrocarbon receptor (AHR) [3]. AHR is a highly conserved transcriptional regulator of the cellular response to environmental stress. Experiments have shown that TCDD over-activates AHR leading to downstream transcription of CYP1A1 and CYP1B1; the latter leads to generation of reactive oxygen species.

The evidence in linking exposure to dioxins to human disease is not confined to laboratory experiments on rodents; there is real world data from actual human exposures through work or through environmental disasters. A chemical reactor exploded in Seveso Italy 1976 and covered a wide area with a vast chemical cloud of TCDD. Many farm animals died; long term follow up of the population exposed to the airborne chemicals revealed a significant increase in the risk of developing malignancy (RR 1.30) [4]. The authors looked at health out comes in three groups with decreasing soil levels of TCBB with increasing distance away from the exploded reactor. A more granular look at the data revealed the following: All cancer incidence did not differ from expectations in any of the contaminated zones. There was an excess of lymphatic/ hematopoietic neoplasms in zones A (four cases; RR, 1.39; 95% CI, 0.52–3.71) and B (29 cases; RR, 1.56; 95% CI, 1.07–2.27); there was also an increased risk of breast cancer was detected in zone A females after 15 years since the accident (five cases, RR, 2.57; 95% CI, 1.07–6.20). Interestingly, there were no cases of soft tissue sarcomas in the most exposed zones (A and B, 1.17 expected) but given the rarity of these tumors it cannot be statistically ruled out.

Fingerhut et al. published a major study in the New England Journal of Medicine in 1991 looking at occupational exposure of workers to TCDD [5]. They performed a retrospective cohort study of mortality among the 5172 workers at 12 plants in the United States that produced chemicals contaminated with TCDD. Occupational exposure was documented by reviewing job descriptions and by measuring TCDD in serum from a sample of 253 workers; the causes of death were taken from death certificates. Interestingly, they found that mortality from several cancers that had been previously associated with TCDD (such as stomach, liver, Hodgkin's disease, and non-Hodgkin's lymphoma) was not significantly elevated in this cohort. Mortality from soft tissue sarcoma and cancers of the respiratory system were elevated however, and there was a small but statistically significant increase in overall cancer mortality in the overall cohort (SMR, 115; 95 percent confidence interval, 102 to 130). They were able to measure TCDD levels in a sample of 253 workers from two plants; exposed worker had a TCDD level of 233 pg per gram of lipid (range, 2 to 3400) versus a mean level of 7 pg per gram was found in the comparison group of 79 unexposed persons. The mean for 119 workers with 1 year or more of exposure was 418 pg per gram.

As a result of medical studies and pressure from Veterans' groups the VA recognizes the following cancers as possibly being caused by exposure to Agent Orange from wartime service: Bladder cancer, Prostate cancer, respiratory cancers, some soft tissues sarcomas, Chronic B-cell leukemia, Hodgkin's disease, Multiple myeloma, Non-Hodgkin's lymphoma.

Scientific Background of Agent Orange/Dioxin

The process by which humans develop cancer is essentially affected by either genetic susceptibility or environmental exposures [6]. Often, it takes many years of exposure to potential carcinogens and a large number of patient cases before enough epidemiological evidence accrues whereby attribution of cause can be safely established. Of course, the establishment of cause can be difficult and is not always a question of scientific data. The paradigmatic example would be the realization that exposure to tobacco smoke was carcinogenic; while there were early concerns of the adverse health effects of smoking, it took years of collecting epidemiologic data as well as countering political and commercial pressures before a more definitive position on causation could be adopted.

The story of Agent Orange, health effects and health advocacy remain complicated to this day. We will first review briefly the historical facts of operation Ranch Hand and then look at data that has emerged in soldiers and Vietnamese citizens exposed to Agent Orange.

Operation Ranch Hand

Between 1961 to 1971, in an attempt to defoliate the landscape and to limit the access of Vietnamese guerillas, (the Viet Cong), to crops to serve as food for their soldiers, the United States and its South Vietnamese allies sprayed nearly seventy-three million liters (over 19 million gallons) of chemical herbicides over two and a half million acres of southern and central Vietnam [7]. The herbicides were not just sprayed from airplanes and helicopters; Navy river boats sprayed herbicides on to river banks and soldiers applied it around the perimeters of camps and airfields. There were a family of different herbicides that were utilized by the military in Vietnam; the code name came from the color of the band around the 55 gallon drum that stored the product [8]. Initially limited in planned use, as the war escalated and as the involvement of American soldiers increased, Agent Orange use was expanded beyond the initial strategy to deprive enemy soldiers of cover towards directly attacking the food supply that enemy combatants relied upon. Even at the time, there was concern that the picture of using herbicidal war against food supplies would be used by North Vietnam and other enemies of the US to paint the US in a poor light; in addition, it was counterproductive to winning the 'hearts and minds' of the Vietnamese people [9].

As the war in south Asia dragged on, it coincided with the dramatic growth of the environmentalism movement in the US, fueled in part by the publication of Rachel Carlson's book 'Silent Spring." Carlson wrote passionately about the harm that pesticides, especially DDT, were causing on the environment and on beloved wildlife such as birds and fish. There was a huge increase in pressure on legislators to safeguard the environment, leading eventually to the first clean air act in 1967 and the National Environmental Policy Act in 1969. The first Earth Day was held April 22, 1970; by the next year, President Nixon had ended Ranch Hand.

The environmental and military questions swirling around the Vietnam War and the use of Agent Orange in herbicidal warfare could not but have roiled many scientists as well. In addition, many scientists in the US, now temporally on the other side of the nuclear strikes on Hiroshima and Nagasaki that resulted from the work of the Manhattan project, were apt to look at technical utilization of scientific advances in the setting of war with a more concerned eye. There was the mistaken tendency to view the use of Agent Orange much the same was that Zyklon B was used in the Nazi extermination camps—although the latter was always used with the plan of genocide whereas the data on human harms of herbicides were far less clear in the early stages of the Vietnam war. Of course, even if humans were not the prime targets of Agent Orange, there is the difficult question of military intention and how trying to deprive the opposing Army of food could possibly not adversely affect civilians [10].

It is also important to note that it was not until 1969, many years in to Operation Ranch Hand, that it was recognized that small amounts of dioxin were found to contaminate Agent Orange.

While herbicides had been used in a military role prior to the war in Vietnam, the unprecedented widespread utilization of herbicides in warfare alarmed major groups of scientists. They were not alone; internal US opposition to the war made bedfellows of the rapidly growing student activists, scientists, left-leaning public intellectuals and Catholic peace advocates such as Dorothy Day and Thomas Merton. In 1970, a committee from the American Association for the Advancement of Science (AAAS) traveled to Vietnam to assess the health and environmental effects of herbicides sprayed by the military. The committee noted that local reports of stillbirths and birth defects might be linked to the dioxin contaminant in Agent Orange. The day after Christmas 1970, the AAAS released "resolution on Chemical defoliants' which called on the US government to cease the utilization of herbicidal warfare in Vietnam. "The key achievement of the AAAS was its ability to publicize the immense chemical destruction of nature in Vietnam as a war crime unjustifiable under any circumstances." [11].

Arthur Galston, the Chairman of the Yale University Botany Department who, termed the term ecocide. Protesting students realized within their own universities there were departments that were part of what they labelled the military-scientific complex. Dow Chemical Company made Agent Orange as well as being the principal supplier of napalm to the US Military.

The 1969 AAAS herbicide assessment commission didn't just focus on the adverse effects of spraying on human health; they reported that 800 thousand civilians were deprived of food crops, there was 500 million dollars in loss of hardwood trees and 1.2 million acres of mangrove forests had been destroyed [12].

Methodological Challenges with Confirming Causality

Clearly, proving unambiguous evidence that exposure to a particular compound is carcinogenic is difficult and prone not just to scientific debate but often political and economic issues affect and cloud assessment of the data. There can often be a very long time from exposure to the development of disease; imperfect records and recall bias can make attribution difficult. In addition, the lack of strong biomarkers to connect and confirm disease stemming from a particular chemical exposures complicates the picture. Nevertheless, even in disease that manifest years to decades after exposure, epidemiological evidence has accrued where certain cancers can be attributed to Agent Orange exposure with a fair degree of confidence.

Ultimately, the Bradford Hill criteria is a useful framework when trying to assess exposure to a toxin and the subsequent development of a disease [13] The criterion include biological plausibility, a biological gradient, temporality, strength, consistency, analogy, specificity, coherence, and experiment (successful prevention of the risk by a specific intervention).

Perhaps the most widely accepted environmental exposure that triggers malignancy in humans is tobacco—both in smokers but in addition, to those exposed to second-hand smoke. Yet it took decades to confirm the association between tobacco smoke and the carcinogens contained within and the development of multiple different types of human cancers. It then should come as no surprise that confirming in a scientifically rigorous fashion without bias the guilt of a particular exposure to an endogenous agent in triggering disease is difficult. Lastly, often these studies can have conflicting data that leave individuals—and courts of law- confused as to the strength of the evidence.

And why attempt to prove causation? The most obvious reason is for those individuals who are suffering the consequences of exposure with death or ongoing disease—whether lung cancer in smokes, malignant mesothelioma in workers exposed to asbestos, and soldiers or civilians exposed to dioxins. But there are additional reasons to probe these effects; developing a better understanding of the pathogenesis of disease, acquiring data for civilians to call on the military or industry to ethically safeguard human wellbeing as much as possible, and to avoid ecological damage that harms other species, plant life and the planet.

Part of the difficulty with ascribing causality is the long latency period between exposure and the manifestation of disease. TCDD has a long half life in humans and in addition, in areas heavily sprayed or near storage facilities, there can be prolonged persistence in the environment and in the local food supply.

Kang et al. performed a health survey of these 1499 Vietnam veterans and a group of 1428 non-Vietnam veterans assigned to chemical operations jobs was conducted using a computer-assisted telephone interview (CATI) system. [14] This study had more data than just recalled exposure to herbicides; they also analyzed serum specimens from a sample of 897 veterans for dioxin. Odds ratios for the development of diabetes, heart disease and COPD were significantly greater for veterans who had sprayed Agent Orange than in veterans who did not.

Large studies have attempted to better gauge the actual risk of cancer from TCDD exposure. Xu et al. performed a systematic analysis of TCDD exposure and cancer incidence [15]. They applied a random-effects model to estimate the pooled odds ratio (OR), risk ratio (RR), standard incidence ratio (SIR) or standard mortality ratio (SMR) for cancer incidence or mortality. 29,605 cancer cases and 3,478,748 participants across 31 studies were included in their analysis. Their conclusions were that higher external exposure level of TCDD was significantly associated with all cancer mortality but not all cancer incidence. In subgroup analysis there was increased mortality ratio of non-Hodgkin's lymphoma in both higher external exposure and blood level of TCDD.

Yi et al. published their analysis of agent orange exposure in Korean veterans who served in Vietnam [16]. From 1 January 1992 to 31 December 2005, 180,639 Korean Vietnam veterans were followed up for vital status and cause of death. They calculated an exposure index for Agent Orange that was based on the proximity of the veteran's unit to AO-sprayed areas, using a geographical information system-based model. The adjusted hazard ratios and 95% confidence intervals were calculated by Cox's proportional hazard model. They found that the all-cause death mortality was elevated in those veterans exposed to Agent Orange. The deaths due to all sites of cancers combined and some specific cancers, including cancers of the stomach, small intestine, liver, larynx, lung, bladder and thyroid gland, as well as chronic myeloid leukemia were positively associated with AO exposure. Interestingly but frustratingly, none of the site-specific cancer outcomes classified by the IOM as being associated with AO exposure was confirmed in this study [17].

While there have been a number of studies that looked at blood levels of TCDD that correlated with recalled exposure to herbicides, ultimately the fact that much of the data on exposure is retrospective and relies upon incomplete data means that there is an unresolvable issue in the middle of any decisive assessment of Agent Orange and cancer development.

Some of the other difficulty in establishing a link between exposure and disease is the difficulty in retrospectively assessing exposure to Agent Orange. Estimates needed to be drawn from flight path coordinates which were sometimes not accurate. Stellman et al. published revised estimates of Agent Orange exposure in 2003 using more complete data [18]. In their work, the estimates of the spray inventory was expanded by more than seven million liters. They also found that millions of Vietnamese citizens were likely to have been directly exposed to sprayed herbicides. There has been some criticisms of the Stellman methodology; however accurately assessing true exposure as best as possible (as opposed to just being in Vietnam during the war) is the only way to truly gauge accurate risks from exposure [19].

Specific Cancer Risk Associated with Agent Orange Exposure

The Veterans Administration acknowledges that there is sufficient evidence that soft tissue sarcoma can be ascribed to TCDD exposure to herbicides while serving in Vietnam. In addition, Hodgkin's disease, non-Hodgkin's lymphoma, chronic lymphocytic leukemia and hairy cell lymphoma are recognized as well. The data tying exposure to Agent Orange to manifestation of a particular diseases is difficult as it is all retrospective and as mentioned above, records are spotty. We will now look at some studies that looked at presumptive exposure and the development of certain malignancies.

Soft Tissue Sarcomas

Soft tissue sarcomas are rare tumors with multiple different histologies that arise from mesenchymal tissue; they represent less than 1% of adult tumors. The small numbers, diverse histologies and biological behaviors, and low numbers of patients enrolled on to multi-institution cooperative trials have hampered efforts to develop effective treatment strategies. While the majority of cases are felt to arise sporadically, there has been clear association between certain exposures such as radiation) and vinyl chloride causing hepatic angiosarcoma. The association between dioxin exposure and the development of soft tissue sarcoma, while acknowledged by the VA, remains disputed on the basis of epidemiological studies.

Sarcomas arising in veterans had been identified as one of the cancers to be associated with exposure to Agent Orange.

While much of the data regarding sarcoma development with exposure history has been mixed, a recent meta-analysis found clearer links. In a large study published in 2021, Edwards et al. reported on a systematic review of the literature and found that exposure to phenoxy herbicides and chlorophenols had a pooled odds ratio of developing sarcoma of 1.85 (95% CI 1.2202.82) from 16 studies with 2254 participants [20]. In the same study, they found that the pooled risk ratio for development of angiosarcoma of the liver and other STS in individuals exposed to vinyl chloride was 19.23 (95% CI 2.03, 182.46) and 2.23 (95 CI 1.55, 3.22) respectively based on 3 cohort studies with 12,816 participants. Exposure to dioxins was also associated with increased mortality from sarcoma; the pooled standardized mortality ratio was 2.56 (95% CI 1.60, 4.10) based on 4 cohort studies with 30,797 participants.

Studying American workers in factories in the US, while still having methodological issues, is somewhat easier than collecting data from soldiers exposed in the fog of war, still, the data continues to yield conflicting information. Fingerhut et al. performed a retrospective study of mortality among a cohort of 5172 workers at 12 plants in the United States that produced chemicals contaminated with TCDD [5]. Exposure was assessed by reviewing job descriptions as well as measuring TCDD in serum from a sample of 253 workers. They found that death from cancers that had previously been associated with TCDD (stomach, liver, and nasal cancers,

Hodgkin's disease, and non-Hodgkin's lymphoma) were not increased in this group. In a sub-cohort of workers with a greater than 1 year exposure and > 20 year latency, mortality was significantly increased for soft-tissue sarcoma (3 deaths). No firm conclusions could be drawn given the small numbers of sarcomas and in addition, there was not a way to evaluate other clinical risk factors (tobacco use? Family history of cancer?)

Saracci et al. performed a historical cohort study of mortality in an international register of 18,910 production workers or sprayers from ten countries [21]. Exposure to dioxins were obtained from questionnaires or factory records. There was no excess seen in all-cause mortality, for all neoplasms, for the most common epithelial cancers, or for lymphomas. There was a statistically non-significant two-fold excess risk, based on 4 observed deaths, was noted for soft-tissue sarcoma. The authors concluded that the excess does not seem to be specifically associated with those herbicides probably contaminated by dioxin.

Myelodysplasia and Leukemia

These devastating hematological malignancies are characterized by various cytopenias and proliferation of malignant blasts. Myelodysplasia (MDS) has been recognized for a long time as being associated with exposures to radiation and benzene. The damaged marrow leads to disordered production and development of cells and there is a risk of progression to acute leukemia. While many of the risk factors for these conditions are patient specific, such as age and male sex for example, it has long been recognized that certain environmental toxins such as benzene and dioxins can initiate the development of these cancers [22]. The fact that often herbicides that were sprayed in Vietnam were mixed with benzene meant that soldiers and citizens were being exposed to two known carcinogens.

Further data from occupational exposures have corroborated the toxicity of exposures to dioxins and marrow toxicity. A retrospective job exposure analysis of workers at a Dow company factory gives additional information regarding exposure to TCDD and hematological cancer risk [23]. When compared with the standard US population, workers had an increased risk of deaths from AML (standardized mortality ration 2.43; 95% confidence interval 1.16–4.46).

MGUS/Multiple Myeloma

The plasma cell disorders are hematological neoplasms that develop from B-lymphocyte lineage cells in the marrow. They can cause symptoms either from local growth and spread in the marrow space causing cytopenias as well as from secretion of monoclonal proteins that can adversely affect kidney function and in extreme cases can cause hyperviscosity syndrome. Multiple myeloma is a cancer of plasma cells in the bone marrow that can lead to punched-out lesions in bone, kidney failure, high calcium and eventually marrow failure. The incidence of multiple myeloma increases with age; while the exact cause is unknown, there are often clear exposure history of radiation exposure and to petroleum products.

Monoclonal gammopathy of undetermined significance (MGUS) is a premalignant condition that often can progress to multiple myeloma. These disorders are seen more commonly in older populations and men develop the disease more than women. There has been data that suggests that dioxins play a rule in increasing the odds of developing these conditions.

Bumma et al. performed a retrospective review of pts. seen at one VA hospital between 2005-2015 with MGUS, smoldering multiple myeloma and multiple myeloma [24]. Of 211 patients, only 11 had documented agent orange exposure. While the risk of transformation to frank myeloma in the exposed group had a significantly larger hazard ratio (11.19), the small numbers of patients make it difficult to draw firm conclusions.

Landgren et al. performed a prospective cohort study in 2013-2015 testing for MGUS in previously stored blood specimens in the Air Force health Study (AFHS) [25]. They looked at 479 Ranch Hand veterans and a matched 479 comparison samples from veterans who served in Viet Nam at the same time but did not take part in herbicide missions. The crude prevalence of MGUS in veterans exposed to Agent Orange was 7.1% and 3.1% in the comparison group.

Non-Hodgkin's Lymphomas and Hodgkin's Disease

Lymphomas are cancers of the lymphatic system that can range from fairly slow-growing, indolent cancers to highly aggressive cancers with some of the fastest tumor doubling times amongst human malignancies. Indolent lymphomas are often observed for a period of time prior to initiating treatment in order to gauge the disease's biological behavior. Ma et al. looked at whether there were differences in the latency period from exposure to dioxins and the manifestation of lymphoma between veterans with exposure to agent orange and those who did not report any [26]. They identified 17,175 veterans with indolent B cell lymphomas; 5233 had agent orange exposure and 11, 942 had none. In general, AO-exposed Veterans had shorter latencies than unexposed veterans but once the lymphoma was diagnosed they did not have worse overall survival.

Some research calls in to question the hypothesis of agent orange exposure and the development of hematological malignancies. Chang et al. performed a systematic critical review of epidemiologic studies that looked at the association of TCDD and the risk of lymphoid malignancies [27]. They looked at three groups: Vietnam veterans who were involved in the spraying and/or handling of Agent Orange or otherwise reported exposure to Agent Orange; workers involved in the production or use of herbicides potentially contaminated by TCDD or of communities exposed to herbicides and dioxin through industrial accidents; and hospitalized patients or general community members with biological measurements of TCDD exposure.

The authors' conclusion was that there was not an association of TCDD exposure with the development of Hodgkin's lymphoma and that the data on TCDD and the development of multiple myeloma is mixed. The small numbers of patients and the difficulty in establishing true exposure classification was listed as confounding reasons for the weak data.

The authors concluded: "Nevertheless, the available epidemiologic evidence does not demonstrate any significant association between valid Agent Orange or TCDD exposure and risk of NHL, HL, or MM among Vietnam veterans."

Prostate Cancer

Prostate cancer is a common malignancy that increases in prevalence as men age. When caught early, prior to spread, there are a number of treatments that can be curative such as radical prostatectomy or radiation therapy. Once it spreads, it often goes to bone where patients can experience a great deal of pain and fractures. Hormonal suppression can often control the disease for many years, but ultimately the disease is able to develop alternative mechanisms of cellular growth and when the disease becomes castration resistant it is much harder to treat.

As the cohort of veterans who served in the Vietnam war entered their 60 s, Chamie et al. looked at whether veterans who claimed exposure to Agent Orange during their service had a greater relative risk of developing prostate cancer than those who were unexposed [28]. They looked at patients at the Northern California veteran Affairs Health System (6214 exposed, 6930 unexposed). Exposure to Agent Orange increased the odds of being diagnosed with prostate cancer (odds ratio 2.19, 95% CI 1.75-2.75). Exposed veterans were diagnosed at a younger age, had a two-fold increase in Gleason 8 diagnoses, were more likely to have developed metastatic disease. Impressively, in the same group smoking history, race and preoperative PSA levels were not found to contribute to the increased risk.

Ansbaugh et al. assessed Agent Orange exposure as a risk factor for the development of prostate cancer in veterans with exposure in Vietnam [29]. In 2720 patients referred for prostate biopsy at the Portland Veterans Affairs Medical center, the authors set out to look at risk factors for developing high grade prostate cancer (prostate cancer is graded by pathologists by a Gleason score, with a higher score reflecting a more aggressive biology and a higher likelihood of the development of metastases). Of the2720 patients who underwent a prostate biopsy, 896 had positive biopsies for prostate cancer (32.9%) and approximately half of those had high grade histology (16.9% of the cohort). Agent Orange exposure was associated with a 52% increase in the risk of detecting prostate cancer (adjusted odds ratio 1.52; 95% confidence interval 1.07–2.13). While exposure did not increase the odds of developing low grade prostate cancer, there was a 75% increase in the risk of high-grade disease (adjusted odds ratio, 1.75, 95% confidence CI 1.12–3.62). Furthermore, exposure to agent orange was associated with a 2.1 increase in the risk of detecting prostate cancer with a Gleason score of >8.

The status of the patients regarding Agent Orange exposure was the classification in the VA's medical record system—defined as exposed or unexposed according to PVAMC standards.

Head and Neck Cancer

Cancer of the head and neck are aggressive cancers that often arise in the oropharynx, larynx, are treated with surgery or chemoradiation and carry a high morbidity.

Mowery et al. looked at data in the Veteran Affairs corporate data Warehouse VA CDW) to assess the presence of head and neck cancers in relation to agent orange exposure [30]. They looked at over eight million veterans in the database, 22% of whom reported agent orange exposure. 54,717 cases of head and neck cancer developed and Agent Orange exposure significantly predicted upper aerodigestive cancers with a RR of 1.16 and nasopharyngeal 1.22. Interestingly, exposure to TCBB did not predict for worse survival and in fact was associated with a superior 10 year survival in the patients with upper aerodigestive cancers.

Other Medical Issues from Exposure to Agent Orange and Dioxins

There are a number of non-malignant diseases associated with exposures to these agents. Confirming causation remains difficult, as proving degree and length of exposure is difficult and often is clouded by faulty patient recall. Still, there remains data that dioxins are associated with a number of other health issues.

Teratogenicity

June 1966 Bionetics Research labs informed the NCI that small amounts of 2,4,5-T injected in to lab rats gave birth in high ratios to offspring with birth defects [31] (ecocide 2543).

Ngo et al. looked at studies that tried to assess exposure to AO and birth defects in the Vietnamese population [32]. The authors reviewed 22 studies including 13 Vietnamese and nine non-Vietnamese studies. The summary relative risk (RR) of birth defects associated with exposure to Agent Orange was 1.95 [95% confidence interval (95% CI) 1.59-2.39], with substantial heterogeneity across studies. Interestingly, Vietnamese studies showed a higher summary RR (RR = 3.00; 95% CI 2.19–4.12) than non-Vietnamese studies (RR = 1.29; 95% CI 1.04–1.59). The magnitude of association seen in subgroup analysis tended to increase with greater degrees of exposure to Agent Orange.

Studies have confirmed the presence of dioxins in breast milk. Wong et al. collected breast milk samples from 137 first time mothers from Hong Kong [33]. They found that older mothers and mothers with longer residency in Hong Kong had higher levels of dioxin in their milk. Exposure to dioxins in utero and in breast milk have been associated with low birth weight and reduced IQ [34]. More recent data has suggested that children born to mothers residing near the former US air base at Da Nang-a known dioxin hot spot—showed evidence of neurodevelopmental abnormalities at age 8 that manifested with sex differences: boys exhibited learning delays and girls ADHD [35].

Dementia

Neurocognitive side effects of exposures to agent orange/dioxins are not limited to children. Martinez et al. performed a cohort study on veterans from 2001 to 2015 with 14 years of follow up [36]. The authors studied a 2% random sample of US veterans of the Vietnam era who received inpatient or outpatient Veterans Health Administration care, and they excluded those patients with dementia at baseline, those without follow-up visits, and those with vague or unclear Agent Orange exposure status. After adjusting for various demographic factors in the cohort, veterans exposed to Agent Orange had a significantly higher risk of developing dementia than those who weren't (hazard ratio of 1.68 (95% CI 1.59–1.77).

Controversies Regarding Attribution of Cause

Clearly, proving unambiguous evidence that exposure to a particular compound is carcinogenic is difficult and prone not just to scientific debate but often political and economic issues affect and cloud assessment of the data. There can often be a very long time from exposure to the development of disease; imperfect records and recall bias can make attribution difficult. In addition, the lack of strong biomarkers to connect and confirm disease stemming from a particular chemical exposures complicates the picture. Nevertheless, even in disease that manifest years to decades after exposure, epidemiological evidence has accrued where certain cancers can be attributed to agent orange exposure with a fair degree of confidence.

Measurement of Dioxin Levels Has Led to Discordant Results

Tai et al. looked at breast milk samples from 520 nursing mothers both in known hot spots as well as sprayed and unsprayed areas revealed differences in concentrations of 2,3,7,8-substituted PCDDs in primiparae mothers living in hot spots (14.10 pg/g lipid), 10.89 pg/g lipid in sprayed and 4.09 in mothers in unsprayed areas [37]. Obviously drawing conclusions are limited but this study suggests that decades after operation Ranch Hand that dioxin products remained in the Vietnamese ecosystem. A slightly later study by Nghi et al. confirmed high levels of 2,3,7,8-tetraCDD in the breast milk of women living near Bien Hoa Air base, another known hot spot [38].

Other studies looking at men more than 40 years after the spraying of herbicides showed a similar story. Manh et al. measured levels of TCDD as well as other dioxins and PCBs in men who lived near hotspots and those who lived in unsprayed areas [39]. The toxic equivalences of PCDDs/PCDFs was 41.7 pg/g lipid in those who lived closest to hot spots while it was 29 pg/g lipid in unsprayed areas.

Schecter et al. found similar results [40]. Using gas chromatography and mass spectroscopy, they assessed adipose tissue, human milk and blood form Vietnamese citizens from sprayed and unsprayed regions. Elevated TCBB levels of 1832 ppt were found in lipid milk in 1970, 103 ppt in 1980. Tissue from Vietnamese subjects in the south had higher dioxin levels than those in the north, as what would expected from where the Ranch Hand missions were performed.

The Role of Physician and Scientific Advocates in Environmental Safety

The 1960s were really a hinge period where many scientists and physicians used their expertise and their visibility to take strong positions on societal issues that they felt fell under their purview.

To paraphrase the historian Amy Hay in her magisterial book "The Defoliation of America" chemical companies almost always claim safety in the absence of overt harm and almost always contest charges of harm [41]. One could say that tobacco companies and Oil companies also take similar stances. This conflicts with the physicians' code to first do no harm. Clearly, certain agents that damage the environment are known ahead of time to pose significant danger to human life and health—atomic weapons for example. But physicians are increasing called upon in areas where there is a conflict between human health and wellbeing and other interests—fossil fuels, air pollution, gun violence, pesticides. It is important that healthcare workers continue to keep the patients' health foremost in mind and lend a rigorous eye to studying potential medical harm so that the public and especially their legislators can be kept up to date on potential harms.

Agent Orange and the ongoing questions of the true magnitude of harm is a vivid example of how public debates regarding ethics, science, interpretation of data and attribution of cause can be difficult and heated. And yet health care workers and scientists cannot avoid becoming involved in these debates and making sure that the data is collected and honestly analyzed; human health and flourishing depend on it.

References

1. Stellman JM, Stellman SD. Agent Orange during the Vietnam war: The lingering issue of its civilian and military health impact. Am J Public Health. 2018;108(6):726–8.
2. Huff JE, Salmon AE, Hopper NK. Long-term carcinogenesis studies on 2,3,7,8-tetrachlorodibenzo-p-dioxin and hexachlorodibenzo-p-dioxins. Cell Biol Toxicol. 1991;7(1):67–94. https://doi.org/10.1007/BF00121331.
3. Baccarelli A, Pesatori AC, Masten SA, et al. Aryl-hydrocarbon receptor-dependent pathway and toxic effects of TCDD in humans: a population-based study in Seveso, Italy. Toxicol Lett. 2004;149(1-3):287–93.
4. Pesatori AC, Consonni D, Rubagotti M, et al. Cancer incidence in the population exposed to dioxin after the "Seveso accident": twenty years of follow-up. Environ Health. 2009;8:29. https://doi.org/10.1186/1476-069X-8-39.
5. Fingerhut MA, Halperin WE, Marlow, et al. Cancer mortality in workers exposed to 2,3,7,8-tetrachlorodibenzo-p-dioxin. N Engl J Med. 1991;324(4):212–8.
6. Walker CL, Ho S. Developmental reprogramming of cancer susceptibility. Nat Rev Cancer. 2012;12(7):479–86. https://doi.org/10.1038/nrc3220.

7. Martini EA. Agent Orange: history, science and the politics of uncertainty. University of Massachusetts Press; 2012.
8. Veterans and Agent Orange Update 2000. National Academies Press; 3rd edition (November 15, 2001). Page 118.
9. Martini, 2012
10. Martini, 2012
11. Zierler D. The invention of ecocide: Agent Orange, Vietnam, and the scientists who changed the way we think about the environment. University of Georgia Press; 2011. p. 148.
12. Hay AM. The defoliation of America: Agent Orange chemicals, citizens, and protests (NEXUS: New Histories of Science, Technology, the Environment, Agriculture, and Medicine). University of Alabama Press; 2012. p. 51.
13. Hill AB. The environment and disease: association or causation? Proc R Soc Med. 1965;58(5):295–300.
14. Kang HK, Dalager NA, Needham LL, et al. Health status of Army Chemical Corps Vietnam veterans who sprayed defoliant in Vietnam. Am J Ind Med. 2006;49(11):875–84. https://doi.org/10.1002/ajim.20385.
15. Xu J, Ye Y, Huang F, et al. Association between dioxin and cancer incidence and mortality: a meta-analysis. Sci Rep. 2016;6:38012. https://doi.org/10.1038/srep38012.
16. Yi S, Or H. Agent Orange exposure and cancer incidence in Korean Vietnam veterans: a prospective cohort study. Cancer. 2014;120:3699–706.
17. Sinks TH. Challenges in investigating the association between Agent Orange and cancer: site-specific cancer risk and accuracy of exposure assessment. Cancer. 2014;120:23. 3595-3597
18. Stellman JM, Stellman SD, Christian R, et al. The extent and patterns of usage of agent Orange and other herbicides in Vietnam. Nature. 2003;422(6933):681–7. https://doi.org/10.1038/nature01537.
19. Stellman JM, Stellman SD. Agent Orange during the Vietnam war: The lingering issue of its civilian and military health impact. Am J Public Health. 2014;108(6):726–8.
20. Edwards D, Voronina A, Attwood K, et al. Association between occupational exposures and sarcoma incidence and mortality: systematic review and meta-analysis. Syst Rev. 2021;10:231. https://doi.org/10.1186/s13643-021-01769-4.
21. Saracci R, Kogevinas M, Bertazzi PA, et al. Cancer mortality in workers exposed to chlorophenoxy herbicides and chlorophenols. Lancet. 1991;338(8774):1027–32. https://doi.org/10.1016/0140-6736(91)91898-5.
22. Shallis RM, Gore SD. Agent Orange and dioxin-induced myeloid leukemia: a weaponized vehicle of leukemogenesis. Leuk Lymphoma. 2022;63(7):1534–43. https://doi.org/10.1080/10428194.2022.2034156. Epub 2022 Feb 2
23. Collins JJ, Ireland B, Buckley CF, et al. Lymphohaematopoeitic cancer mortality among workers with benzene exposure. Occup Environ Med. 2003;60:676–9.
24. Bumma N, Nagasaka M, Hemingway G, et al. Effect of exposure to Agent Orange on the risk of monoclonal gammopathy and subsequent transformation to multiple myeloma: a single-Center experience from the Veterans Affairs Hospital, Detroit. Clin Lymphoma Myeloma Leuk. 2020;20(5):305–11.
25. Landgren O, Shim YK, Michalek J, et al. Agent Orange exposure and monoclonal gammopathy of undetermined significance: an operation ranch hand veteran cohort study. JAMA Oncol. 2015;1(8):1061–8. https://doi.org/10.1001/jamaoncol.2015.2938.
26. Ma H, Ricks-Oddie J, Gupta P, et al. IBCL-349 veterans with Agent Orange exposure have a shorter latency from exposure to diagnosis of indolent B-cell lymphoid malignancies compared to unexposed veterans. Clin Lymphoma Myeloma Leuk. 2022;22(Supplement 2):S389–90.
27. Chang ET, Boffetta P, Adami HO, et al. A critical review of the epidemiology of agent Orange or 2,3,7,8-tetrachlorodibenzo-p-dioxin and lymphoid malignancies. Ann Epidemiol. 2015;25(4):275–292.e30. https://doi.org/10.1016/j.annepidem.2015.01.002.
28. Chamie K, White RW, Lee D, et al. Agent Orange exposure, Vietnam War veterans, and the risk of prostate cancer. Cancer. 2008;113(9):2464–70.

29. Ansbaugh N, Shannon J, Mori M, et al. Agent Orange as a risk factor for high-grade prostate cancer. Cancer. 2013;119(13):2399–404. https://doi.org/10.1002/cncr.27941. Epub 2013 May 13
30. Mowery A, Conlin M, Clayburgh D, et al. Increased risk of head and neck cancer in Agent Orange exposed Vietnam Era veterans. Oral Oncol. 2020;100:104483. https://doi.org/10.1016/j.oraloncology.2019.104483. Epub 2019 Dec 3
31. Zierler 2011
32. Ngo AD, Taylor R, Roberts CL, et al. Association between Agent Orange and birth defects: systematic review and meta-analysis. Int J Epidemiol. 2006;35(5):1220–30. https://doi.org/10.1093/ije/dyl038. Epub 2006 Mar 16
33. Wong TW, Wong AHS, Nelson EA, et al. Levels of PCDDs, PCDFs, and dioxin-like PCBs in human milk among Hong Kong mothers. Sci Total Environ. 2013;463-464:1230–8. https://doi.org/10.1016/j.scitotenv.2012.07.097.
34. Lundqvist C, Zuurbier M, Leijs M, et al. The effects of PCBs and dioxins on child health. Acta Paediatr Suppl. 2006;95(453):55–64. https://doi.org/10.1080/08035320600886257.
35. Tran NN, Pham-The T, Pham TN, et al. Neurodevelopmental effects of perinatal TCDD exposure differ from those of other PCDD/Fs in Vietnamese children living near the former US Air Base in Da Nang, Vietnam. Toxics. 2023;11(2):103. https://doi.org/10.3390/toxics11020103.
36. Martinez S, Yaffe K, Li Y, et al. Agent Orange exposure and dementia diagnosis in US veterans of the Vietnam era. JAMA Neurol. 2021;78(4):473–7. https://doi.org/10.1001/jamaneurol.2020.5011.
37. Tai PT, Nishijo M, Kido T, et al. Dioxin concentrations in breast milk of Vietnamese nursing mothers: a survey four decades after the herbicide spraying. Environ Sci Technol. 2011;45(15):6625–32. https://doi.org/10.1021/es201666d. Epub 2011 Jul 14
38. Nghi TT, Muneko Nishijo HD, Manh. Dioxins and Nonortho PCBs in breast milk of Vietnamese mothers living in the largest hot spot of dioxin contamination. Environ Sci Technol. 2015;49(9):5732–42. https://doi.org/10.1021/es506211p. Epub 2015 Apr 21
39. Manh HD, Kido T, Okamoto R, et al. Serum dioxin levels in Vietnamese men more than 40 years after herbicide spraying. Environ Sci Technol. 2014;48(6):3496–503. https://doi.org/10.1021/es404853h. Epub 2014 Mar 4
40. Schecter A, Dai LC, Thuy LT, et al. Agent Orange and the Vietnamese: the persistence of elevated dioxin levels in human tissues. Am J Public Health. 1995;85(4):516–22.
41. Hay, 2021

Nuclear Weapons and Cancer

13

Kevin T. Tran and Andrew M. Farach

In August of 1945 the first and second nuclear weapons were used in warfare; a uranium bomb was dropped on Hiroshima on August 6, 1945 and 3 days later, a plutonium bomb was used against Nagasaki. The destructive forces of these bombs devastated the two cities, resulting in the estimated deaths of 90,000 to 120,000 of the 330,000 civilian population of Hiroshima and 60,000 to 80,000 out of the 280,000 in Nagasaki [1, 2]. On October 12, 1945 the Unites States, in cooperation with Japanese scientists, formed the Joint Commission for the Investigation of the Effects of the Atomic Bomb in Japan and looked at the acute effects of the bomb. As a result of these early studies, a" long-range, continuing study of the biological and medical effects of the atomic bomb on man" was approved by President Truman on November 26, 1946. [3] Much of the medical knowledge we know about in utero effects of radiation, heritable effects, and carcinogenesis comes from these studies.

The atomic bomb survivors of Hiroshima and Nagasaki have been followed through the Life Span Study (LSS) cohort by the Radiation Effects Research Foundation (RERF) and the Atomic Bomb Casualty Commission (ABCC), looking at a variety of late effects from radiation exposure. The LSS cohort subjects were recruited through the 1950 Japanese National Census which included a questionnaire about atomic bomb exposures as well as through two surveys by the Atomic Bomb Casualty Commission in 1950 and 1951, and resident surveys by Hiroshima in 1953 and Nagasaki in 1950. The final cohort includes 120,321 subjects (82,214 from Hiroshima and 38,107 from Nagasaki) who were within 2.5 km of the

K. T. Tran
Department of Radiation Oncology, The University of Texas Medical Branch, Galveston, TX, USA
e-mail: kettran@UTMB.EDU

A. M. Farach (✉)
Department of Radiation Oncology, Houston Methodist Hospital, Houston, TX, USA
e-mail: amfarach@houstonmethodist.org

E. H. Bernicker (ed.), *Environmental Oncology*,
https://doi.org/10.1007/978-3-031-33750-5_13

hypocenters at the time of the bombing as well as an age- and sex-matched cohort of people who were between 2.5 and 10 km from the hypocenters (see Fig. 13.1). Follow-up of these people started on October 1, 1950 [2].

Dose estimations have been calculated and have taken into account many factors including the distance and elevation in which subjects were from the hypocenter as well as terrain shielding, in what is known as the Dosimetry System 2002 (DS02). This system includes calculated doses for 15 organ sites and uses a linear model to predict excess relative risk (ERR) for solid cancers and major causes of death, while using a linear-quadratic model for ERR of leukemias [4, 5].

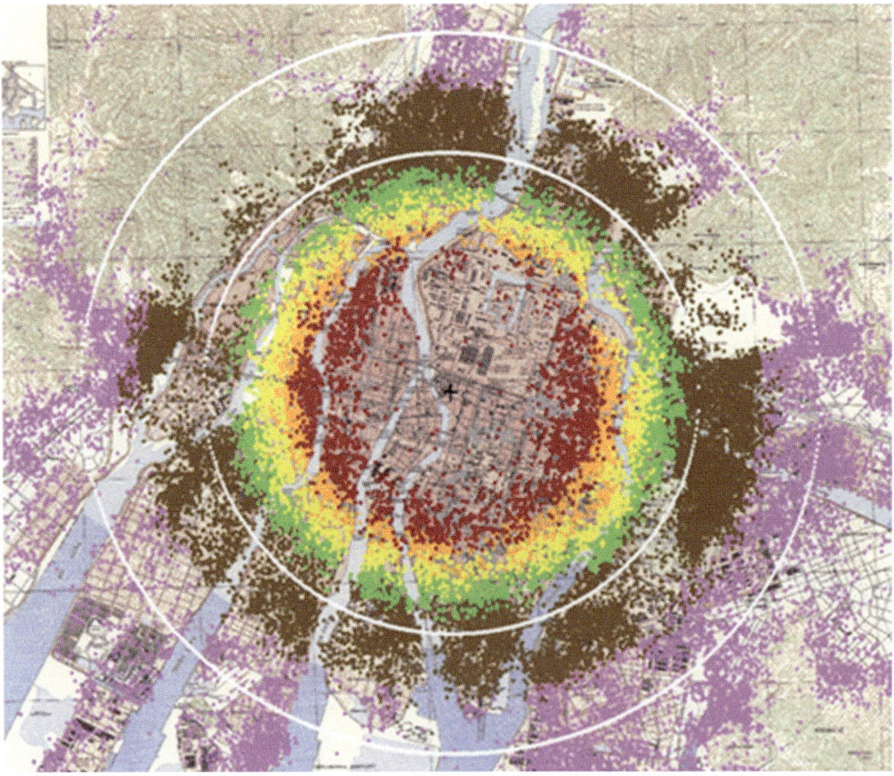

Fig. 13.1 Location of individuals in the superimposed on a map of the city with color denoting estimated radiation dose ranges (gray = unknown; red = >1000 mGy; orange = 500–1000 mGy; yellow = 200–500 mGy; brown = 5–100 mGy; pink = <5 mGy). Rings represent 2- and 3-km distances from the hypocenter (+). From: Douple: Disaster med. Public health prep., Volume 5(20110300) Supplement 1.March 2011.S122

Cancer Mortality

The LSS has published 14 reports looking at the mortality of atomic bomb survivors, the most recent looking at data from 1950–2003. The most important finding from the LSS is the increased risk of cancer mortality throughout life. The current data shows a proportional increase in risk of death from solid cancers with increasing radiation dose, with an excess death rate of 26/10,000 person-years per Gy of radiation. It is also seen that there is a higher relative risk for cancer death for those exposed at younger ages and that relative risk declined with increasing attained age.

Though this is the largest cohort with the longest follow-up after exposure to ionizing radiation, there are several limitations to this study. These subjects were all survivors of the atomic bombings. This excludes those who succumbed to the physical injuries and burns from the bombings as well as those that expired in the post-war setting where the infrastructure was lacking, and nutrition and hygiene were poor. This may have excluded a population of subjects vulnerable to radiation that died before 1950 [2, 5].

The following sections will look more specifically at the different cancer subtypes and the excess risk seen in the LSS cohort.

Hematologic Malignancies

The incidence and deaths from leukemia were first reported in atomic bomb survivors in the years 1948, 1949, and 1950. Initial studies showed an increase in both the incidence and death rate from leukemia in the exposed populations of Hiroshima and Nagasaki as compared to the non-exposed population. It was also seen that those closer to the hypocenters had a higher incidence of leukemia. Of these patients, excess incidence of acute (AML) and chronic myeloid leukemia (CML) was observed, while chronic lymphocytic leukemia (CLL) did not seem to be related to radiation exposure. Later LSS reports estimated the ERR at 1 Gy for leukemia was 3.1. They also showed ERR per Gy diminishes in magnitude with increasing age of exposure through age 30. For those less than 30 years at age of exposure, the peak ERR per Gy was estimated to be at 7 years after exposure [6–8].

When looking at the incidence of all leukemia, the preferred model for ERR was linear quadratic. However when looking at specific leukemias, it was seen that there were better fitting models; acute lymphoblastic leukemia (ALL) and CML both had linear non-threshold models while AML had a best fit with a purely quadratic model. Essentially, exposure to ionizing radiation has shown an increase in excessive risk in all types of leukemia except for CLL. However specific subtypes are best fit with different models making it difficult to accurately create a single a dose-response curve for all leukemia [7, 8].

Adult T-cell leukemia (ATL) is known to be endemic to the Nagasaki population due to prevalence human T-cell lymphotropic virus type-1 (HTLV-1). Through the LSS, there was no indication of excess risk of developing ATL though it was noted that ATL as an entity was not coded until the International Classification of Diseases 10 (ICD 10), almost 35 years after the atomic bombings [7–9].

There is limited evidence to suggest increased risk of incidence or death from lymphoma associated with ionizing radiation from the atomic bombs. Hodgkin's disease was rare among the LSS cohort and when looking at the Non-Hodgkin's disease population, there was no ERR seen. [7]

Similarly to CLL, multiple myeloma was not shown to be associated with radiation exposure through the LSS cohort. [7]

Thyroid Cancer

The thyroid gland is seen to be highly sensitive to radiation exposure, showing both malignant and non-malignant radiation-related changes. There has been links to autoimmune thyroid disease including hypoparathyroidism and hyperparathyroidism, with increases in antithyroid antibodies detected in the serum of atomic bomb survivors. When looking at thyroid cancers specifically, most cases were histologically papillary carcinomas and crude incidence rates increased in response to radiation dose among those who were children or adolescents at the time of exposure. Thyroid cancer followed a linear dose response model with the pooled ERR per Gy of about 1.28, however the highest excess risk was seen in the group younger than 9 years old at time of exposure. [10–12]

Though the thyroid is sensitive to radiation, non-malignant disease can often be managed with medications while malignant disease can be treated successfully with surgery or radioactive iodine. These account for the low mortality rate in the LSS cohort due to thyroid cancers [5].

Breast Cancer

We know that there is a radiation-related risk for breast cancer from both the atomic bomb studies as well as childhood Hodgkin's lymphoma patients treated with chest radiation therapy.

Though doses to the breast for survivors of the atomic bomb were much lower than those children treated with radiation for Hodgkin's lymphoma, both showed an increase risk of developing radiation-related breast cancer. In the Hodgkin's lymphoma group, the highest risk was seen in female patients from age 10–16 [13].

In the LSS the highest excess risk for breast cancer was in the female population exposed before the age of 20 while the lowest was seen in the over 40 age group [14–17]. Interestingly, no particular age or age range within those less than 20 years of age at exposure stood out as having higher incidences of breast cancer; the data did not support previous hypotheses that radiation-related breast cancer may be exceedingly high in the population around the age of menarche or thelarche [18]. The overall ERR per Gy for the cohort was 1.6 and the data fit a linear dose-response curve. When looking at the data of time from exposure until the appearance of radiation-related excess risk, a best fit model suggested a latent period of about 12 years. It was also that 34 breast cancer cases were cancers of the opposite breasts, diagnosed either simultaneously or months to years after initial diagnosis [16].

In summary the data show a linear dose-response to increasing radiation dose for the excess incidence of breast cancer, particularly in younger women with a latency period of about 12 years.

Lung Cancer

While lung cancer is strongly associated with smoking, radiation exposure has also been shown to increase the risks of lung cancer. Lung cancer is the second most common cancer among the LSS cohort with an estimated ERR per Gy of 0.81 and follows a linear dose response. Data on smoking was available for approximately 60% of the LSS cohort, with roughly 85% of men and 18% of women indicating that they had ever smoked. When looking at the joint effect of smoking and radiation, a complex interaction was observed; for light smokers, less than a pack of cigarettes a day, a super-multiplicative interaction was seen, while for heavy smokers an additive interaction was seen. This is to say that there was a large radiation-related risk in light smokers but little radiation-related risk in heavy smokers [19, 20].

Interestingly, ERR per Gy for lung cancer was seen to increase with age of exposure, while in other cancer sites, it decreases.

When looking at the histological subtypes of lung cancer, smoking conveyed a higher risk for small cell carcinoma and squamous cell carcinoma while conveying a lower risk for adenocarcinoma. A clear interaction between development of a specific histological subtype and radiation-smoking effects was not clearly seen; it was suggested that radiation effects may share a complex dependence on smoking intensity [19, 21].

Bladder Cancer

There are many factors associated with urothelial carcinoma of the bladder cancer including smoking, exposures to chemicals such as aromatic amines, excessive use of analgesics, and ionizing radiation. Some factors such as alcohol use have less of a clear association with bladder cancer and some lifestyle factors, such as high fruit and vegetable diets are thought to be protective against bladder cancer. In the LSS, bladder cancer was seen to have one of the highest ERR per Gy however the initial studies had difficulties quantifying how much of the excess risk was due to ionizing radiation rather than the other associated factors. [22]

Later studies took into account lifestyle factors and found an ERR per Gy adjusted for all lifestyle factors of 0.96 and follow a linear dose-response curve.

When looking at different lifestyle factors, only smoking was significantly associated with urothelial carcinoma. A high female to male ERR ratio was also seen though it is noted that in the general Japanese population, the incidence bladder cancer is four times as common in men than women. ERR per Gy was not seen to vary with age at exposure or attained age. This all suggests that though smoking is clearly associated with bladder cancer, the bladder is sensitive to ionizing radiation [22, 23].

Skin Cancer

It is known that exposure to non-ionizing ultraviolet radiation has an increased risk of non-melanoma skin cancers, and the LSS shows that similarly there is an excess risk from exposure to ionizing radiation. Particularly, when looking at histology, it was seen that basal cell carcinoma had an increased ERR per Gy of 1.9 and follows a linear-dose response while there was no significant risk related to squamous cell carcinoma. There were too few cases of melanoma in the LSS cohort for adequate analysis. It is noted that only about 2% of all solid cancers from the LSS cohort were skin cancers, and that the incidence of skin cancers in Asian populations is much lower than in Caucasian populations. They also noted that the squamous cell carcinoma occurs much for frequently at baseline than basal cell carcinoma; even with the increased excessive risk of developing basal cell carcinoma, the incidence of basal cell carcinoma did not surpass the baseline incidence of squamous cell carcinoma [22, 24, 25].

Colon and Rectal Cancer

Since the 1980s, the incidence of colon cancer rates in Hiroshima and Nagasaki have dramatically increased, likely to the increased emphasis on surveillance and screening as well as the change in coding for the tumor registries. This rate has seen to increase among men more rapidly as compared to women, and age-standardized colon cancer rates in Japan are similar to those in the United States. Despite increasing baseline rates of colon cancer in Japan, an excess risk was observed among those exposed to the atomic bombs. The ERR per Gy for women was observed to be 0.34 while in men it was 0.73 and followed a linear dose-response. Age at exposure was not seen to have effect on the ERR [5, 22].

Rectal cancers also saw increases in the baseline incidence over the past several decades, albeit at a slower pace than that of colon cancer. Similar to colon cancer, rectal cancer incidences were higher in men than in women however a higher incidence was seen in the cohort born in the years following the bombings. Under a linear dose-response model, the ERR per Gy was estimated to be 0.19 and found not to be statistically significant. Otherwise with no effect from age at exposure, attained age, or gender [22].

Summary of Cancer Risks

Leukemia deaths were the first major radiation-related health effect seen in the LSS with solid-cancer deaths following soon after. The LSS has shown increased risk of cancers for many organ sites, with the highest risks seen in the bladder, female breast, and lungs. Relatively high risks were seen in the brain, ovary, thyroid, colon, and esophagus, while there was no association seen in rectal, prostate, or melanoma.

In-Utero Changes

Of the 120,000 atomic bomb survivors in the LSS, a small subset of 15,899 who were younger than the age of 6 years old at the time of exposure, as well as 3268 subjects who were in utero at the time of exposure were followed closely. Follow-up of these patients started in 1958 when the tumor registries were started. In this full cohort, 1216 solid cancers were recorded through the end of 1999, of which 743 were first primary cancers among those with dosimetric information. Solid cancer and leukemia incidence rates were seen to be higher among the exposed subjects. Of these, only eight solid cancers were seen in these groups between ages 14 and 19, only one of which was in the in-utero groups (a Wilms tumor). It was seen that incidence rate started to rise significantly after age 40, however when looking at the excess relative risk, it was seen to decrease with attained age. This decreased excess risk is thought to be more marked in the in-utero cohort compared to the early childhood cohort [26].

No solid cancer deaths were recorded for ages less than 17 though both solid cancer and leukemia mortality rates were elevated in both the in-utero group and early childhood group. The ERR per Gy for solid cancer mortality was 2.4 and most of the deaths were in females and attributed to cancers of the breast, digestive tract, and reproductive system. There was an absence of leukemia deaths among the in-utero cohort, whereas a dose response was seen in the early childhood group [27].

Severe intellectual disability (ID) was shown to be related to in-utero exposure to radiation at gestational ages of 8–15 weeks as well as 16–25 weeks. Gestational age is thought to be the most important factor in determining radiation related damage to the brain; the thought is that during the fetal stage (after week 8 of gestation), development of the cerebral cortex progresses rapidly and is highly susceptible to radiation damage due to the actively dividing cells. In weeks 0–7, precursor neurons and neuroganglia start to emerge and are mitotically active. During weeks 8–15 these neurons rapidly increase in number and migrate to the cortex. In weeks 16–25 differentiation increases, and synaptogenesis occurs. In the last period, > = 26 weeks, continued differentiation occurs. [28, 29]

Of the patients with dosimetric data available, there were 30 cases of severe ID, five of which were excluded as they were likely non-radiation-related (Down's syndrome, ID in a sibling, and encephalitis). Of note, the ABCC identified severe ID through clinical assessment by one or more pediatricians, rather than by IQ scores.

Gestational age was estimated based on the date of birth, and assuming the mean gestational time being 280 days. There were 17 observed cases in the group of 8–15 weeks gestational age, while no cases were seen between 0–7 weeks or > = 26 weeks. Looking at the dose response, a threshold as low as 0.30 Gy is seen to increase the risk of ID [28, 29].

In 1955 and 1956, the ABCC collected IQ data in prenatally exposed survivors and non-exposed residents who were at that time 10–11 years old. Of the 1673 examinees, the average IQ for the whole group was 107.8, while the average IQ for those with small head size and ID was 63.8, those with regular head size and ID was 68.9, and those with small head but no ID was 96.4. It was seen that among children exposed to radiation prior to 8 weeks or after 26 weeks of gestation, there was no effect on intelligence. Of those with decreased IQ, there was marked diminution if exposed between 8–15 weeks, though there was still an observed decrease if exposed between 16–25 weeks. The decreased IQ followed a linear dose-response to absorbed uterine dose in mothers and was seen to be ~25–29 points per Gy [28].

Small head size, which includes microcephaly and craniostenosis, is described as having a head circumference two or more standard deviations below expected sex- and age-specific head size. In this cohort, 62 were described as having small head size between the ages of 9–19 years. Of the reported 30 cases with ID, 26 were included in the small head size cohort. Most of the individuals with small head size were exposed during the first or second trimester (up to ~week 24) and on dose-response models shows zero threshold. [28, 30, 31]

The F1 Cohort

Early on, it was recognized that following the offspring of atomic bomb survivors was important, which led to the formation of the F1 cohort. This group included approximately 77,000 subjects born between the time of the bombings and 1984. [2]

Some of the earliest studies from this cohort tried looking for genetic aberrations. When looking at 185 children, some of which were conceived before the bombing and some after the bombing, karyotype analysis showed 182 normal subjects [32, 33]. When performing electrophoresis to identify loss of enzyme activity mutations, only one mutation in the children of proximally exposed parents was seen in 60,529 locus products, similar to the control group. That is to say, the rate of mutations between the two groups was identical [33, 34].

When looking at cancer incidence rates in the F1 cohort, about 31,000 children of exposed parents (with average conjoint gonad exposure of 0.43 Sv) as well as a cohort of 41,000 children in a control group were followed until age 20. In this group there were 43 malignant tumors identified in the children of exposed parents while 49 were diagnosed in the control. Essentially, an increase in cancer incidence among those born to exposed parents was not seen. [35, 36]

Mortality rate was another area of interest when following the F1 cohort. Mortality data was followed through 1999 and during this time 314 cancer deaths and 1125 non-cancer deaths were observed. Of these, 33 cancer deaths and 741 non-cancer deaths occurred before the age of 20. The general trend for deaths was that infection, respiratory disease, or digestive disease was more common earlier, however as the population aged, cancer and circulatory disease became more common. When compared to the control group, the mortality rate from cancer deaths and non-cancer deaths of the children of exposed parents was not seen to be higher. [37]

Epilogue

The catastrophic nuclear events that transpired in Japan following the bombings of Hiroshima and Nagasaki, while unspeakable, provided a unique opportunity for the study of the long-term effects of radiation exposure and carcinogenesis. The studies stemming from the dropping of the atomic bomb have helped establish acceptable limits for occupational radiation exposure, and to safely harness the power of radioactivity for the greater good of humanity in such things as radiotherapeutics and imaging. We also learned how to minimize harmful in utero effects of radiation exposure. Albert Einstein once said, "I know not with what weapons World War III will be fought, but World War IV will be fought with sticks and stones." In an uncertain world with the omnipresent threat of nuclear warfare it is our hope never to be presented such a learning opportunity again.

References

1. A nuclear shadow from Hiroshima and Nagasaki to Fukushima. Lancet. 2015;386(9992):403. https://doi.org/10.1016/S0140-6736(15)61429-5.
2. Douple EB, Mabuchi K, Cullings HM, et al. Long-term radiation-related health effects in a unique human population: lessons learned from the atomic bomb survivors of Hiroshima and Nagasaki. Disaster Med Public Health Prep. 2011;5(SUPPL:1–S133. https://doi.org/10.1001/dmp.2011.21.
3. Beebe GW. Reflections on the work of the atomic bomb casualty commission in Japan. Epidemiol Rev. 1979;1(1):184–210. https://doi.org/10.1093/oxfordjournals.epirev.a036210.
4. Cullings HM, Grant EJ, Egbert SD, et al. DS02R1: improvements to atomic bomb survivors' input data and implementation of dosimetry system 2002 (DS02) and resulting changes in estimated doses. Health Phys. 2017;112(1):56–97. https://doi.org/10.1097/HP.0000000000000598.
5. Ozasa K, Shimizu Y, Suyama A, et al. Studies of the mortality of atomic bomb survivors, report 14, 1950-2003: an overview of cancer and noncancer diseases. Radiat Res. 2012;177(3):229–43. https://doi.org/10.1667/RR2629.1.
6. Folley JH, Borges W, Yamawaki T. Incidence of leukemia in survivors of the atomic bomb in Hiroshima and Nagasaki, Japan. Am J Med. 1952;13(3):311–21. https://doi.org/10.1016/0002-9343(52)90285-4.
7. Preston DL, Kusumi S, Tomonaga M, et al. Cancer incidence in atomic bomb survivors. Part III: leukemia, lymphoma and multiple myeloma, 1950-1987. Radiat Res. 1994;137(2 SUPPL):S68. https://doi.org/10.2307/3578893.
8. Richardson D, Sugiyama H, Nishi N, et al. Ionizing radiation and leukemia mortality among Japanese atomic bomb survivors, 19502000. Radiat Res. 2009;172(3):368–82. https://doi.org/10.1667/RR1801.1.

9. Arisawa K, Soda M, Akahoshi M, et al. Human T-cell lymphotropic virus type-1 infection and risk of cancer: 15.4 year longitudinal study among atomic bomb survivors in Nagasaki, Japan. Cancer Sci. 2006;97(6):535–9. https://doi.org/10.1111/j.1349-7006.2006.00212.x.
10. Furukawa K, Preston D, Funamoto S, et al. Long-term trend of thyroid cancer risk among Japanese atomic-bomb survivors: 60 years after exposure. Int J Cancer. 2013;132(5):1222–6. https://doi.org/10.1002/ijc.27749.
11. Yoshimoto Y, Ezaki H, Etoh R, Hiraoka T, Akiba S. Prevalence rate of thyroid diseases among autopsy cases of the atomic bomb survivors in Hiroshima, 1951-1985. Radiat Res. 1995;141(3):278–86. https://doi.org/10.2307/3579004.
12. Imaizumi M, Usa T, Tominaga T, et al. Radiation dose-response relationships for thyroid nodules and autoimmune thyroid diseases in Hiroshima and Nagasaki atomic bomb survivors 55-58 years after radiation exposure. JAMA. 2006;295(9):1011–22. https://doi.org/10.1001/jama.295.9.1011.
13. Holmqvist AS, Chen Y, Berano Teh J, et al. Risk of solid subsequent malignant neoplasms after childhood Hodgkin lymphoma—identification of high-risk populations to guide surveillance: a report from the late effects study group. Cancer. 2019;125(8):1373–83. https://doi.org/10.1002/cncr.31807.
14. Tokunaga M, Land CE, Tokuoka S, Nishimori I, Soda M, Akiba S. Incidence of female breast cancer among atomic bomb survivors, 1950-1985. Radiat Res. 1994;138(2):209–23. https://doi.org/10.2307/3578591.
15. Tokunaga M, Land CE, Yamamoto T, et al. Incidence of female breast cancer among atomic bomb survivors, Hiroshima and Nagasaki, 1950–1980. Radiat Res. 1987;112(2) https://doi.org/10.2307/3577254.
16. Land CE, Tokunaga M, Koyama K, et al. Incidence of female breast cancer among atomic bomb survivors, Hiroshima and Nagasaki, 1950–1990. Radiat Res. 2003;160(6) https://doi.org/10.1667/RR3082.
17. McGregor DH, Land CE, Choi K, et al. Breast cancer incidence among atomic bomb survivors, Hiroshima and Nagasaki, 1950-69. J Natl Cancer Inst. 1977;59(3):799–811. https://doi.org/10.1093/jnci/59.3.799.
18. Russo J, Russo IH. Toward a unified concept of mammary carcinogenesis. Prog Clin Biol Res. 1997;396:1–16.
19. Cahoon EK, Preston DL, Pierce DA, et al. Lung, laryngeal and other respiratory cancer incidence among Japanese atomic bomb survivors: an updated analysis from 1958 through 2009. Radiat Res. 2017;187(5):538–48. https://doi.org/10.1667/RR14583.1.
20. Furukawa K, Preston DL, Lönn S, et al. Radiation and smoking effects on lung cancer incidence among atomic bomb survivors. Radiat Res. 2010;174(1):72–82. https://doi.org/10.1667/RR2083.1.
21. Egawa H, Furukawa K, Preston D, et al. Radiation and smoking effects on lung cancer incidence by histological types among atomic bomb survivors. Radiat Res. 2012;178(3):191–201. https://doi.org/10.1667/RR2819.1.
22. Preston DL, Ron E, Tokuoka S, et al. Solid cancer incidence in atomic bomb survivors: 1958-1998. Radiat Res. 2007;168(1):1–64. https://doi.org/10.1667/RR0763.1.
23. Grant EJ, Ozasa K, Preston DL, et al. Effects of radiation and lifestyle factors on risks of urothelial carcinoma in the life span study of atomic bomb survivors. Radiat Res. 2012;178(1):86–98. https://doi.org/10.1667/RR2841.1.
24. Ron E, Preston DL, Kishikawa M, et al. Skin tumor risk among atomic-bomb survivors in Japan. Cancer Causes Control. 1998;9(4):393–401. https://doi.org/10.1023/A:1008867617415.
25. Kishikawa M, Koyama K, Iseki M, et al. Histologic characteristics of skin cancer in Hiroshima and Nagasaki: background incidence and radiation effects. Int J Cancer. 2005;117(3):363–9. https://doi.org/10.1002/ijc.21156.
26. Preston DL, Cullings H, Suyama A, et al. Solid cancer incidence in atomic bomb survivors exposed in utero or as young children. J Natl Cancer Inst. 2008;100(6):428–36. https://doi.org/10.1093/jnci/djn045.

27. Delongchamp RR, Mabuchi K, Yoshimoto Y, Preston DL. Cancer mortality among atomic bomb survivors exposed in utero or as young children, October 1950–May 1992. Radiat Res. 1997;147(3). https://doi.org/10.2307/3579348.
28. Otake M, Schull WJ. Review: radiation-related brain damage and growth retardation among the prenatally exposed atomic bomb survivors. Int J Radiat Biol. 1998;74(2):159–71. https://doi.org/10.1080/095530098141555.
29. Otake M, Schull WJ, Lee S. Threshold for radiation-related severe mental retardation in prenatally exposed A-bomb survivors: a re-analysis. Int J Radiat Biol. 1996;70(6):755–63. https://doi.org/10.1080/095530096144644.
30. Otake M, Schull WJ. Radiation-related small head sizes among prenatally exposed a-bomb survivors. Int J Radiat Biol. 1993;63(2):255–70. https://doi.org/10.1080/09553009314550341.
31. Wood JW, Johnson KG, Omori Y. In utero exposure to the Hiroshima atomic bomb. An evaluation of head size and mental retardation: twenty years later. Pediatrics. 1967;39(3):385–92.
32. Awa AA, Bloom AD, Yoshida MC, Neriishi S, Archer PG. Cytogenetic study of the offspring of atom bomb survivors (17). Nature. 1968;218(5139):367–8. https://doi.org/10.1038/218367a0.
33. Nakamura N. Genetic effects of radiation in atomic-bomb survivors and their children: past, present and future. J Radiat Res. 2006;47(SUPPL. B):B67–73. https://doi.org/10.1269/jrr.47.B67.
34. Neel JV, Satoh C, Goriki K, et al. Search for mutations altering protein charge and/or function in children of atomic bomb survivors: final report. Am J Hum Genet. 1988;42(5):663–76.
35. Yoshimoto Y, Neel JV, Schull WJ, et al. Malignant tumors during the first 2 decades of life in the offspring of atomic bomb survivors. Am J Hum Genet. 1990;46(6):1041–52.
36. Izumi S, Koyama K, Soda M, Suyama A. Cancer incidence in children and young adults did not increase relative to parental exposure to atomic bombs. Br J Cancer. 2003;89(9):1709–13. https://doi.org/10.1038/sj.bjc.6601322.
37. Izumi S, Suyama A, Koyama K. Radiation-related mortality among offspring of atomic bomb survivors: a half-century of follow-up. Int J Cancer. 2003;107(2):292–7. https://doi.org/10.1002/ijc.11400.

Index

E. H. Bernicker (ed.), *Environmental Oncology*,
https://doi.org/10.1007/978-3-031-33750-5

MIX
Papier aus verantwortungsvollen Quellen
Paper from responsible sources
FSC® C105338

If you have any concerns about our products,
you can contact us on
ProductSafety@springernature.com

In case Publisher is established outside the EU,
the EU authorized representative is:
Springer Nature Customer Service Center GmbH
Europaplatz 3, 69115 Heidelberg, Germany

Printed by Libri Plureos GmbH
in Hamburg, Germany